T0137874

Lecture Notes in Computer Science 13813

More information about this series at https://link.springer.com/bookseries/558

Adam Krzyzak · Ching Y. Suen ·
Andrea Torsello · Nicola Nobile (Eds.)

Structural, Syntactic, and Statistical Pattern Recognition

Joint IAPR International Workshops, S+SSPR 2022
Montreal, QC, Canada, August 26–27, 2022
Proceedings

Editors
Adam Krzyzak (ID)
Concordia University
Montreal, QC, Canada

Ching Y. Suen
Concordia University
Montreal, QC, Canada

Andrea Torsello
University Ca' Foscari of Venice
Venice, Italy

Nicola Nobile (ID)
Concordia University
Montreal, QC, Canada

ISSN 0302-9743   ISSN 1611-3349 (electronic)
Lecture Notes in Computer Science
ISBN 978-3-031-23027-1   ISBN 978-3-031-23028-8 (eBook)
https://doi.org/10.1007/978-3-031-23028-8

This Springer imprint is published by the registered company Springer Nature Switzerland AG
The registered company address is: Gewerbestrasse 11, 6330 Cham, Switzerland

Preface

This volume contains the papers presented at the joint IAPR International Workshops on Structural and Syntactic Pattern Recognition (SSPR 2022) and Statistical Techniques in Pattern Recognition (SPR 2022). S+SSPR is a joint event organized by Technical Committee 1 (Statistical Pattern Recognition Technique) and Technical Committee 2 (Structural and Syntactical Pattern Recognition) of the International Association of Pattern Recognition (IAPR). Following the tradition of holding 10 S +SSPR events in different cities around the world over the past 20 years, S+SSPR 2022 was held during August 26–27 right after the International Conference on Pattern Recognition (ICPR) in Montreal.

S+SSPR 2022 received 50 submissions. The reviewing process was single blind, where each submission was reviewed by at least two and usually three Program Committee members. Papers by PC chairs or members where were handled by the other PC chairs and assigned to PC members is a way to avoid conflicts of interest. In the end 32 papers were accepted for presentation at the workshops. The accepted papers cover topics of current interest in pattern recognition, including graphical analysis, machine learning, neural networks, classification techniques, deep learning, and various applications.

We were delighted to have four prominent keynote speakers: Yoshua Bengio (Turing Prize Winner) who was the recipient of the IAPR TC1 Pierre Devijver Award for 2022, Linda Shapiro, Mohamed Cheriet, and Patia Radeva.

We would like to thank all the Program Committee members for their help in the review process and express our appreciation to Springer for publishing this volume. More information about the workshops and organization can be found on the website: https://iapr.org/ssspr2022/.

November 2022

Adam Krzyzak
Ching Y. Suen
Andrea Torsello
Nicola Nobile

Organization

Honorary Chair

Edwin Hancock   University of York, UK

Chairs

Adam Krzyzak   Concordia University, Canada
Ching Y. Suen   Concordia University, Canada
Andrea Torsello   University Ca' Foscari of Venice, Italy

SPR Chair

Simone Scardapane   Sapienza University, Italy

SSPR Chair

Xiao Bai   Beihang University, China

Secretariat and Web Design

Nicola Nobile   Concordia University, Canada

Program Chairs

Adam Krzyzak   Concordia University, Canada
Ching Y. Suen   Concordia University, Canada
Andrea Torsello   University Ca' Foscari of Venice, Italy
Nicola Nobile   Concordia University, Canada

Program Committee

Najla Al-Qawasmeh   Concordia University, Canada
Mrouj Almuhajri   Concordia University, Canada
Ethem Alpaydin   Ozyegin University, Turkey
Juan Humberto Sossa   Instituto Politécnico Nacional, Mexico
  Azuela
Silvia Biasotti   IMATI-CNR, Italy
Luc Brun   GREYC, CNRS, ENSICAEN, France
Jeongik Cho   Concordia University, Canada
Ambra Demontis   University of Cagliari, Italy

Alessio Devoto	Sapienza Università di Roma, Italy
Pasi Fränti	University of Eastern Finland, Finland
Giorgio Fumera	University of Cagliari, Italy
Eleonora Grassucci	Sapienza Università di Roma, Italy
Daniele Grattarola	École Polytechnique Fédérale de Lausanne, France
Michal Haindl	Institute of Information Theory and Automation, Czech Republic
Paul Honeine	University of Rouen Normandy, France
Donato Impedovo	Università degli studi di Bari "Aldo Moro", Italy
Muna Khayyat	Morgan Stanley, Canada
Saeed Khazaee	Concordia University, Canada
Adam Krzyzak	Concordia University, Canada
Valerio Marsocci	Sapienza Università di Roma, Italy
Andrea Mastropietro	Sapienza Università di Roma, Italy
Nicola Nobile	Concordia University, Canada
Mauricio Orozco-Alzate	Universidad Nacional de Colombia, Colombia
Maryam Sharifi Rad	Concordia University, Canada
Jairo Rocha	University of the Balearic Islands, Spain
Luca Rossi	Queen Mary University of London, UK
Simone Scardapane	Sapienza University of Rome, Italy
Indro Spinelli	Sapienza Università di Roma, Italy
Ching Y. Suen	Concordia University, Canada
Kar-Ann Toh	Yonsei University, South Korea
Alessio Verdone	Sapienza Università di Roma, Italy
Richard Wilson	University of York, UK
Terry Windeatt	University of Surrey, UK
Michal Wozniak	Wroclaw University of Science and Technology, Poland
Jing-Hao Xue	University College London, UK
Jun Zhou	Griffith University, Australia

Sponsors

S+SSPR 2022 is proud to be sponsored by CENPARMI and other scientific, technological, and industrial partners.

Contents

Realization of Autoencoders by Kernel Methods

Shumpei Morishita[1]([⊠]) [iD], Mineichi Kudo[1] [iD], Keigo Kimura[1] [iD], and Lu Sun[2] [iD]

[1] Division of Computer Science and Information Technology, Graduate School of Information Science and Technology, Hokkaido University, Sapporo 060-0814, Japan
{morishita_s,mine,kimura5}@ist.hokudai.ac.jp
[2] School of Information Science and Technology, ShanghaiTech University, Shanghai 201210, China
sunlu1@shanghaitech.edu.cn

Abstract. An autoencoder is a neural network to realize an identity mapping with hidden layers of a relatively small number of nodes. However, the role of the hidden layers is not clear because they are automatically determined through the learning process. We propose to realize autoencoders by a set of linear combinations of kernels instead of neural networks. In this framework, the roles of the encoder and/or decoder, are explicitly determined by a user. We show that it is possible to replace almost every type of autoencoders realized by neural networks with this approach. We compare the pros and cons of this kernel approach and the neural network approach.

1 Introduction

An autoencoder is a neural network trained to attempt to copy its input to its output. Most autoencoders have a hidden layer with a small number of nodes, for extracting the latent characteristics of the original input vector. The majority are three-layer neural networks, but there exist neural networks with deep structures and convolution operators. Autoencoders have been used for a variety of goals: denoising [16], artificial sample generation [8], classification, anomaly detection [3], etc. The key point of autoencoders is the efficiency and semantics of information compression realized in the hidden layer. The functional behavior of an autoencoder can be divided into two mappings separated before and after the hidden layer. The first mapping, called an *encoder*, transforms an input vector x into a compressed vector z in the latent space. The second mapping, called a *decoder*, transforms z back to x in the original space as precisely as possible. Both are learned by backpropagation in neural networks. The largest merit is that the features (the way of encoding) are found automatically in hidden layers, as they are often better than human-crafted features for classification. Such features found by autoencoders are sometimes used even for manifold learning and for visualization [17]. However, we cannot predict, in advance and even after learning, what features are extracted and how effective they are (a black-box mechanism). To cope with this problem, in this paper, we present an alternative

A. Krzyzak et al. (Eds.): S+SSPR 2022, LNCS 13813, pp. 1–10, 2022.
https://doi.org/10.1007/978-3-031-23028-8_1

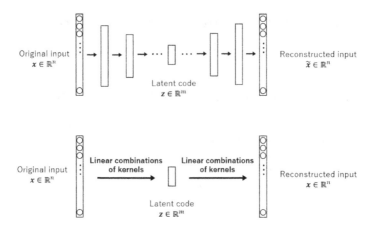

Fig. 1. Models of autoencoder realized in a neural network (above) and a set of linear combinations of kernels (below).

way to realize another type of autoencoder, and then show its applicability for several applications. We realize the autoencoders using a set of linear combinations of kernels associated with training samples both in the encoder and in the decoder (Fig. 1). It enables us to simulate perfectly any input-output relationship under an easily-satisfied condition imposed to kernels. In other words, we can control the role of the encoder and thus that of the decoder. With its high flexibility, we show that the kernel approach is available for almost all applications that have been made by neural network based autoencoders.

2   Related Work

We will use kernels for realizing autoencoders. However, such a trial is not novel. Autoencoders realized by kernels have been studied already. In [6], the encoder is realized by linear combinations of kernels associated with training samples, as we do, but the latent variables are binary and the decoder is a linear transformation of the latent vectors. The latent vectors are designed to enhance class separability. Another study [11] gives a more general framework with strong theoretical analyses, and proposes even multi-layer kernel mappings. In both studies, however, the latent variables are learned so as to optimize the mapping from input x to output $\tilde{x}$ to meet a criterion, so that they are learned from data as the same as neural net autoencoders. In other words, they are not semantically designed. Compared with those approaches, in our model, we determine the latent variables z explicitly and semantically, and then use them as the output in the encoder and as the input in the decoder. The encoder is learned from a finite set of input-output pairs $\{(x_i, z_i)\}_{i=1}^n$, and the decoder is learned from the set of the reversed pairs $\{(z_i, x_i)\}_{i=1}^n$. Note that z_i's are arbitrary given by a user as the complemental input. For example, z_i can be the mapped point of x_i by some manifold learning [12], or can be the jth standard basis e_j if x_i belongs to the

jth class in supervised learning. Unlike the above kernel based approaches, we can simulate an identity mapping perfectly over all the training pairs. In summary, our approach is different from both neural network based and previously proposed kernel based approaches in the sense that the latent variables are given explicitly and the autoencoder performs an identity mapping exactly. This gives us a large amount of flexibility and a clear semantic interpretation in the design of the latent space as a feature extractor.

3 Autoencoders by Kernel Methods

3.1 Encoder and Decoder

In our study, an encoder and a decoder are both realized as non-linear mappings using a reproducing kernel $k(\cdot, \cdot)$ [2]. The concept is as follows. We map the sample $\boldsymbol{x}$ from the original space $\mathbb{R}^M$ to a higher-dimensional space $\mathbb{R}^{\mathcal{M}}$, typically an infinite-dimensional space, using a non-linear mapping $\phi(\boldsymbol{x})$ and then map $\phi(\boldsymbol{x})$ to a lower-dimensional space $\mathbb{R}^m$, using a linear mapping :

$$\boldsymbol{x}(\in \mathbb{R}^M) \xrightarrow{\text{non-linear}} \phi(\boldsymbol{x})(\in \mathbb{R}^{\mathcal{M}}) \xrightarrow{\text{linear}} \boldsymbol{z}(\in \mathbb{R}^m).$$

Typically, $m < M$, but $m \geq M$ is also possible in this transformation, as we will do in the decoder. Specifically, using a reproducing kernel $k(\cdot, \cdot)$, we map $\boldsymbol{x}$ to a function $\phi(\boldsymbol{x}) = k(\boldsymbol{x}, \cdot)$ in the reproducing kernel Hilbert space (RKHS) $\mathcal{H}_k$ where the inner product is given by $\phi(\boldsymbol{x})^T \phi(\boldsymbol{y}) = k(\boldsymbol{x}, \boldsymbol{y})$ (kernel trick) (T denotes the transpose). We limit ourselves into $\mathrm{Span}(k(\cdot, \boldsymbol{x}_1), k(\cdot, \boldsymbol{x}_2), ..., k(\cdot, \boldsymbol{x}_n)) \subseteq \mathcal{H}_k$ and consider a function f on the subspace as a linear combination $f(\boldsymbol{x}) = \sum_{i=1}^{n} c_i k(\boldsymbol{x}, \boldsymbol{x}_i)$, where $\boldsymbol{x}_i$ is the ith training sample. Then, a collection of m such functions enables us to map $\boldsymbol{x}$ to $\boldsymbol{z} = (f_1(\boldsymbol{x}), f_2(\boldsymbol{x}), ..., f_m(\boldsymbol{x}))^T$. With a coefficient matrix $C \in \mathbb{R}^{m \times n}$ and a kernel vector $K(\boldsymbol{x}) = (k(\boldsymbol{x}, \boldsymbol{x}_1), ..., k(\boldsymbol{x}, \boldsymbol{x}_n))^T \in \mathbb{R}^n$, this can be written in a matrix form as

$$\boldsymbol{z} = CK(\boldsymbol{x}) = \begin{pmatrix} \mathbb{c}_1^T \\ \mathbb{c}_2^T \\ \vdots \\ \mathbb{c}_m^T \end{pmatrix} \begin{pmatrix} k(\boldsymbol{x}, \boldsymbol{x}_1) \\ k(\boldsymbol{x}, \boldsymbol{x}_2) \\ \vdots \\ k(\boldsymbol{x}, \boldsymbol{x}_n) \end{pmatrix} \in \mathbb{R}^m, \tag{1}$$

where $\mathbb{c}_j^T$ is the jth row in C. The coefficient matrix C determines the mapping of the training data $\{\boldsymbol{x}_i\}_{i=1}^n$ into $\{\boldsymbol{z}_i\}_{i=1}^n$:

$$Z = (\boldsymbol{z}_1, \boldsymbol{z}_2, ..., \boldsymbol{z}_n) = C\mathbb{G}_X, \tag{2}$$

where $\mathbb{G}_X$ is so-called the *Gram matrix* of the kernel:

$$\mathbb{G}_X = (K(\boldsymbol{x}_1), K(\boldsymbol{x}_2), ..., K(\boldsymbol{x}_n)) = (k(\boldsymbol{x}_i, \boldsymbol{x}_j)) \in \mathbb{R}^{n \times n}.$$

3.2   Fundamental Mapping Without Loss

Equation (1) shows a mapping from $\mathbb{R}^M$ to $\mathbb{R}^m$ given C, and Eq. (2) shows the positions where all the training samples are mapped by it. Now, suppose that $\mathbb{G}_X$ is invertible. Then, from (2), we can construct C optimally from n input-output pairs $\{(\boldsymbol{x}_i, \boldsymbol{z}_i)\}_{i=1}^n$ by $C = Z\mathbb{G}_X^{-1}$. This realizes a perfect simulation over training samples. This is formally stated as a theorem:

Theorem 1. *Given the training data $\{\boldsymbol{x}_i \in \mathbb{R}^M\}_{i=1}^n$ with their mapping points $\{\boldsymbol{z}_i \in \mathbb{R}^m\}_{i=1}^n$, if the Gram matrix of a reproducing kernel over $\{\boldsymbol{x}_i\}_{i=1}^n$ is non-singular, then a function can be realized in linear combinations of kernels (1) so as to map $\boldsymbol{x}_i$ into $\boldsymbol{z}_i$ perfectly for every $i \in \{1, 2, ..., n\}$.*

Proof. It is realized by setting $C = Z\mathbb{G}_X^{-1}$ in (2). Then, $C\mathbb{G}_X = Z\mathbb{G}_X^{-1}\mathbb{G}_X = Z$.

Fortunately, for a Gaussian kernel, it is known that $\mathbb{G}_X$ is always non-singular if all data are distinct [9], so that we can simulate perfectly any implicit function or embedding that produced $\boldsymbol{z}_i$'s from $\boldsymbol{x}_i$'s.

3.3   Kernelized Autoencoder

We exploit Theorem 1 for realizing both an encoder and a decoder :

$$\text{(Encoding)} \quad \boldsymbol{z} = C_e K_X(\boldsymbol{x}) \quad \text{where} \quad C_e = Z\mathbb{G}_X^{-1}, \tag{3}$$

$$\text{(Decoding)} \quad \tilde{\boldsymbol{x}} = C_d K_Z(\boldsymbol{z}) \quad \text{where} \quad C_d = X\mathbb{G}_Z^{-1}. \tag{4}$$

Here, $\mathbb{G}_X$ is the Gram matrix of a reproducing kernel over $\{\boldsymbol{x}_i\}_{i=1}^n$ and $\mathbb{G}_Z$ is the Gram matrix of a reproducing kernel over $\{\boldsymbol{z}_i\}_{i=1}^n$. Also, $K_X(\cdot)$ and $K_Z(\cdot)$ show the column vectors $K_X(\boldsymbol{x}) = (k(\boldsymbol{x}_1, \boldsymbol{x}), k(\boldsymbol{x}_2, \boldsymbol{x}), ..., k(\boldsymbol{x}_n, \boldsymbol{x}))^T$ and $K_Z(\boldsymbol{z}) = (k(\boldsymbol{z}_1, \boldsymbol{z}), k(\boldsymbol{z}_2, \boldsymbol{z}), ..., k(\boldsymbol{z}_n, \boldsymbol{z}))^T$, respectively. Different kernels can be used between the encoder and the decoder. The coefficient matrix C_e for the encoder is learned from $\{(\boldsymbol{x}_i, \boldsymbol{z}_i)\}_{i=1}^n$, and the coefficient matrix C_d for the decoder is learned from $\{(\boldsymbol{z}_i, \boldsymbol{x}_i)\}_{i=1}^n$. From Theorem 1, as long as $\{\boldsymbol{x}_i\}_{i=1}^n$ and $\{\boldsymbol{z}_i\}_{i=1}^n$ are distinct, respectively, the successive two mappings of (3) and (4) with the nonsingular Gram matrices realize an identity mapping perfectly over $\{(\boldsymbol{x}_i, \boldsymbol{z}_i)\}_{i=1}^n$. As a result, we have a $kernel - AE$:

$$\hat{\boldsymbol{x}} = C_d K_Z(C_e K_X(\boldsymbol{x})) \tag{5}$$

that copies an input to the output perfectly if the input $\boldsymbol{x}$ is one of training samples.

4   Comparison with Neural Networks

We have investigated the difference between neural network based autoencoders (NNet-AEs) and the proposed kernel-based autoencoders (kernel-AEs) (Table 1). The major difference is the way by which the latent vectors are determined.

Table 1. Comparison of NNet-AE and Kernel-AE. In complexity, n is the number of samples, m is the average number of nodes per layer, l is the number of layers, and t is the number of iterations.

Viewpoint	NNet-AEs	Kernel-AE (proposal)
Latent vectors	Determined by training process	Given by a user
Feature extraction	Learned automatically	Simulation of an embedding
Reconstruction	Close but not perfect	Perfect over training data
Generalization ability	Unknown	Unknown but estimated as high
Time complexity	$O(nm^2lt)$	$O(n^3)$
Amount of data for training	Large	Just as given
Generalization evaluation	None	Fast LOO

NNet-AE learns the latent vectors in an end-to-end learning manner, while kernel-AE requires them as input. A feature extractor is also learned in NNet-AE, while it is given as a simulator in kernel-AE. For the latent vectors in kernel-AE, any embedding or dimensionality reduction method are able to be used, such as PCA [1], kernel PCA [13], t-SNE [12], unsupervised or supervised Laplacian Eigenmaps (LEs) [4,15]. An unsupervised LE method [4] is used in the experiments.

The objective of realizing an identity mapping is the same, but NNet-AE realizes it approximately, while kernel-AE realizes it perfectly for training samples (Theorem 1). In the viewpoint of the model complexity, NNet-AE has a complexity proportional to the number of the edges, while kernel-AE has a complexity proportional to the number of samples. Therefore, conventional analysis would warn the risk of over-fitting for NNet-AEs with a large-scale architecture, although recent some studies deny over-fitting. Kernel-AE does not seem to over fit data because the dimensionality of the function space is the same as the number of samples as seen in (1). In addition, in kernel-AE, a fast leave-one out (LOO) technique can be used for generalization evaluation in the encoder and in the decoder [10], respectively.

The time complexity is closely related to the model complexity. In the training phase, NNet-AE needs to train all the weights in $\bar{O}(nm^2lt)$, where n is the number of samples, m is the average number of nodes per layer, l is the number of layers, and t is the number of iterations. Kernel-AE needs to calculate the inverse of the Gram matrix in $\bar{O}(n^3)$. In the testing phase, NNet-AE forwards the calculation results layer to layer, so that it needs $\bar{O}(m^2l)$ in total, while kernel-AE needs $\bar{O}(n)$ for $K(\boldsymbol{x})$ and $\bar{O}(nm)$ for the linear mapping, both in the encoder and decoder parts. In general, to draw the maximum performance from a neural network, a large number of training samples is necessary, so an NNet-AE is. In contrast, kernel-AE does work even for a small number of samples.

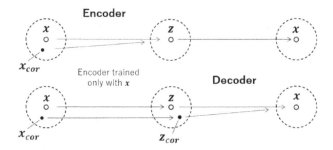

Fig. 2. Two types in kernel-DAE. Denoising is realized in encoding (above) or in decoding (below). x_{cor} is a corrupted x.

5   Applications

Two popular applications of NNet-AEs were carried out by the proposed kernel-AEs.

5.1   Denoising Autoencoders

Denoising autoencoder (DAE) receives a corrupted data and outputs the clean original data. In NNet-DAE, the encoder and decoder are learned at the same time from a set of input-output pairs $\{(x_{\text{cor},i}, x_i)\}_{i=1}^n$. On the other hand, in kernel-DAE, we can consider two types of kernel-DAEs (Fig. 2): denoising in the encoder or denoising in the decoder. In either realization, z_i's (or $z_{\text{cor},i}$'s or both) are arbitrary determined by a user.

We conducted experiments to compare NNet-DAE with those two types of kernel-DAEs. For kernel-DAE, we used a Gaussian kernel:

$$k(x, y) = \exp\left(-\frac{||x - y||^2}{\sigma^2}\right),$$

where σ^2 is the variance parameter. The parameter σ^2 is set to be proportional to the squared distance of the closest pair in the encoder, while is chosen in the decoder so as to be small but to avoid the rank deterioration of $\mathbb{G}_Z$.

The kernel-AE is learned as follows. First, we map the clear samples $\{x_i\}_{i=1}^n$ into $\{z_i \in \mathbb{R}^m\}_{i=1}^n$ by LE. In encoder-based model, we give $2n$ pairs of $\{(x_i, z_i), (x_{\text{cor},i}, z_i)\}_{i=1}^n$ to train the encoder by (3), and the n reversed pairs of $\{(z_i, x_i)\}_{i=1}^n$ to train the decoder in (4). In decoder-based model, we apply LE to the set of x_i and $x_{\text{cor},i}$ to have the set of z_i and $z_{\text{cor},i}$. Then we train the encoder by (3) with those $2n$ pairs $\{(x_i, z_i), (x_{\text{cor},i}, z_{\text{cor},i})\}_{i=1}^n$ and the decoder by (4) with $\{(z_i, x_i), (z_{\text{cor},i}, x_i)\}_{i=1}^n$.

We used MNIST image dataset [5], a collection of hand-written digits (0–9) of size 28×28 taking a pixel value in $[0, 255]$. We added a noise to each pixel randomly according to Gaussian with mean zero and standard deviation of 50. For kernel-AE, $10,000$ images and their corrupted $10,000$ images were

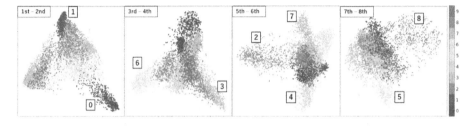

Fig. 3. Results of dimensionality reduction of MNIST to a 10-dimensional space using LE. Shown is scatter plots in some pairs of axes.

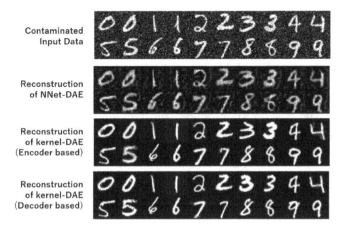

Fig. 4. Denoisng of test samples by NNet-DAE and kernel-DAEs. The images of test data are added Gaussian noise with standard deviation of 50 pixel-wisely.

used for training. For NNet-DAE, 60, 000 pairs were used for training. The used NNet-DAE is a convolutional 8 layers deep neural network [3].

The dimension m is often chosen experimentally in NNet. In kernel-AE, we set m to 10 so as to separate classes in the latent space as seen in Fig. 3. We can see that one to three classes are well separated from the other classes in each pair of features.

The result of denoising is shown in Fig. 4. The NNet-DAE outputs are blurred, whereas the kernel-DAE outputs are sharp. Therefore, the kernel-DAE is superior to the NNet-DAE in noise removal. This is because kernel-AE outputs a linear combination of the non-corrupted samples as seen in (4). On the other hand, the character form is a little different from the original character. The difference between types was ignorable in kernel-DAE.

5.2 Generative Autoencoders

Next, we constructed generative autoencoders (GAE) for generating new artificial but realistic data. Generative models aiming at this goal have gathered a lot of attention in the field of artificial intelligence in recent years. Typically we design an encoder so as to compress data in the latent space in such a way that they obey a pre-chosen probability distribution. Then we generate a random latent vector according to the distribution, and pass it to the pre-designed decoder for producing realistic data. Two typical generative models are Variational Bayes autoencoders (VAEs) and Generative Adversarial Networks (GANs) [7]. VAE designs an encoder to produce a distribution close to a specified distribution $p(z)$, typically, a Gaussian distribution. In kernel-VAE, we transform the latent vectors so as to obey a specified distribution. Specifically, we first map samples x's into z's in a 2D space by LE, then transform the distribution of z's into a standard Gaussian distribution of $\tilde{z}$'s. After that, we train the decoder such that $\tilde{z}$ is mapped into the original x. Transformation of z to $\tilde{z}$ is made by partitioning the bounding box of z's into several pieces such that each piece has almost equal probability in the radius direction. We generated some digits in MNIST dataset by NNet-VAE [8] and by kernel-VAE.

The results are shown in Fig. 5. For NNet-VAE, we adopted the architecture of five layers with the third hidden layer of two nodes ($m = 2$). Similar to the case of DAE, the NNet-VAE produced blurred samples, whereas kernel-VAE produced sharp samples. We see also more variety in the thickness and in the character form in kernel-VAE. We have also confirmed that any sample generated by kernel-VAE is out of training samples.

6   Discussion

The flexibility of kernel-AEs comes from the fact that we can arbitrarily choose latent vectors, while it can be a burden for the designer. A rough guideline of design is as follows. If the aim is classification, we can design the latent vectors such that they are separated class by class. If the aim is a generation of images, we can design the latent vectors in such a way that no blank space is left and those vectors are distributed uniformly. If the aim is not a realization of an identity mapping, such as DAE or VAE, we can do that by changing the output in the encoder or the input in the decoder or both. Since kernel-AE is a continuous mapping, we can expect that the nature of latent vectors is kept mostly in the outputs.

Latent vectors might be naturally given in some applications. For example, the discovery of new compounds from combinations of elements has been gathering a great deal of attention in material science [14]. Then, a mixture ratio vector of the possible elements is connected to the performance of the produced compound. We might be able to use such pairs in kernel-VAE with the vector representations of compounds.

The kernel-AE is not learned from data only, because it needs the corresponding latent vectors as well. Thus, in that sense, kernel-AE might not be an *auto*-encoder, rather it might be a *semi-auto*-encoder.

In this paper, we have simulated the input-output relationship by a linear transformation using kernels in (1), but it is easy to extend to an affine

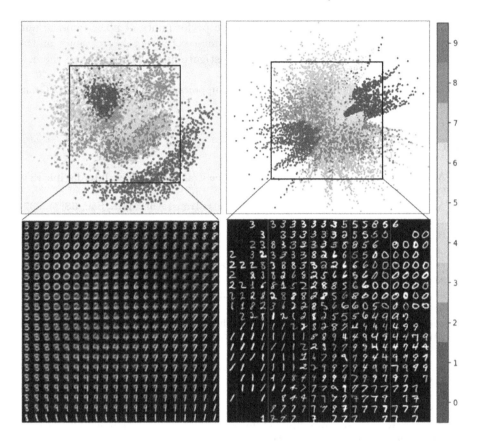

Fig. 5. NNet-VAE (left) and kernel-VAEs (right) on MNIST. In kernel-VAEs, the distributions of latent vectors (upper row) are transformed so as to obey a Gaussian (right). Artificial data is generated at grid points inside a solid square through a decoder (lower row). Only part of the region is used for generation of artificial samples.

transformation: $z = CK(x) + c_0 \, (c_0 \in \mathbb{R}^m)$. This is made by replacing Z with $Z - c_0 1^T$ (where 1 is the vector of all one's). In addition, we can relax the criterion from the (perfect) simulation to a regularized error minimization such as $\|Z - C\mathbb{G}_X\| + \lambda C\mathbb{G}C^T$. Then it suffices to replace $\mathbb{G}_X^{-1}$ with $(\mathbb{G}_X + \lambda)^{-1}$ in (3) and (4). Another merit is that we have an efficient LOO evaluation algorithm for each encoder and decoder [10].

7   Conclusion

We have realized kernel-autoencoders (kernel-AEs) by a set of linear combinations of kernels instead of neural networks. The kernel-AEs can perfectly reconstruct the training data. We compared kernel-AEs with neural network autoencoders (NNet-AEs) in denoising and in data generation. In the experiments, kernel-VAE was superior to NNet-AE in the performance of noise removal and sharp character generation. Since the performance of kernel-AEs depends on the choice of the latent vectors, we need to investigate more how to choose them.

Acknowledgment. This work was partially supported by JSPS KAKENHI (Grant Number 19H04128).

References

1. Abdi, H., Williams, L.J.: Principal component analysis. Wiley Interdiscipl. Rev. Comput. Stat. **2**(4), 433–459 (2010)
2. Aronszajn, N.: Theory of reproducing kernels. Trans. Am. Math. Soc. **68**(3), 337–404 (1950)
3. Bank, D., Koenigstein, N., Giryes, R.: Autoencoders. arXiv preprint arXiv:2003.05991 (2020)
4. Belkin, M., Niyogi, P.: Laplacian eigenmaps for dimensionality reduction and data representation. Neural Comput. **15**(6), 1373–1396 (2003)
5. Deng, L.: The mnist database of handwritten digit images for machine learning research [best of the web]. IEEE Signal Process. Mag. **29**(6), 141–142 (2012)
6. Gholami, B., Hajisami, A.: Kernel auto-encoder for semi-supervised hashing. In: 2016 IEEE Winter Conference on Applications ofCComputer Vision (WACV), pp. 1–8. IEEE (2016)
7. Goodfellow, I.J., et al.: Generative adversarial networks. arXiv preprint arXiv:1406.2661 (2014)
8. Kingma, D.P., Welling, M.: Auto-encoding variational bayes. arXiv preprint arXiv:1312.6114 (2013)
9. Knaf, H.: Kernel fisher discriminant functions - a concise and rigorous introduction. Tech. Rep. 117, Fraunhofer (ITWM) (2007)
10. Kudo, M., et alEfficient leave-one-out evaluation of kernelized implicit mappings. Accepted in S+SSPR (2022)
11. Laforgue, P., Clémençon, S., d'Alché Buc, F.: Autoencoding any data through kernel autoencoders. In: The 22nd International Conference on Artificial Intelligence and Statistics, pp. 1061–1069. PMLR (2019)
12. Van der Maaten, L., Hinton, G.: Visualizing data using t-SNE. J. Mach. Learn. Res. **9**(11), 2579-=-2605 (2008)
13. Mika, S., et al.: Kernel PCA and de-noising in feature spaces. In: NIPS, vol. 11, pp. 536–542 (1998)
14. Noh, J., et al.: Machine-enabled inverse design of inorganic solid materials: promises and challenges. Chem. Sci. **11**(19), 4871–4881 (2020)
15. Tai, M., et al.: Kernelized supervised laplacian eigenmap for visualization and classification of multi-label data. Pattern Recogn. **123**, 108399 (2022)
16. Vincent, P., et al.: Extracting and composing robust features with denoising autoencoders. In: Proceedings of the 25th International Conference on Machine Learning, pp. 1096–1103. ICML2008, Association for Computing Machinery, New York, NY, USA (2008)
17. Wang, W., et al.: Generalized autoencoder: a neural network framework for dimensionality reduction. In: Proceedings of the IEEE Conference on Computer Vision and Pattern Recognition (CVPR) Workshops (2014)

Maximal Independent Vertex Set Applied to Graph Pooling

Stevan Stanovic[1]($\boxtimes$) ⓘ, Benoit Gaüzère[2] ⓘ, and Luc Brun[1] ⓘ

[1] Normandie University, ENSICAEN, CNRS, UNICAEN,
GREYC UMR 6072, 14000 Caen, France
{stevan.stanovic,luc.brun}@ensicaen.fr
[2] Normandie University, INSA de Rouen, University of Rouen, University of Le Havre,
LITIS EA 4108, 76800 Saint-Étienne-du-Rouvray, France
benoit.gauzere@insa-rouen.fr

Abstract. Convolutional neural networks (CNN) have enabled major advances in image classification through convolution and pooling. In particular, image pooling transforms a connected discrete grid into a reduced grid with the same connectivity and allows reduction functions to take into account all the pixels of an image. However, a pooling satisfying such properties does not exist for graphs. Indeed, some methods are based on a vertex selection step which induces an important loss of information. Other methods learn a fuzzy clustering of vertex sets which induces almost complete reduced graphs. We propose to overcome both problems using a new pooling method, named MIVSPool. This method is based on a selection of vertices called surviving vertices using a Maximal Independent Vertex Set (MIVS) and an assignment of the remaining vertices to the survivors. Consequently, our method does not discard any vertex information nor artificially increase the density of the graph. Experimental results show an increase in accuracy for graph classification on various standard datasets.

Keywords: Graph Neural Networks · Graph pooling · Graph classification · Maximal independant vertex set

1 Introduction

Convolutional neural networks (CNN) have enabled major advances in image classification. An image can be defined as a connected discrete grid and this property enables to define efficient convolution and pooling operations. Nevertheless, social networks, molecules or traffic infrastructures are not represented by a grid but by graphs. Convolution and pooling have been adapted to these structured data into Graph Neural Networks (GNN) [11]. The adaptation of convolution to graphs can be performed by using learned aggregation functions which combine the value of each vertex with the ones of its neighborhood [6,7]. For graph pooling, the adaptation is mainly performed by selecting vertices [2,5,8,15] or by learning a fuzzy clustering [13].

The work reported in the paper was supported by French ANR grant #ANR-21-CE23-0025 CoDeGNN.

A. Krzyzak et al. (Eds.): S+SSPR 2022, LNCS 13813, pp. 11–21, 2022.
https://doi.org/10.1007/978-3-031-23028-8_2

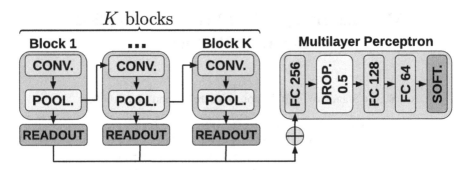

Fig. 1. General architecture of our GNN. Each block is composed of a convolution layer followed by a pooling layer. Features learned after each block are combined to have several levels of description of the graph.

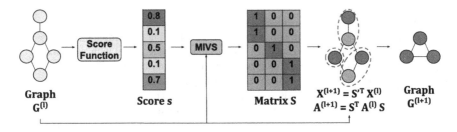

Fig. 2. Proposed graph pooling. Each node is associated to a score (Sect. 3.2). Based on this score, a MIVS is computed from which a reduction matrix S is derived. Applying S to both feature and structure leads to a reduced graph $G^{(l+1)}$.

In this paper, we detail graph convolution and pooling as well as the construction of the reduced graph in Sect. 2. Section 3 presents our method (Fig. 2) and explains its differences from other pooling methods. We present in Sect. 4 our experiments where we study different heuristics to select surviving vertices and the mean complexity of our pooling algorithm. We finally compare our method to other pooling strategies on standard datasets using an unified architecture (Fig. 1) in Sect. 4.

2   Related Work

GNNs are neural networks applied to graphs. Inspired by CNNs, they are like the latter based on convolution and pooling operations. Let us note that convolution operation should be permutation equivariant. Intuitively, this condition states that a convolution operation may be permuted with a permutation of the graph's nodes.

Convolution operations diffuse the information by a message passing mechanism which allows to learn a representation for each node by aggregating the information of their n-hop neighborhood. Optimized according to a particular task, the resulting nodes' features can be used as input of a node classifier or regressor. According to [1], current aggregation functions correspond mainly to low-pass filters.

2.1   Graph Pooling

Considering graph level tasks like graph classification, we need to compute a graph representation as a fixed sized vector by summing node's representation. This operation is called global pooling. Global pooling realizes an aggregation of all the vertices of a graph and summarizes it in a single vector. Such aggregation must be permutation invariant and is usually performed using basic operators like a sum, mean, maximum or more complex ones [14].

However, global pooling performed on all the vertices of a graph leads to sum up an unbounded amount of information (proportional to the graph's size) into a fixed size vector and thus potentially induces a large amount of information loss. Many authors [8,13] have proposed to decompose GNNs in several steps alternating convolution and pooling to reduce the graph size while averaging vertices values. Such methods are called hierarchical pooling. The choice of the number of vertices for the reduced graph can be fixed or adapted according to the original graph's size with a ratio. The hierarchical pooling, like the convolution operation should be permutation equivariant.

Notations. For any pooling layer l, we consider a graph $\mathcal{G}^{(l)} = (\mathcal{V}^{(l)}, \mathcal{E}^{(l)})$ where $\mathcal{V}^{(l)}$ and $\mathcal{E}^{(l)}$ are respectively the set of vertices and the set of edges of the graph. Let $n_l = |\mathcal{V}^{(l)}|$, we can define $\mathbf{A}^{(l)} \in \mathbb{R}^{n_l \times n_l}$ the adjacency matrix associated to the graph $\mathcal{G}^{(l)}$ where $\mathbf{A}_{ij}^{(l)} = 1$ if it exists an edge between the vertices i and j, 0 otherwise. We also note $\mathbf{X}^{(l)} \in \mathbb{R}^{n_l \times f_l}$ the feature matrix of the graph $\mathcal{G}^{(l)}$ where f_l is the dimension of nodes' attributes.

Construction of the Set of Surviving Vertices. It exists multiple methods to reduce the size of a graph within the GNN framework. However, most of these methods lead to the construction of a reduction matrix $S^{(l)} \in \mathbb{R}^{n_l \times n_{l+1}}$ where n_l and n_{l+1} are respectively the sizes of the original and the reduced graph. This matrix is used to define the attributes and the adjacency matrix of the reduced graph. Each surviving vertex i contributing to its own cluster, we suppose that $S_{i,i}^{(l)} = 1$.

Construction of Attributes and Adjacency Matrix. The reduction matrix $S^{(l)}$ is the basis of the construction of the reduced graph. Based on $S^{(l)}$, the following two equations allow this construction:

$$X^{(l+1)} = S^{(l)\top} X^{(l)} \tag{1}$$

This last equation defines the attribute of each surviving vertex v_i as a weighted sum of the attributes of the vertices v_j of $G^{(l)}$ such that $S_{ji}^{(l)} \neq 0$.

$$A^{(l+1)} = S^{(l)\top} A^{(l)} S^{(l)} \tag{2}$$

Equation (2) can be rewritten as follows for any pair of surviving vertex (i, j):

$$(S^{(l)\top} A^{(l)} S^{(l)})_{i,j} = \sum_{k,l}^{n_l} A_{k,l}^{(l)} S_{k,i}^{(l)} S_{l,j}^{(l)} \tag{3}$$

Two surviving vertices are therefore adjacent in the reduced graph if they are adjacent in the initial graph ($A_{i,j}^{(l)} = S_{i,i}^{(l)} = S_{j,j}^{(l)} = 1$). Moreover, surviving vertices i and j are adjacent in the reduced graph if it exists a pair of non-surviving adjacent vertices (k, l) assigned respectively to i and j ($A_{k,l}^{(l)} = S_{k,i}^{(l)} = S_{l,j}^{(l)} = 1$).

Families of Methods. Pooling methods can be divided in two families. First family consists of methods based on the selection of surviving vertices on a given criteria. This criteria can be the result of a combinatorial algorithm [2] or a learning step like in Top-k methods [5,8]. The second family regroups methods based on a node's clustering as in DiffPool [13]. Each cluster is associated to a surviving vertex.

Methods of the second group use a fixed number of clusters. Hence, learning $S^{(l)}$ does not allow these methods to take into account the variable size and topology of the graphs. Moreover, due to training, such methods produce dense matrices $S^{(l)}$ with few nonzero values. Equation (3) shows that the matrix $A^{(l+1)}$ is then dense and the corresponding graph has a density close to 1 (i.e. will be a complete graph or almost complete graph). Consequently, the structure of the graph is not respected. We say that $S^{(l)}$ is *complete*.

For Top-k methods, we define the reduction matrix $S^{(l)}$ as the restriction of the identity matrix I_{n_l} to the column indices idx corresponding to the surviving vertices:

$$S^{(l)} = [I_{n_l}]_{:,idx} \tag{4}$$

Given a surviving vertex i in the original graph, we have thus $S_{i,\phi(i)}^{(l)} = 1$, where $\phi(i)$ denotes the column index of i in the reduced matrix $S^{(l)}$. All other entries of $S^{(l)}$ are set to 0. The matrix $S^{(l)}$ is called *selective* since it selects the attributes of the surviving vertices and removes those of non-surviving vertices. Moreover, in this case, rows of $S^{(l)}$ corresponding to non-surviving vertices are equal to 0 and two surviving vertices will be adjacent if and only if they were adjacent before the reduction. This last point may induce the creation of disconnected reduced graphs. Moreover, the drop of non-survivors' features leads to an important loss of information. Let us note that MVPool [15] increases the density of the graph by considering power 2 or 3 of the adjacency matrix in order to limit the disconnections of the reduced graph. It additionally adds edges to the reduced graph through an additional layer called Structure Learning.

An alternative solution consists to drop Eq. (2) and to perform a Kron reduction [2] in order to connect all pairs of surviving vertices adjacent to a same removed vertex. This reduction increases the density of the graph and a sparsification step is required. This last point can creates a disconnected reduced graph. Moreover, the time complexity of the Kron reduction is approximately $\mathcal{O}((\frac{n_l}{2})^3)$, due to the inversion of a part of the Laplacian matrix of $\mathcal{G}^{(l)}$.

Let us finally note that it exists a graph pooling method which uses an approximation of a Maximum Independent Vertex Set called MEWIS [10]. The layer used to compute this approximation significantly increases the complexity of the whole GNN. Moreover, the approximated result provided by this layer does not guarantee that the resulting set of selected vertices is even maximal.

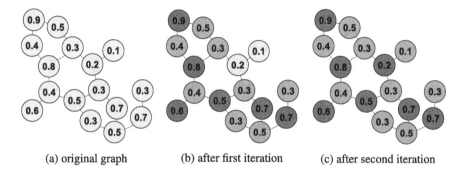

(a) original graph   (b) after first iteration   (c) after second iteration

Fig. 3. Evolution of MIVS algorithm on a graph from the NCI1 dataset (Sect. 4.1). Number inside each vertex corresponds to its score s. Candidate, survivor and non-survivor vertices are respectively denoted by: ○, ●, ◕.

Unlike other methods, we propose to preserve the structure of the original graph as well as the attribute information. We satisfy these properties thanks to a selection of surviving vertices distributed equally on the graph and with an assignment of non-surviving vertices to surviving vertices. We named our pooling method MIVSPool.

3 Proposed Method

3.1 Maximal Independent Vertex Set (MIVS)

Before describing the algorithm, we introduce the notion of Maximal Independent Set and show how this notion can be applied to graph vertices.

Maximal Independent Set. Let $\mathcal{X}$ be a finite set and $\mathcal{N}$ a neighborhood function. A subset $\mathcal{J}$ of $\mathcal{X}$ is independent if:

$$\forall (x, y) \in \mathcal{J}^2 : x \notin \mathcal{N}(y) \tag{5}$$

$\mathcal{J}$ is a subset of $\mathcal{X}$ such that for any $(x, y) \in \mathcal{J}^2$, x and y are not neighbors. The elements of $\mathcal{J}$ are called the surviving elements.

An independent set $\mathcal{J}$ is said to be maximal when no element can be added to it without breaking the independence property, i.e., when we have:

$$\forall x \in \mathcal{X} - \mathcal{J}, \exists y \in \mathcal{J} : x \in \mathcal{N}(y) \tag{6}$$

Equation (6) states that each non-surviving element has to be in the neighborhood of at least one element of $\mathcal{J}$. The elements of $\mathcal{J}$ are denoted as survivors.

Using Eqs. (5) and (6), we have a Maximal Independent Set. We note that it is a maximal but not necessarily a maximum. Indeed, our Maximal Independent Set $\mathcal{J}$ is not necessarily those whose cardinality is maximum with respect to all Maximal Independent Sets of $\mathcal{X}$.

If we interpret the construction of a Maximal Independent Set as a subsampling operation, Eq. (5) can be interpreted as a condition preventing the oversampling (two adjacent vertices cannot be simultaneously selected) which thus guarantees an uniform distribution of survivors. Conversely, Eq. (6) prevents subsampling: Any non-selected vertex is at a distance 1 from a surviving vertex.

Maximal Independent Vertex Set. A Maximal Independent Vertex Set (MIVS) [9] of a graph $\mathcal{G}^{(l)} = (\mathcal{V}^{(l)}, \mathcal{E}^{(l)})$ is therefore a Maximal Independent Set where the neighborhood is deduced from the edge set $\mathcal{E}^{(l)}$. Adapting Eqs. (5) and (6), we select surviving vertices $\mathcal{V}^{(l+1)}$ as described below:

$$\forall (v, v') \in (\mathcal{V}^{(l+1)})^2 : (v, v') \notin \mathcal{E}^{(l)} \tag{7}$$

$$\forall v \in \mathcal{V}^{(l)} - \mathcal{V}^{(l+1)}, \exists v' \in \mathcal{V}^{(l+1)} : (v, v') \in \mathcal{E}^{(l)} \tag{8}$$

Equations (7) and (8) define the MIVS procedure and state that two surviving vertices cannot be neighbors and a non-surviving vertex must have at least one survivor in its neighborhood. Nevertheless, these two equations don't explain how to select these vertices.

Meer's Algorithm. A simple method has been proposed by Meer [9] using an iterative procedure assigning a uniform random variable $\mathcal{U}([0, 1])$ to each vertex. In this procedure, each surviving vertex corresponds at a local maximum of the random variable with respect to its neighbors. According to Eq. (7), vertices adjacent to these survivors are labeled as non-survivors. Other vertices (not yet labeled) are labeled candidates and the algorithm iterates the selection of surviving and non-surviving vertices on this reduced vertex set (Fig. 3). The algorithm convergence is guaranteed since, at each iteration, at least one candidate is labeled as a survivor. For each vertex v_i, its random variable is denoted by x_i while Booleans p_i and q_i are $True$ if v_i is respectively a survivor and a candidate. The stopping criterion of the algorithm is obtained when all q_i are $False$, i.e. when Eq. (8) is satisfied. At the initialisation of the algorithm, all p_i are $False$ and all q_i are $True$ and, for each iteration k, variables $p_i^{(k)}$ and $q_i^{(k)}$ are updated by:

$$p_i^{(k+1)} = p_i^{(k)} \vee (q_i^{(k)} \wedge (x_i = \max\{x_j | (v_i, v_j) \in \mathcal{E}^{(l)} \wedge q_j^{(k)}\}))$$
$$q_i^{(k+1)} = \wedge_{j | (v_i, v_j) \in \mathcal{E}^{(l)}} \overline{p}_j^{(k+1)} \tag{9}$$

In other words, a survivor at iteration $k + 1$ is a survivor at iteration k or is a candidate whose random variable (x_i) is greater than the ones of its candidate neighbors. A vertex is candidate at iteration $k + 1$ if it is not adjacent to a survivor. We note that the neighborhood includes the central vertex. This procedure only involves local computations and is therefore parallelizable.

3.2   Adaptation of MIVS to Deep Learning

Scoring System. Using Meer's algorithm [9], the uniform random variable $\mathcal{U}([0, 1])$ plays an important role in the selection of the surviving vertex set. We propose to modify this variable so that x_i used in the algorithm is learnt and represents the relevance of vertex v_i. We obtain this last property by using an attention mechanism like in Sag-Pool [8] where a score vector $s \in \mathbb{R}^{n_l \times 1}$ is returned by a GCN [7]. Using this score in our MIVS, we select a set of surviving vertices $\mathcal{V}^{(l+1)}$ corresponding to local maxima of the function of interest encoded by s. An uniform distribution of vertices on the graph is guaranteed by the computation of the MIVS.

Assignment of Non-surviving Vertices. In order to construct our reduction matrix $S^{(l)}$ and take all vertices information into account, we need to assign non-surviving vertices. As a reminder, Eq. (8) states that each non-survivor has at least a surviving neighbor. Assuming that the score function encodes the relevance of each vertex, we assign each non-surviving vertex to its surviving neighbor with the highest score. We obtain a reduction matrix $S^{(l)}$ with n_{l+1} clusters corresponding to surviving vertices:

$$\forall v_j \in \mathcal{V}^{(l)} - \mathcal{V}^{(l+1)}, \exists! v_i \in \mathcal{V}^{(l+1)} | S^{(l)}_{ji} = 1$$
$$\text{with } i = argmax(s_k, (v_k, v_j) \in \mathcal{E}^{(l)} \wedge p^N_k) \tag{10}$$

where N is the number of iterations required to have a MIVS and s the score vector.

Construction of Reduced Attributes and Adjacency Matrix. The construction of attributes of the reduced graph $\mathcal{G}^{(l+1)}$ is obtained thanks to an average weighted by s from the set of aggregated vertices to the surviving vertices:

$$X^{(l+1)}_i = \frac{1}{\sum_{j|S^{(l)}_{ji}=1} s_j} \sum_{j|S^{(l)}_{ji}=1} s_j X^{(l)}_j \tag{11}$$

Equation (11) allows to take into account the importance of each vertex in the computation of the attributes of the reduced graph and put more attention on vertices with a high score. As a consequence the learnt vector s can be interpreted as a relevance value. This operation can be achieved by a transformation similar to Eq. (1) by substituting $S^{(l)}$ by the matrix $S^{(l)'} = D_2 D_1 S^{(l)}$ with $D_1 = diag(s)$ and $D_2 = diag(\frac{1}{1^\top D_1 S^{(l)}})$. The construction of the reduced adjacency matrix $A^{(l+1)}$ is obtained thanks to Eq. (2).

Relaxation of MIVSPool. On some highly dense graphs, MIVSPool may provide a decimation ratio lower than 0.5. In order to correct this point, we additionally introduce MIVSPool$_{comp.}$ which is a MIVSPool where we force the addition of surviving vertices using a Top-k so that the pooling ratio remains always equal to 0.5. Note that Eq. (7) is then no more valid while Eq. (8) still holds.

4 Experiments

4.1 Datasets

To evaluate our MIVSPool, we test it on a benchmark of four standard datasets: D&D [4], PROTEINS [3,4], NCI1 [12] and ENZYMES [3]. The statistics of datasets are reported on Table 1. D&D and PROTEINS describe proteins and the aim is to classify them as enzyme or non-enzyme. Nodes represent the amino acids and two nodes are connected by an edge if they are less than 6 Ångström apart. NCI1 describes molecules and the purpose is to classify them as cancerous or non-cancerous. Each vertex stands for an atom and edges between vertices represent bonds between atoms. ENZYMES is a dataset of protein tertiary structures obtained from the BRENDA enzyme database.

Table 1. Statistics of datasets

Dataset	#Graphs	#Classes	Avg $|\mathcal{V}|$	Avg $|\mathcal{E}|$
D&D	1178	2	284 ± 272	715 ± 694
PROTEINS	1113	2	39 ± 46	72 ± 84
NCI1	4110	2	29 ± 13	32 ± 14
ENZYMES	600	6	33 ± 15	62 ± 26

4.2 Model Architecture and Training Procedure

The model architecture consists of K blocks made up of a GCN [7] convolution followed by a graph pooling. A Readout layer is applied after each block using a concatenation of the average and the maximum of vertices' features matrix $X^{(l)}$. At the end of our network, the K Readout are summed and the result is sent to a Multilayer Perceptron. The latter is composed by three fully connected layers (respectively 256, 128 and 64 for the number of hidden neurons) and a droupout of 0.5 is applied between the first two. The classification is obtained by a Softmax layer (see Fig. 1).

For the training procedure, we use Pytorch Geometric and we evaluate our neural network with a 10-fold cross validation. We repeat this procedure ten times without setting the seed. The dataset is split in three parts: 80% for the training set, 10% for the validation set and 10% for the test set.

For the hyperparameters, we use the Adam optimizer, set the dimension of node representation, the batch size and the number of epochs respectively at 128, 512 and 1000 and an early stopping is applied if the validation loss did not improved after 100 epochs. A grid search is used for the learning rate within $\{1e^{-3}, 1e^{-4}, 1e^{-5}\}$, the weight decay within $\{1e^{-3}, 1e^{-4}, 1e^{-5}\}$ and the number of blocks K in $[3, 5]$ to find the best configuration.

4.3 Ablation Studies

Scoring Function. encodes the relevance of each vertex and as a direct influence on the set of selected surviving vertices by MIVSPool. We vary it using a uniform random variable $\mathcal{U}([0, 1])$ like in Meer [9] (MIVSPool$_{rand}$), a trainable normalized projection vector on the features $X^{(l)}$ like in [5] (MIVSPool$_{Top-k}$), a self-attention mechanism thanks to a graph convolution like in SagPool [8] (MIVSPool$_{SagPool}$) and finally a

Table 2. Study of MIVSPool according to the score function

Score function	MIVSPool$_{rand}$	MIVSPool$_{Top-k}$	MIVSPool$_{SagPool}$	MIVSPool$_{MVPool}$
D&D	77.04 ± 0.63	77.10 ± 1.00	75.10 ± 0.76	$\mathbf{77.38 \pm 0.94}$
PROTEINS	75.15 ± 0.44	75.36 ± 0.60	75.11 ± 0.74	$\mathbf{75.62 \pm 0.47}$
NCI1	72.16 ± 0.55	$\mathbf{73.82 \pm 0.94}$	73.72 ± 0.71	72.97 ± 0.71
ENZYMES	45.80 ± 1.35	37.55 ± 1.94	38.68 ± 2.81	$\mathbf{46.80 \pm 1.53}$

Table 3. Averaged number of iteration of MIVS for each pooling step.

Dataset	Pooling 1	Pooling 2	Pooling 3	Pooling 4	Pooling 5
D&D	4.1 ± 0.6	3.6 ± 0.6	3.2 ± 0.6	-	-
PROTEINS	3.1 ± 0.7	2.7 ± 0.6	2.3 ± 0.5	-	-
NCI1	3.2 ± 0.5	2.9 ± 0.5	2.5 ± 0.5	2.3 ± 0.5	2.1 ± 0.3
ENZYMES	3.2 ± 0.6	2.7 ± 0.6	2.4 ± 0.5	-	-

trainable multi-view system across structure and features graph information like in MVPool [15] (MIVSPool$_{MVPool}$). For trainable variations, Eq.(11) allows to propagate the training to the next convolution and, therefore, our pooling method is end-to-end trainable. The results reported Table 2 show that the choice of the score function has a minor influence on the accuracy. Note that, since Top-k and SagPool trainable score functions don't consider structural information, they consequently perform poorly on ENZYMES. However, the use of the MVPool trainable score function allows to obtain the best accuracy on three datasets with comparable standard deviations to other heuristics. In the rest of our article, we choose to take MIVSPool$_{MVPool}$ as a reference that we denote MIVSPool.

Average Number of Iterations Required by MIVSPool for Each Pooling Step.
Table 3 presents the mean number of iterations for each dataset and each pooling step computed over 10 epochs. Note that the number of pooling steps is determined using K-folds for each dataset. Despite an important difference between graph sizes among datasets (Table 1), we note that the number of iterations is comparable between them and less than 5. Knowing that an iteration of MIVS on a graph $\mathcal{G} = (\mathcal{V}, \mathcal{E})$ requires about $|\mathcal{V}|d_{max}$ computations, where d_{max} is the maximum degree of $\mathcal{G}$, the computation of MIVS on one of the graphs of our three datasets is bounded by $5|\mathcal{V}|d_{max}$. This complexity is less than the one needed by the Kron transformation (Sect. 2.1).

4.4 Comparison of MIVSPool According to Other Methods

We compare our method to three state-of-art methods: gPool [5], SagPool [8] and MVPool-SL [15]. Results in Table 4 show that MIVSPool$_{comp.}$ and MIVSPool obtain the highest or second highest accuracies on D&D, PROTEINS and ENZYMES. This point confirms the efficiency of MIVSPool and the fact that some configurations (highly connected graphs) may induce low decimation rate by MIVSPool which slightly decreases the accuracy. For NCI1, the highest accuracy is obtained by MVPool-SL [15], MIVSPool$_{comp.}$ being ranked second. Let us note that the mean vertex's degree in this dataset is 2.15 ± 0.11 with many graphs being almost linear. On such simplified topology considering two hops has done by MVPool-SL allows to recover the structure of the reduced graph.

Table 4. Comparison of MIVSPool according to other hierarchical methods. Highest and second highest accuracies are respectively in **bold** and blue.

Dataset	gPool [5]	SagPool [8]	MVPool-SL [15]	MIVSPool	MIVSPool$_{comp.}$
D&D	75.76 ± 0.82	75.92 ± 0.92	77.26 ± 0.37	77.38 ± 0.94	**77.88 ± 0.73**
PROTEINS	73.49 ± 1.44	74.30 ± 0.62	75.04 ± 0.67	75.62 ± 0.47	**75.81 ± 0.75**
NCI1	71.66 ± 1.04	72.86 ± 0.68	**74.79 ± 0.58**	72.97 ± 0.71	73.44 ± 0.68
ENZYMES	36.93 ± 2.45	35.30 ± 1.80	39.45 ± 2.57	**46.80 ± 1.53**	45.60 ± 2.37

5   Conclusion

Our graph pooling method MIVSPool is based on a selection of surviving vertices thanks to a Maximal Independent Vertex Set (MIVS) and an assignment of non-surviving vertices to surviving ones. Unlike state-of-art methods, our method allows to preserve the totality of graph information during its reduction.

Acknowledgements. The work was performed using computing resources of CRIANN (Normandy, France).

References

1. Balcilar, M., Renton, G., Héroux, P., Gaüzère, B., Adam, S., Honeine, P.: Analyzing the expressive power of graph neural networks in a spectral perspective. In: International Conference on Learning Representations (2021)
2. Bianchi, F.M., Grattarola, D., Livi, L., Alippi, C.: Hierarchical representation learning in graph neural networks with node decimation pooling. IEEE Trans. Neural Networks Learn. Syst. **33**(5), 2195–2207 (2022)
3. Borgwardt, K.M., Ong, C.S., Schönauer, S., Vishwanathan, S., Smola, A.J., Kriegel, H.P.: Protein function prediction via graph kernels. Bioinformatics **21**(suppl_1), 47–56 (2005)
4. Dobson, P.D., Doig, A.J.: Distinguishing enzyme structures from non-enzymes without alignments. J. Mol. Biol. **330**(4), 771–783 (2003)
5. Gao, H., Ji, S.: Graph u-nets. In: International Conference on Machine Learning, pp. 2083–2092. PMLR (2019)
6. Hamilton, W., Ying, Z., Leskovec, J.: Inductive representation learning on large graphs. Adv. Neural. Inf. Process. Syst. **30**, 1024–1034 (2017)
7. Kipf, T.N., Welling, M.: Semi-supervised classification with graph convolutional networks. In: International Conference on Learning Representations (ICLR) (2017)
8. Lee, J., Lee, I., Kang, J.: Self-attention graph pooling. In: International Conference on Machine Learning, pp. 3734–3743. PMLR (2019)
9. Meer, P.: Stochastic image pyramids. Compu. Vis. Graphics Image Process. **45**(3), 269–294 (1989)
10. Nouranizadeh, A., Matinkia, M., Rahmati, M., Safabakhsh, R.: Maximum entropy weighted independent set pooling for graph neural networks. ArXiv abs/2107.01410 (2021)
11. Scarselli, F., Gori, M., Tsoi, A.C., Hagenbuchner, M., Monfardini, G.: The graph neural network model. IEEE Trans. Neural Netw. **20**(1), 61–80 (2008)
12. Wale, N., Watson, I.A., Karypis, G.: Comparison of descriptor spaces for chemical compound retrieval and classification. Knowl. Inf. Syst. **14**(3), 347–375 (2008)

13. Ying, Z., et al.: Hierarchical graph representation learning with differentiable pooling. Adv. Neural. Inf. Process. Syst. **31**, 4805–4815 (2018)
14. Zhang, M., Cui, Z., Neumann, M., Chen, Y.: An end-to-end deep learning architecture for graph classification. Proc. AAAI Conf. Artif. Intell. **32**(1), 4438–4445 (2018)
15. Zhang, Z., et al.: Hierarchical multi-view graph pooling with structure learning. IEEE Trans. Knowl. Data Eng. 35, 545–559 (2021)

Annotation-Free Keyword Spotting in Historical Vietnamese Manuscripts Using Graph Matching

Anna Scius-Bertrand[1,2,3(✉)], Linda Studer[1,2], Andreas Fischer[1,2], and Marc Bui[3]

[1] iCoSys, HES-SO, Fribourg, Switzerland
{anna.scius-bertrand,linda.studer,andreas.fischer}@hefr.ch
[2] DIVA, University of Fribourg, Fribourg, Switzerland
[3] Ecole Pratique des Hautes Etudes, Paris, France
marc.bui@ephe.psl.eu

Abstract. Finding key terms in scanned historical manuscripts is invaluable for accessing our written cultural heritage. While keyword spotting (KWS) approaches based on machine learning achieve the best spotting results in the current state of the art, they are limited by the fact that annotated learning samples are needed to infer the writing style of a particular manuscript collection. In this paper, we propose an annotation-free KWS method that does not require any labeled handwriting sample but learns from a printed font instead. First, we train a deep convolutional character detection system on synthetic pages using printed characters. Afterwards, the structure of the detected characters is modeled by means of graphs and is compared with search terms using graph matching. We evaluate our method for spotting logographic Chu Nom characters on the newly introduced Kieu database, which is a historical Vietnamese manuscripts containing 719 scanned pages of the famous Tale of Kieu. Our results show that search terms can be found with promising precision both when providing handwritten samples (query by example) as well as printed characters (query by string).

Keywords: Historical documents · Keyword spotting · Annotation-free · Kieu database · Chu Nom characters · Character detection · Handwriting graphs · Hausdorff edit distance

1 Introduction

Despite strong progress in the past two decades, automated reading of historical handwriting remains a challenging open problem in the field of pattern recognition [5]. One of the main obstacles is the large variety of scripts, languages, writing instruments, and writing materials that need to be modeled. The current state of the art follows the paradigm of learning by examples using machine learning techniques, leading to high accuracy for automatic transcription under

A. Krzyzak et al. (Eds.): S+SSPR 2022, LNCS 13813, pp. 22–32, 2022.
https://doi.org/10.1007/978-3-031-23028-8_3

the condition that a large amount of annotated handwriting samples are available for some manuscript collections. Especially for ancient languages, which are only known by few human experts, obtaining annotated samples is time-consuming and thus hinders an automatic transcription of historical documents at large scale.

As an alternative to producing a full transcription, keyword spotting (KWS) has been proposed to find specific search terms in historical manuscripts [6]. Although the number of annotated learning samples can be reduced when focusing only on a few search terms, it has become evident over the past decade that the learning by examples paradigm is still needed to achieve high precision for KWS. Prominent examples include the use of annotated word images for training word embeddings with deep convolutional neural networks (CNN) [13] and the use of annotated text line images for training character models with hidden Markov models (HMM) or long short-term memory networks (LSTM) [15].

For historical Vietnamese manuscripts, transfer learning from printed fonts to handwritten Chu Nom characters has recently been shown to be feasible for automatic transcription [9], reducing the amount of annotated handwriting samples required during training. For the tasks of character detection (localizing characters, without recognition) and transcription alignment (localizing characters of a known transcription), it was even possible to completely replace handwritten learning samples with printed ones by means of self-calibration and unsupervised clustering algorithms, leading to fully annotation-free detection [11] and alignment [12] methods, respectively.

In this paper, we go a step further and investigate to what extent annotation-free KWS is feasible for historical Vietnamese manuscripts when considering only printed fonts during training. The proposed method consists of two components. First, a deep convolutional neural network is trained on synthetic printed pages to detect characters. Secondly, a handwritten sample of the search term (query by example) or a printed sample (query by string) is compared with the detected characters to retrieve similar instances. Such an approach is highly valuable for an initial exploration of historical document collections, because it operates directly on the scanned page images and only requires printed fonts of the Chu Nom characters to make the handwritten content searchable. Furthermore, it allows to gather basic statistics about the document collection (number of columns, characters, etc.) and facilitates ground truth creation, especially the annotation of handwritten characters.

To compare images of the logographic Chu Nom characters, we leverage methods from structural pattern recognition [2,14] to focus on the core structure of the writing and thus supporting the transfer from printed to handwritten characters. Specifically, we consider keypoint graphs extracted from character skeleton images [3] and compare them efficiently by means of the Hausdorff edit distance (HED) [4], a quadratic-time approximation (lower bound) of the graph edit distance. A similar approach has already proven successful for signature verification [8] and for spotting handwritten words in Latin script [1]. When compared with [1], the proposed method has two main advantages. First, it is segmentation-free, i.e. it does not require pre-segmented word images. Secondly,

Fig. 1. Kieu database

it is fully annotation-free, i.e. even the parameters of graph extraction and HED are optimized on a set of printed characters instead of using a human-annotated validation set. It is also noteworthy that it is the first time that this structural approach is investigated for non-Latin characters. Hopefully, the publication of the Chu Nom character dataset from the books of Kieu used in this work will be a valuable contribution to the community.

The remainder of the paper is structured as follows. Section 2 describes the Kieu database, a newly introduced research dataset used for experimental evaluation. Section 3 details the proposed KWS method, Sect. 4 reports the experimental results, and Sect. 5 draws some conclusions.

2 Kieu Database

The Kieu database consists of five books with versions of the Tale of Kieu, one of the most famous stories of Vietnam. The versions date from 1866, 1870, 1871, 1872 and 1902, respectively, and are written in Chu Nom. This logographic script was used in Vietnam from the 10th to the 20th century before it was replaced with a Latin-based alphabet. Figure 1 illustrates an example page for each of the five books. The images and transcriptions used to create the database were kindly provided by the Vietnamese Nom Preservation Foundation[1].

We used the transcription alignment method from [12] to align the transcriptions with the page images and manually verified the results, discarding pages that contained errors. The resulting Kieu database[2] consists of 719 pages with a total of 97,152 characters annotations, i.e. bounding boxes and unicode encodings, for 13,207 unique characters. The images retrieved from the webpage have a relatively low resolution, introducing an additional challenge for KWS when combined with the large number of unique characters.

3 Annotation-Free Keyword Spotting (KWS)

An overview of the proposed annotation-free KWS method is provided in Fig. 2. First, a YOLO-based deep convolutional character detection system is trained

---

[1] http://www.nomfoundation.org.

[2] Available here: https://github.com/asciusb/Kieu-database.

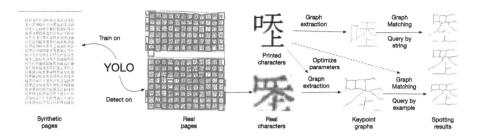

Fig. 2. Overview of the workflow of our proposed keyword spotting method

on synthetic pages with printed Chu Nom characters. Therefore, the training does not require human annotations of real page images. The trained network is then applied to real page images to extract characters. Typical errors at this stage include missing character parts and false positives, especially in the border regions as illustrated in Fig. 2.

Next, the character images are represented by means of keypoint graphs, and HED-based graph matching with query graphs is performed to retrieve the most similar samples. Both the parameters of graph extraction and graph matching are optimized with validation experiments that take only graphs from printed Chu Nom characters into account. Therefore, the parameter optimization does not require human annotations of real character images.

The query graphs can be obtained either from real character images (query by example) or printed characters (query by string). The former approach requires the user to manually select one or several templates of the query term on real page images. The latter allows to enter the search term in form of plain text and is thus completely independent of the scanned manuscripts.

In the following, we provide more details on the different steps of our method.

3.1 Synthetic Dataset Creation

We use five fonts to generate images of printed characters, namely Nom Na Tong Light, Han Nom A et B, Han Nom Gothic, Han Nom Minh, and Han Nom Kai. The synthetic training pages are created by writing columns of random printed Chu Nom characters on a white background surrounded by a black border. Afterwards, a series of image transformations are applied for data augmentation, including changes in brightness, applying Gaussian blur, and adding salt and pepper noise. An example is illustrated in Fig. 2.

3.2 Character Detection

For character detection, we consider the YOLO [10] (You Only Look Once) architecture for one-stage object detection with deep convolutional neural networks. More specifically, we employ the latest version YOLOv5 [7], which integrates several improvements over the original architecture that are important for character detection, including a higher resolution, multi-scale features, and multi-scale anchor boxes that allow to detect also small characters in low-resolution images.

3.3 Graph Extraction

Following the procedure proposed in [3], keypoint graphs are extracted from the detected character images, real and synthetic, as illustrated in Fig. 2. First, a local edge enhancement by means of a Difference of Gaussians (DoG) is performed before applying a global threshold for binarization. Afterwards, the binary image is thinned to one pixel width and three types of keypoints are located on the skeleton image: endpoints, junction points, and a random point on circular structures. To obtain a labeled keypoint graph $g = (V, E)$, all keypoints are added to the set of nodes V with coordinate labels (x, y). Afterwards, the skeleton is sampled at distance D pixels between the keypoints, adding further nodes to V. Unlabeled edges are added to the set of edges E for each pair of nodes that is directly connected on the skeleton.

We introduce another parameter for graph extraction to cope with low-resolution images, i.e. a scaling factor $S > 1$, which is applied to resize (upscale) the character images at the beginning of the process, before applying DoG and binarization. This super-resolution allows to insert even more nodes between two keypoints than the number of pixels in the original image, thus emphasizing the structure even for very small strokes, which are important to distinguish similar Chu Nom characters.

3.4 Graph Matching

We use the graph edit distance (GED) to compare Chu Nom characters based on their keypoint graphs. GED calculates the minimum transformation cost between two graphs with respect to cost of node deletion ($u \rightarrow \epsilon$), node insertion ($\epsilon \rightarrow v$), node label substitution ($u \rightarrow v$), edge deletion ($s \rightarrow \epsilon$), and edge insertion ($\epsilon \rightarrow t$). To overcome the computational constraints of exact the GED, which is NP-complete, we use the Hausdorff edit distance (HED) [4] to compute an approximation (lower bound) in quadratic time:

$$HED_c(g_1, g_2) = \sum_{u \in V_1} \min_{v \in V_2 \cup \{\epsilon\}} f_c(u, v) + \sum_{v \in V_2} \min_{u \in V_1 \cup \{\epsilon\}} f_c(u, v) \tag{1}$$

where c is the cost function for the edit operations and $f_c(u, v)$ the cost for assigning node u to node v, taking into account their adjacent edges as well.

We use the Euclidean cost function, i.e. constant costs c_V and c_E

$$\begin{aligned} c(u \rightarrow \epsilon) = c(\epsilon \rightarrow v) = c_V \\ c(s \rightarrow \epsilon) = c(\epsilon \rightarrow t) = c_E \end{aligned} \tag{2}$$

for node and edge deletion and insertion, and the Euclidean distance

$$c(u \rightarrow v) = ||(x_u, y_u) - (x_v, y_v)|| \tag{3}$$

for node label substitution.

3.5 Keyword Spotting (KWS)

To create a keyword, one or several template graphs $\mathcal{T} = \{t_1, \ldots, t_m\}$ are extracted either from real character images or from synthetic images using printed fonts. Then, for spotting a keyword, the template graphs are compared with each character graph $c_1, \ldots, c_n$ in the document collection, and scored according to the minimum HED, such that the most similar character have the lowest score.

$$score(c_i) = \min_{t \in \mathcal{T}} d(c_i, t) \tag{4}$$

4 Experimental Evaluation

To evaluate the proposed annotation-free KWS method, we conduct a series of experimental evaluations on the five books of the Kieu database. In a first step, we optimize the different meta-parameters without using human annotations. In a second step, we perform an an ablation study to explore the limitations of the method by gradually rendering the task more difficult, while keeping the parameters fixed.

For the performance evaluation, the character graphs of the document collection are sorted according to Eq. 4 to compute recall and precision for each possible score threshold. For each keyword, the average precision is computed (AP) and the mean average precision (mAP) is considered as the final performance measure when spotting N keywords:

$$mAP = \frac{1}{N} \sum_{i=1}^{N} AP_i. \tag{5}$$

4.1 Task Setup and Parameter Optimization

We compare the following three scenarios in terms of what data (real vs. synthetic font characters) is used for parameter and model selection:

– **Annotated-only.** As a baseline approach, we use the first 5 pages of each book to select keyword templates, the next 5 pages as a validation set to optimize the parameters, and the next 5 pages as a test set to evaluate the spotting performance with the best parameter configuration. All keywords are considered that have at least 3 templates and that appear at least once in the validation and test set, respectively. Table 1 shows the size of the test sets for each book and the number of keywords that are spotted.
– **Font-validation.** To avoid the need for expert annotations for parameter optimization, we replace the validation set with synthetic printed characters. 20 random characters are printed in 5 Chu Nom fonts to obtain keyword templates. Another 900 random non-keyword characters are printed with a random font, leading to a synthetic validation set with a total of 1,000 characters.

Table 1. Overview of the test sets of the Kieu database.

	1866	1870	1871	1872	1902
Pages	5	5	5	5	5
Characters	840	490	840	700	700
Keywords	80	36	93	68	63

Table 2. Chosen meta-parameters after optimization.

Parameter	Annotated-only	Font-validation
Scale	3	4
Norm	z-score	z-score
Node cost c_V	1.0	0.1
Edge cost c_E	1.0	2.0

- **Annotation-free.** To fully avoid expert annotations, we use the best parameters obtained by font-validation and perform an automatic character detection on the 5 test pages instead of using the ground truth bounding boxes. This setup corresponds to the KWS method proposed in Sect. 3.

The parameters optimization and selection is performed as follows:

- **Character detection parameters.** One fixed setup is tested for YOLO-based character detection. We use the default configuration[3] of the medium-sized YOLOv5m model with COCO-pretrained weights, and an initial learning rate of 0.0032. A total of 30,000 synthetic pages using printed Chu Nom characters (see Sect. 3.1 and 3.2) are used for training over 25 epochs until convergence.
- **Graph extraction parameters.** Six scaling factors $S \in \{1, 2, \ldots, 6\}$ are tested in combination with a fixed node distance of $D = 3$ pixels to explore different degrees of super-resolution (see Sect. 3.3). Furthermore, three setups are tested for normalizing the node labels, i.e. using the raw coordinates, centering to zero mean, and normalizing to zero mean and unit variance (z-score).
- **Graph matching parameters.** 25 combinations of node and edge costs (c_V, c_E) are tested (see Sect. 3.4). For the raw and centered coordinates, we investigate the range of $c_V, c_E \in \{1.0, 5.0, 10.0, 15.0, 20.0\}$ and for the z-score normalized ones we consider $c_V, c_E \in \{0.1, 0.5, 1.0, 1.5, 2.0\}$.

4.2 Results

Table 2 indicates the optimal parameter values obtained with respect to the mAP on the annotated validation set (annotation only), and on the synthetic

---

[3] github.com/ultralytics/yolov5, commit cc03c1d5727e178438e9f0ce0450fa6bdbbe1ea7.

Table 3. Mean average precition (mAP) on the test set after parameter optimization. The best results are highlighted in bold font.

	1866	1870	1871	1872	1902	Average
Annotated-only	**0.79**	0.93	0.67	0.68	**0.74**	0.76
Font-validation	**0.79**	**0.97**	**0.73**	**0.70**	**0.74**	**0.78**
Annotation-free	**0.79**	0.94	**0.73**	0.66	0.71	0.77

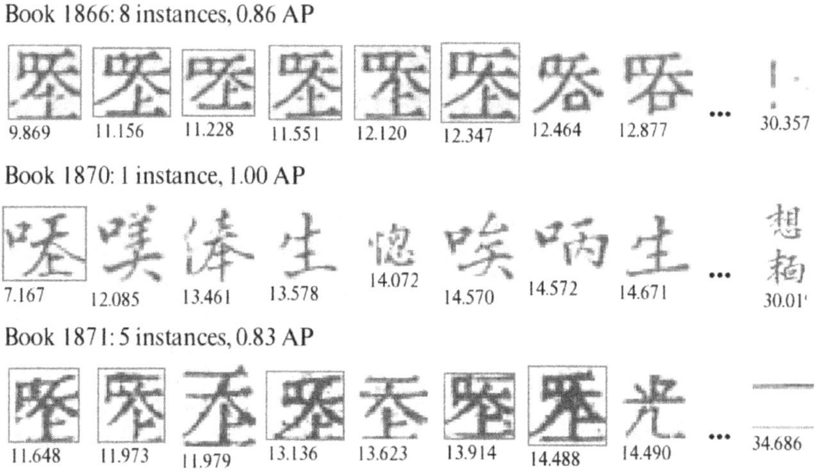

Book 1866: 8 instances, 0.86 AP

Book 1870: 1 instance, 1.00 AP

Book 1871: 5 instances, 0.83 AP

Fig. 3. Exemplary spotting results for the character "word". Correct retrieval results are marked in green. The top-8 results as well as the character with the worst score are shown, ranked according to the Hausdorff edit distance (indicated below). (Color figure online)

validation set with printed characters (font-validation). In both cases, a super-resolution is preferred, highlighting details of small strokes in the Chu Nom characters by inserting additional nodes. Also, normalizing the coordinates to zero mean and unit variance is beneficial (z-score), removing small variations in position and scale among the characters. The optimal values for node and edge deletion and insertion are a different when optimizing on printed characters instead of real ones. The font-validation suggests an emphasis on the edges, thus increasing the importance of the character structure.

Table 3 shows the mAP results achieved on the test set for each book individually, and on average over all books. For annotated-only, we used the 5 annotated validation pages of book 1866 for parameter optimization. When compared with font-validation, we observe a slight overfitting to this book, whereas the font-validation parameters generalize better to all five books, achieving 0.78 mAP. When using the same parameters as for font-validation, but detecting the characters automatically instead of using the ground truth bounding boxes, we report

Table 4. Ablation study. Mean average precision (mAP) on the test set for different keyword spotting tasks with increasing difficulty.

Query by	# Pages	# Templ.	Book version					Average
			1866	870	1871	1872	1902	
Example	5	3	0.79	0.94	0.73	0.66	0.71	0.77
	10	3	0.77	0.92	0.73	0.60	0.74	0.75
	20	3	0.73	0.92	0.70	0.56	0.74	0.73
	5	3	0.77	0.93	0.71	0.63	0.66	0.74
	5	1	0.70	0.92	0.64	0.54	0.62	0.68
String	5	1	0.65	0.73	0.59	0.61	0.59	0.63

between 0.66 mAP (book 1872) and 0.94 mAP (book 1870), resulting in 0.77 mAP on average for fully annotation-free KWS.

Figure 3 illustrates exemplary spotting results on the test sets when searching for the character "word". For each book, the number of instances of the character and the achieved average precision (AP) is reported. For example, in book 1870, the only present instance is retrieved with the smallest HED, thus leading to a perfect spotting result of 1.0 AP. The retrieved character with the lowest HED is typically a detection error, such as elements of the page border, partial characters, merged characters, or page background. Thus, errors at the character detection stage do not have a negative impact on the KWS.

4.3   Ablation Study

After obtaining strong results with the initial setup of three real keyword templates, we gradually increase the difficulty of the KWS task in an ablation study. Table 4 shows the results. When increasing the number of test pages from 5 to 20, we include more characters, which may be similar to the keywords. Nevertheless, the spotting performance is only reduced by 0.04 mAP on average. When using only a single keyword template, however, the decrease of 0.09 mAP is already more significant.

Remarkably, our proposed method is also able to perform query by string KWS using printed templates of the keyword, highlighting the effectiveness of the structural representation. However, there is a decrease in mAP of 0.14, illustrating the need to model the variability of the handwriting in order to achieve the best results.

5   Conclusions

We have introduced a new approach to spot keywords in historical Vietnamese manuscripts, which is directly applicable to a collection of scanned page images without requiring human annotations. The structural pattern recognition method for KWS is able to achieve a promising performance of 0.77 mAP for query by example and 0.63 mAP for query by string on the Kieu database. Thus, our method is ideally suited for an initial exploration of manuscript collections, especially because of its ability to perform knowledge transfer from printed to handwritten characters.

Future research includes improvements of the spotting method by means of geometric deep learning for graph-based representations, data augmentation for modeling variations in the handwriting, and a possible generalization of the method to other scripts and languages.

References

1. Ameri, M.R., Stauffer, M., Riesen, K., Bui, T.D., Fischer, A.: Graph-based keyword spotting in historical manuscripts using Hausdorff edit distance. Pattern Recogn. Lett. **121**, 61–67 (2019)
2. Conte, D., Foggia, P., Sansone, C., Vento, M.: Thirty years of graph matching in pattern recognition. Int. J. Pattern Recogn. Artif. Intell. **18**(3), 265–298 (2004)
3. Fischer, A., Riesen, K., Bunke, H.: Graph similarity features for HMM-based handwriting recognition in historical documents. In: Proceedings of the International Conference on Frontiers in Handwriting Recognition, pp. 253–258 (2010)
4. Fischer, A., Suen, C.Y., Frinken, V., Riesen, K., Bunke, H.: Approximation of graph edit distance based on Hausdorff matching. Pattern Recogn. **48**(2), 331–343 (2015)
5. Fischer, A., Liwicki, M., Ingold, R. (eds.): Handwritten Historical Document Analysis, Recognition, and Retrieval - State of the Art and Future Trends. World Scientific, Singapore (2020)
6. Giotis, A.P., Sfikas, G., Gatos, B., Nikou, C.: A survey of document image word spotting techniques. Pattern Recogn. **68**, 310–332 (2017)
7. Jocher, G., et al.: ultralytics/yolov5: v4.0 - nn.SILU() activations, weights & biases logging, PyTorch hub integration (2021). https://doi.org/10.5281/ZENODO.4418161
8. Maergner, P., et al.: Combining graph edit distance and triplet networks for offline signature verification. Pattern Recogn. Lett. **125**, 527–533 (2019)
9. Nguyen, K.C., Nguyen, C.T., Nakagawa, M.: Nom document digitalization by deep convolution neural networks. Pattern Recogn. Lett. **133**, 8–16 (2020)
10. Redmon, J., Divvala, S., Girshick, R., Farhadi, A.: You only look once: unified, real-time object detection. In: Proceedings of the International Conference on Computer Vision and Pattern Recognition (CVPR), pp. 779–788 (2016)
11. Scius-Bertrand, A., Jungo, M., Wolf, B., Fischer, A., Bui, M.: Annotation-free character detection in historical Vietnamese stele images. In: Proceedings of the 16th International Conference on Document Analysis and Recognition (ICDAR), pp. 432–447 (2021)
12. Scius-Bertrand, A., Jungo, M., Wolf, B., Fischer, A., Bui, M.: Transcription alignment of historical Vietnamese manuscripts without human-annotated learning samples. Appl. Sci. **11**(11), 4894 (2021)
13. Sudholt, S., Fink, G.A.: PHOCNet: a deep convolutional neural network for word spotting in handwritten documents. In: Proceedings of the 15th International Conference on Frontiers in Handwriting Recognition, pp. 277–282 (2016)

14. Vento, M.: A one hour trip in the world of graphs, looking at the papers of the last ten years. In: Proceedings of the International Workshop on Graph-Based Representation, pp. 1–10 (2013)
15. Vidal, E., Toselli, A.H., Puigcerver, J.: A probabilistic framework for lexicon-based keyword spotting in handwritten text images. CoRR abs/2104.04556 (2021)

Interactive Generalized Dirichlet Mixture Allocation Model

Kamal Maanicshah[1](✉)📧, Manar Amayri[2]📧, and Nizar Bouguila[1]📧

[1] Concordia Institute of Information and Systems Engineering, Concordia University,
1455 Boulevard de Maisonneuve O, Montreal, Quebec H3G 1M8, Canada
k_mathin@live.concordia.ca, nizar.bouguila@concordia.ca
[2] G-SCOP Lab, Grenoble Institute of Technology, 8185 Grenoble, France
manar.amayri@grenoble-inp.fr

Abstract. A lot of efforts have been put in recent times for research in the field of natural language processing. Extracting topics is undoubtedly one of the most important tasks in this area of research. Latent Dirichlet allocation (LDA) is a widely used model that can perform this task in an unsupervised manner efficiently. It has been proved recently that using priors other than Dirichlet can be advantageous in extracting better quality topics from the data. Hence, in our paper we introduce the interactive latent generalized Dirichlet allocation model to extract topics. The model infers better topics using little information provided by the users through interactive learning. We use a variational algorithm for efficient inference. The model is validated against text datasets based on extracting topics related to news categories and types of emotions to test its efficiency.

Keywords: Topic models · Interactive learning · Generalized dirichlet allocation · Mixture models · Variational learning

1 Introduction

LDA is one of the most important models for topic learning tasks in the field of natural language processing [2]. The model helps us to learn the composition of documents or sentences to gain insights on the topics representing that particular document. Classical algorithms like LDA and mixture models are still widely used owing to the interpretability of these models. Deep learning models which have been considered to give optimal results in recent years lack this feature.

Knowing that all data cannot be properly modelled with the assumption of Dirichlet prior for the topics proportions research has been conducted replacing it with other alternatives. Generalized Dirichlet prior is one such alternative explored in recent research [1]. The generalized Dirichlet (GD) distribution has a general covariance structure as opposed to a negative one in the case of Dirichlet. This gives GD distribution a better capability to fit the data. Motivated by the advantages and results obtained in earlier research, in our paper we use the generalized Dirichlet allocation model.

A. Krzyzak et al. (Eds.): S+SSPR 2022, LNCS 13813, pp. 33–42, 2022.
https://doi.org/10.1007/978-3-031-23028-8_4

In [4], the authors introduce the concept of a mixture of LDA, giving rise to latent Dirichlet mixture allocation (LDMA) which gives a better approximation of the topics. The presence of mixture model gives additional support to the topics proportions which leads to the extraction of more relevant topics. Hence, following in the footsteps of this paper, we use the latent generalized Dirichlet mixture model (LGDMA) in our work. The quality of topics learned by this approach is further improved by integrating an interactive algorithm similar to [8]. The algorithm gives partial control to the user to modify the topic proportions based on coherence score. In our case we use flexible criteria that can be tailored to the dataset. This human knowledge helps us to derive more accurate topics from the data.

To infer the parameters of our model, we follow an atypical approach of variational inference used in [6] as opposed to the method followed in [2] which is generally used. The variational approach gives us an edge over pure Bayesian methods like Gibbs sampling as convergence is guaranteed in the former. It also helps us to steer away from the problem of getting stuck at local minima induced by maximum likelihood estimation (MLE).

The rest of the paper is organized as follows: Sect. 2 introduces out interactive latent generalized Dirichlet mixture allocation (iLGDMA) model. The variational inference approach for estimating the parameters are explained in Sect. 3 followed by the interactive learning algorithm in Sect. 4. The experiments performed and the results obtained are discussed in Sect. 5. We conclude our findings in Sect. 6.

2 Model Description

Consider a corpus containing D documents. For each document d, we have a word vector $\vec{w}_d = (w_{d1}, w_{d2}, ..., w_{dN_d})$, where N_d is the number of words in document d. Every word n can also be represented by a V dimensional vector representing the number of words in the vocabulary with the condition $w_{dnv} = 1$ when w_{dn} is the vth word in the vocabulary and $w_{dnv} = 0$ otherwise. The affinity that a word n from a document d is associated with a particular topic k is represented by a latent variable $\mathcal{Z} = \{\vec{z}_d\}$, where $\vec{Z}_d = (\vec{z}_{d1}, \vec{z}_{d2}, ..., \vec{z}_{dN_d})$. For K topics, $\vec{z}_{dn}$ is in turn a K dimensional vector with $z_{dnk} = 1$ if the word is identified to be from topic k and 0 otherwise. The topic probabilities concerning each word in the vocabulary to a topic k is given by the vector $\vec{\beta}_k = (\beta_{k1}, \beta_{k2}, ..., \beta_{kV})$. This variable will be modified to suit our interactive learning algorithm in Sect. 4. The prior for the topic proportions $\vec{\theta}_d$ in our case is a mixture of generalized Dirichlet components with parameters $\vec{\sigma} = \{\sigma_l\} = \{\sigma_{lk}\}$ and $\vec{\tau} = \{\tau_l\} = \{\tau_{lk}\}$ where, $l = 1, 2, ..., L$ indicates one of the L components of the mixture model. $\mathcal{Y} = (\vec{y}_1, \vec{y}_2, ..., \vec{y}_D)$ is the indicator matrix which represents the component to which the document might belong to. Hence, $\vec{y}_d = (y_{d1}, y_{d2}, ..., y_{dL})$ which is a one hot encoded vector with 1 wherever the document d belongs to the cluster L and 0 if not. The mixing parameters of the mixture model is given by, $\vec{\pi} = (\pi_1, \pi_2, ..., \pi_L)$ obeying the rules; $\sum_l^L \pi_l = 1$ and $0 \leq \pi_l \leq 1$. Suppose, we are given a dataset $W = \{\vec{w}_d\}$ containing the word vectors of D documents, we can write the marginal likelihood of our model as,

$$p(W \mid \vec{\pi}, \vec{\sigma}, \vec{\tau}, \vec{\beta}) = \prod_{d=1}^{D} \int \left[\left(\sum_{y_d} p(\vec{\theta}_d \mid y_d, \vec{\sigma}, \vec{\tau}) p(\vec{y}_d \mid \vec{\pi}) \right) \right.$$
$$\left. \times \prod_{n=1}^{N_d} \sum_{z_{dn}} p(w_{dn} \mid z_{dn}, \vec{\beta}) p(z_{dn} \mid \vec{\theta}_d) \right] d\vec{\theta}_d \qquad (1)$$

In Eq. 1, the terms corresponding to the mixture model is given by,

$$p(\vec{\theta}_d \mid \vec{y}_d, \vec{\sigma}, \vec{\tau}) p(\vec{y}_d \mid \vec{\pi}) = \prod_{l=1}^{L} \left[\pi_l \prod_{k=1}^{K} \frac{\Gamma(\tau_{lk} + \sigma_{lk})}{\Gamma(\tau_{lk})\Gamma(\sigma_{lk})} \theta_{dk}^{\sigma_{lk}-1} \left(1 - \sum_{j=1}^{k} \theta_{dj} \right)^{\gamma_{lk}} \right]^{y_{dl}} \qquad (2)$$

Here, $\gamma_{lk} = \tau_{lk} - \tau_{l(k+1)} - \sigma_{l(k+1)}$ for $k = 1, 2, ..., K-1$ and $\gamma_{lk} = \sigma_{lk} - 1$ for $k = K$. Since, $\vec{\beta}$ is the term concerning the words belonging to a particular topic, to integrate our interactive algorithm, we split it into two separate terms, $\vec{\beta}_o$ which is the probability generated by our iLGDMA model and $\vec{\beta}_u$ is the user defined probability obtained through our algorithm. The algorithm is described in detail in Sect. 4. However, since $\vec{\beta}_o$ is the only variable to be calculated, we continue our definition of the model describing the combined probability as $\vec{\beta}$ for clarity. By definition, we can write $p(w_{dn} \mid z_{dn}, \vec{\beta})$ and $p(z_{dn} \mid \vec{\theta}_d)$ as,

$$p(w_{dn} \mid z_{dn}, \vec{\beta}) = \prod_{k=1}^{K} \left(\prod_{v=1}^{V} \beta_{kv}^{w_{dnv}} \right)^{z_{dnk}}; \qquad p(z_{dn} \mid \vec{\theta}_d) = \prod_{k=1}^{K} \theta_{dk}^{z_{dnk}} \qquad (3)$$

Introducing Gamma priors has been found to be an efficient method to provide better fit to the data [7]. Since GD does not have a conjugate prior, using Gamma priors would be a wise choice [7]. The Gamma prior for the parameter $\vec{\sigma}$ is given by,

$$p(\sigma_{lk}) = \mathcal{G}(\sigma_{lk} \mid \upsilon_{lk}, \nu_{lk}) = \frac{\nu_{lk}^{\upsilon_{lk}}}{\Gamma(\upsilon_{lk})} \sigma_{lk}^{\upsilon_{lk}-1} e^{-\nu_{lk}\sigma_{lk}} \qquad (4)$$

where $\mathcal{G}(\cdot)$ indicates a Gamma distribution. Similarly, we can write the prior for $\vec{\tau}$ as $p(\tau_{lk}) = \mathcal{G}(\tau_{lk} \mid s_{lk}, t_{lk})$. Following the suggestion from [2], we follow variational smoothing to handle sparse data by assuming a Dirichlet prior over the parameter $\vec{\beta}$, given by,

$$p(\vec{\beta}_k \mid \vec{\lambda}_k) = \frac{\Gamma(\sum_{v=1}^{V} \lambda_{kv})}{\prod_{v=1}^{V} \Gamma(\lambda_{kv})} \prod_{v=1}^{V} \beta_{kv}^{\lambda_{kv}-1} \qquad (5)$$

Assuming another GD distribution over $\vec{\theta}$ as a variational prior helps us to simplify the derivations. Hence, we define,

$$p(\vec{\theta}_d \mid \vec{g}_d, \vec{h}_d) = \prod_{k=1}^{K} \frac{\Gamma(g_{dk} + h_{dk})}{\Gamma(g_{dk})\Gamma(h_{dk})} \theta_{dk}^{g_{dk}-1} \left(1 - \sum_{j=1}^{k} \theta_{dj} \right)^{\zeta_{dk}} \qquad (6)$$

provided, $\zeta_{dk} = h_{dk} - g_{d(k-1)} - h_{d(k-1)}$ while $k \leq K - 1$ and $\zeta_{dk} = h_{dk} - 1$ when $k = K$ respectively. Following these assumptions and consolidating the corresponding parameters of the model as $\Theta = \{\mathcal{Z}, \vec{\beta}, \vec{\theta}, \vec{\sigma}, \vec{\tau}, \vec{y}\}$ we can write the posterior distribution $p(W, \Theta)$ as,

$$
\begin{aligned}
p(W, \Theta) &= p(W \mid \mathcal{Z}, \vec{\beta}, \vec{\theta}, \vec{\sigma}, \vec{\tau}, \vec{y}) \\
&= p(\vec{W} \mid \mathcal{Z}, \vec{\beta}) p(\vec{z} \mid \vec{\theta}) p(\vec{\theta} \mid \vec{\sigma}, \vec{\tau}, \vec{y}) p(\vec{y} \mid \vec{\pi}) p(\vec{\theta} \mid \vec{g}, \vec{h}) p(\vec{\beta} \mid \vec{\lambda}) \\
&\quad \times p(\vec{\sigma} \mid \vec{v}, \vec{\nu}) p(\vec{\tau} \mid \vec{s}, \vec{t})
\end{aligned}
\tag{7}
$$

3   Variational Inference

The rudimentary principle behind variational inference is to assume a variational posterior distribution $Q(\Theta)$ and approximate it to be almost equal to the real posterior distribution given by $p(W \mid \Theta)$. We can accomplish this by calculating the Kullback-Leibler (KL) divergence between the two distributions and minimizing it until the value reaches 0. The closer the value is to 0, more similar the two distributions are to each other. So, the KL divergence between the two distributions $Q(\Theta)$ and $p(W \mid \Theta)$ is given by,

$$
KL(Q \parallel P) = -\int Q(\Theta) \ln\left(\frac{p(W \mid \Theta)}{Q(\Theta)}\right) d\Theta = \ln p(W) - \mathcal{L}(Q)
\tag{8}
$$

where, $\mathcal{L}(Q) = \int Q(\Theta) \ln\left(\frac{p(W,\Theta)}{Q(\Theta)}\right) d\Theta$ is the lower bound. The KL divergence moves closer to zero with increasing lower bound. So our objective is to find optimal variational solutions which maximize the lower bound. To make the inference easier, we use the mean field theory [11] to arrive at a tractable posterior with the assumption that the parameters involved are independent and identically distributed. This gives $Q(\Theta) = \prod_{j=1}^{J} \Theta_j$, provided $j = 1, 2, ..., J$ represents the parameters of the model. Now, the variational solution for each parameter Θ_j can be found by using the expectations of all other parameters excluding Θ_j with the equation:

$$
Q_j(\Theta_j) = \frac{\exp\langle \ln p(W, \Theta)\rangle_{\neq j}}{\int \exp\langle \ln p(W, \Theta)\rangle_{\neq j} d\Theta}
\tag{9}
$$

The variational solutions thus found are:

$$
Q(\vec{y}) = \prod_{d=1}^{D} \prod_{l=1}^{L} r_{dl}^{y_{dl}}, \quad Q(\mathcal{Z}) = \prod_{d=1}^{D} \prod_{N=1}^{N_d} \prod_{k=1}^{K} \phi_{dnk}^{z_{dnk}}
\tag{10}
$$

$$
Q(\vec{\sigma}) = \prod_{l=1}^{L} \prod_{k=1}^{K} \frac{\nu_{lk}^{* v_{lk}^*}}{\Gamma(v_{lk}^*)} \sigma_{lk}^{v_{lk}^* - 1} e^{-\nu_{lk}^* \sigma_{lk}}
\tag{11}
$$

$$Q(\vec{\tau}) = \prod_{l=1}^{L} \prod_{k=1}^{K} \frac{t_{lk}^{*s_{lk}^{*}}}{\Gamma(s_{lk}^{*})} \tau_{lk}^{s_{lk}^{*}-1} e^{-t_{lk}^{*}\tau_{lk}} \tag{12}$$

$$Q(\vec{\beta}) = \prod_{k=1}^{K} \prod_{v=1}^{V} \frac{\Gamma(\sum_{v=1}^{V} \lambda_{kv}^{*})}{\prod_{v=1}^{V} \Gamma(\lambda_{kv}^{*})} \beta_{kv}^{\lambda_{kv}^{*}-1} \tag{13}$$

$$Q(\vec{\theta}) = \prod_{d=1}^{D} \prod_{k=1}^{K} \frac{\Gamma(g_{dk}^{*}+h_{dk}^{*})}{\Gamma(g_{dk}^{*})\Gamma(h_{lk}^{*})} \theta_{dk}^{g_{dk}^{*}-1} \left(1 - \sum_{j=1}^{k} \theta_{dj}\right)^{\zeta_{dk}^{*}} \tag{14}$$

where,

$$r_{dl} = \frac{\rho_{dl}}{\sum_{l=1}^{L} \rho_{dl}}, \phi_{dnk} = \frac{\delta_{dnk}}{\sum_{k=1}^{K} \delta_{dnk}}, \pi_l = \frac{1}{D} \sum_{d=1}^{D} r_{dl} \tag{15}$$

$$\rho_{dl} = exp\left\{ \ln \pi_l + \mathcal{R}_l + \sum_{k=1}^{K} (\sigma_{lk}-1)\langle \ln \theta_{dk} \rangle + \gamma_{lk}\left\langle 1 - \sum_{j=1}^{k} \theta_{dj} \right\rangle \right\} \tag{16}$$

$$\delta_{dnk} = exp(\langle \ln \beta_{kv} \rangle + \langle \ln \theta_{dk} \rangle) \tag{17}$$

Here, $\mathcal{R}$ is the taylor series approximations of $\left\langle \ln \frac{\Gamma(\sigma+\tau)}{\Gamma(\sigma)\Gamma(\tau)} \right\rangle$ since it is intractable [5] and is given by,

$$\begin{aligned}
\mathcal{R} = & \ln \frac{\Gamma(\bar{\sigma}+\bar{\tau})}{\Gamma(\bar{\sigma})\Gamma(\bar{\tau})} + \bar{\sigma}\left[\Psi(\bar{\sigma}+\bar{\tau}) - \Psi(\bar{\sigma})\right](\langle \ln \sigma \rangle - \ln \bar{\sigma}) \\
& + \bar{\tau}\left[\Psi(\bar{\sigma}+\bar{\tau}) - \Psi(\bar{\tau})\right](\langle \ln \tau \rangle - \ln \bar{\tau}) \\
& + 0.5\bar{\sigma}^2\left[\Psi'(\bar{\sigma}+\bar{\tau}) - \Psi'(\bar{\sigma})\right]\langle(\ln \sigma - \ln \bar{\sigma})^2\rangle \\
& + 0.5\bar{\tau}^2\left[\Psi'(\bar{\sigma}+\bar{\tau}) - \Psi'(\bar{\tau})\right]\langle(\ln \tau - \ln \bar{\tau})^2\rangle \\
& + \bar{\sigma}\bar{\tau}\Psi'(\bar{\sigma}+\bar{\tau})(\langle \ln \sigma \rangle - \ln \bar{\sigma})(\langle \ln \tau \rangle - \ln \bar{\tau})
\end{aligned} \tag{18}$$

$$\begin{aligned}
\upsilon_{lk}^{*} = & \upsilon_{lk} + \sum_{d=1}^{D} \langle y_{dl} \rangle \left[\Psi(\bar{\sigma}_{lk} + \bar{\tau}_{lk}) - \Psi(\bar{\sigma}_{lk}) \right. \\
& \left. + \bar{\tau}_{lk}\Psi'(\bar{\sigma}_{lk} + \bar{\tau}_{lk})(\langle \ln \tau_{lk} \rangle - \ln \bar{\tau}_{lk}) \right] \bar{\sigma}_{lk}
\end{aligned} \tag{19}$$

$$\begin{aligned}
s_{lk}^{*} = & s_{lk} + \sum_{d=1}^{D} \langle y_{dl} \rangle \left[\Psi(\bar{\tau}_{lk} + \bar{\sigma}_{lk}) - \Psi(\bar{\tau}_{lk}) \right. \\
& \left. + \bar{\sigma}_{lk}\Psi'(\bar{\tau}_{lk} + \bar{\sigma}_{lk})(\langle \ln \sigma_{lk} \rangle - \ln \bar{\sigma}_{lk}) \right] \bar{\tau}_{lk}
\end{aligned} \tag{20}$$

$$\upsilon_{lk}^{*} = \upsilon_{lk} - \sum_{d=1}^{D} \langle y_{dl} \rangle \langle \ln \theta_{dk} \rangle \tag{21}$$

$$t_{lk}^* = t_{lk} - \sum_{d=1}^{D} \langle y_{dl} \rangle \left\langle \ln \left[1 - \sum_{j=1}^{K} \theta_{dj} \right] \right\rangle \tag{22}$$

$$g_{dk}^* = g_{dk} + \sum_{n=1}^{N_d} \langle z_{dnk} \rangle + \sum_{l=1}^{L} \langle y_{dl} \rangle \sigma_{lk} \tag{23}$$

$$h_{dk}^* = h_{dk} + \sum_{l=1}^{L} \langle y_{dl} \rangle \tau_{lk} + \sum_{kk=k+1}^{K} \phi_{dn(kk)} \tag{24}$$

$$\lambda_{kv}^* = \lambda_{kv} + \sum_{d=1}^{D} \sum_{n=1}^{N_d} \sum_{v=1}^{V} \phi_{dnk} w_{dnv} \tag{25}$$

$$\pi_l = \frac{1}{D} \sum_{d=1}^{D} r_{dl} \tag{26}$$

$\langle \cdot \rangle$ in these set of equations implies expectation of that particular variable and $(\bar{\cdot})$ is its corresponding mean. The corresponding values of these expectations [3] are given by,

$$\langle \ln \theta_{dk} \rangle = \sum_{j=1}^{k} (\Psi(g_{dk}) - \Psi(g_{dk} + h_{dk})) \tag{27}$$

$$\left\langle 1 - \sum_{j=1}^{k} \theta_{dj} \right\rangle = \sum_{j=1}^{k} (\Psi(h_{dk}) - \Psi(g_{dk} + h_{dk})) \tag{28}$$

$$\bar{\sigma}_{lk} = \frac{v_{lk}^*}{\nu_{lk}^*}, \langle \ln \sigma_{lk} \rangle = \Psi(v_{lk}^*) - \ln \nu_{lk}^* \tag{29}$$

$$\langle (\ln \sigma_{lk} - \ln \bar{\sigma}_{lk})^2 \rangle = [\Psi(v_{lk}^*) - \ln v_{lk}^*]^2 + \Psi'(v_{lk}^*) \tag{30}$$

$$\bar{\tau}_{lk} = \frac{s_{lk}^*}{t_{lk}^*}, \langle \ln \tau_{lk} \rangle = \Psi(s_{lk}^*) - \ln t_{lk}^* \tag{31}$$

$$\langle (\ln \tau_{lk} - \ln \bar{\tau}_{lk})^2 \rangle = [\Psi(s_{lk}^*) - \ln s_{lk}^*]^2 + \Psi'(s_{lk}^*) \tag{32}$$

$$\langle z_{dnk} \rangle = \phi_{dnk}, \langle y_{dl} \rangle = r_{dl}, \langle \ln \beta_{kv} \rangle = \Psi(\lambda_{kv}) - \Psi(\sum_{f=1}^{V} \lambda_{kf}) \tag{33}$$

$\Psi(\cdot)$ and $\Psi(\cdot)'$ indicate digamma and trigamma functions respectively. Generally the variational algorithm works by iterating through Eqs. 10–14 until the values don't change considerably. However, we tweak this algorithm to suit our needs for an interactive algorithm in the next section.

4   Interactive Learning Algorithm

The primal motive to incorporate an interactive learning algorithm is to enhance topic quality, by asking the users to decide the probability of words within the topic. Using the definitions of $\vec{\beta}_o$ and $\vec{\beta}_u$ mentioned earlier and adding weights to this objective and subjective probabilities, $\vec{\beta}$ can be defined as.

$$\vec{\beta} = \eta_1 \vec{\beta}_o + \eta_2 \vec{\beta}_u \tag{34}$$

where, η_1 and η_2 are the weights given to the objective and subjective probabilities respectively. The intention of using this values is to control the effect of user input as it is probable that in some cases the user might not be well versed in the topic under consideration. So assigning a lower value to η_2 in this case helps us to reduce the impact of user defined input and vice-versa. The only criteria to be taken care of here is that $\eta_1 + \eta_2 = 1$. Another important question is to identify when to prompt users for input. The criteria can be custom-defined based on our needs and the problem we are trying to solve. For example, the criteria could be the number of iterations or when the coherence score beyond a certain value or when the classification accuracy with the topics is above a certain level. So first we run the variational algorithm until our criteria is reached and then we calculate the UMass coherence score [10] for each of the K topics with the formula,

$$score_{UMass}(k) = \sum_{i=2}^{M_k} \sum_{j=1}^{M_k-1} \log \frac{p(w_i, w_j) + 1}{p(w_i)} \tag{35}$$

where M_k is the set of words in a topic with w_i and w_j being the ith and j^{th} word in the topic. We can then prompt the users to modify the probabilities of T_k words within each topic. It would be better to show the topics with low coherence score in case of large number of topics. If $\{p_{o1}^k, p_{o2}^k, ..., p_{oT}^k\}$ are the probabilities inferred by our model and $\{p_{u1}^k, p_{u2}^k, ..., p_{uT'}^k\}$ is the set of probabilities of T_k' words that the user choose to modify, then the new probabilities for this set of words is obtained by $\beta_{ut}^K = p_{ot}^k * p_{ut}^k$. The probabilities that had been reduced is distributed among the rest of the words proportionally by using the equation:

$$\beta_{ut}^k = p_{ot} * \left(1 + \frac{p_r}{\sum_t p_{ot}}\right); \forall t \notin T_k' \tag{36}$$

Here, $p_r = \sum_t^{T_k'} p_{ot}^k * (1 - p_{ut}^k)$. The value of $\vec{\beta}_u$ can be substituted in Eq. 34. The variational algorithm is again carried out to attain the new set of words in each topic.

5   Experimental Results

To see how our model performs with real data, we use two real world datasets namely, BBC news[1] and twitter emotions [12]. The former consisted of 2325

---

[1] http://mlg.ucd.ie/datasets/bbc.html.

documents from 5 categories specifically business (610), entertainment (386), politics (417), sports (511) and technology (401). On the other hand, the later entails a huge corpus 416,809 tweets representing emotions related to anger, joy, fear, love, sadness and surprise. To keep things simple we choose 2000 samples form each emotion category to test our model. The criteria in our case will be, to pause when the accuracy of a supervised version of LGDMA model following [9] attains a threshold accuracy which can be varied. At this point the user will be prompted to enter new probabilities for the words in each topic. After removing the stop words and words less than four letters we create a bag of words model with 1800 and 1000 most frequent words as the vocabulary. UMass score explained in the previous section is used as a metric to record the performance of our model.

We compare our model with vanilla LGDMA, latent generalized Dirichlet allocation (LGDA) and LDA models respectively. Our main comparison is between iLGDMA and LGDMA since the interactive version is a direct improvement over LGDMA. For both the experiments the value of L was found to give better results when set to 3. The rest of the parameters are randomly initiated. The coherence score of the two models when the value of K is set as 10, 15, 20 and 25 is shown in Table 1 and Table 2 for the two datasets respectively. For both the cases, we see that iLGDMA achieves a better coherence score than rest of the bunch.

Table 1. Average coherence score of all topics for BBC news data

Model	$K = 10$	$K = 15$	$K = 20$	$K = 25$
iLGDMA	−1.04	−1.16	−1.20	−1.56
LGDMA	−1.20	−1.18	−1.36	−1.73
LGDA	−1.30	−1.27	−1.38	−1.77
LDA	−1.22	−1.40	−2.08	−1.87

Table 2. Average coherence score of all topics for emotions data

Model	$K = 10$	$K = 15$	$K = 20$	$K = 25$
iLGDMA	−4.65	−4.87	−4.67	−5.52
LGDMA	−4.86	−4.99	−5.90	−6.30
LGDA	−4.99	−5.09	−5.94	−7.48
LDA	−5.13	−6.17	−6.30	−7.67

However, it would be more appropriate to compare what percentage of increase in topic coherence our model provides compared to the rest, especially LGDMA. Figure 1 shows the percentage increase in coherence score using our iLGDMA with the other models for BBC news data. The figure also provides a

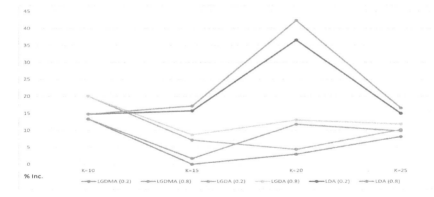

Fig. 1. % Increase in topic quality for BBC news

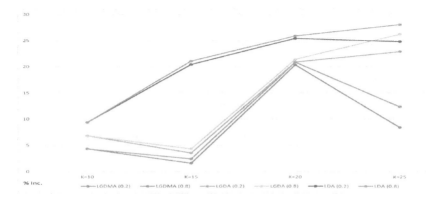

Fig. 2. % Increase in topic quality for emotions dataset

fine comparison between the percentage increase achieved by varying the weights of user defined probabilities. The experiments were conducted with objective and subjective properties set to 0.2 and 0.8 to simulate a well versed user and vice-versa in case of a user with low subject knowledge. We can see a clear improvement when the user is someone who knows the subject as opposed to a mundane user. This was replicated by lowering the probability of a few correctly identified words in a topic in case of a common user. The figure shows that even in this case we can still see considerable improvements over the other models. This shows the robustness of our model.

We can observe similar results in the case of tweets labelled with emotions. Another notable observation from the two experiments is that, though the weights were varied when the value of K was set to 10, the coherence score remained the same. This is because, most of the words in the topics were already closely related and little information from the user was more than enough to improve the coherence to the best possible value. The figures also show that our model outputs better topics almost with a percentage increase of at least about 25% with respect to LDA for certain K values.

6   Conclusion

Our paper presents an unique model to learn topics from data with an interactive approach which is flexible and can be custom-fit for the user. The trend we observed in our experiments establishes the efficiency of our model to extract improved quality topics from data. The advantage of using weights to define the impact of user inputs is also clearly visible from our experiments. Using a mixture of GD priors instead of a commonly used prior tends to help in the learning of better topics which is seen in our comparison with other models. The percentage increase in the coherence score of the derived topics is especially very high when compared to LDA which is the baseline model for unsupervised learning of topics. Future research could involve using Beta-Liouville allocation models which might be an efficient replacement for Dirichlet as well.

References

1. Bakhtiari, A.S., Bouguila, N.: A variational Bayes model for count data learning and classification. Eng. Appl. Artif. Intell. **35**, 176–186 (2014)
2. Blei, D.M., Ng, A.Y., Jordan, M.I.: Latent dirichlet allocation. J. Mach. Learn. Res. **3**, 993–1022 (2003)
3. Bouguila, N.: Deriving kernels from generalized dirichlet mixture models and applications. Inf. Process. Manage. **49**(1), 123–137 (2013)
4. Chien, J.T., Lee, C.H., Tan, Z.H.: Latent dirichlet mixture model. Neurocomputing **278**, 12–22 (2018)
5. Fan, W., Bouguila, N.: Online variational learning of generalized dirichlet mixture models with feature selection. Neurocomputing **126**, 166–179 (2014)
6. Fan, W., Bouguila, N.: Topic novelty detection using infinite variational inverted dirichlet mixture models. In: 2015 IEEE 14th International Conference on Machine Learning and Applications (ICMLA), pp. 70–75. IEEE (2015)
7. Fan, W., Bouguila, N., Ziou, D.: Variational learning for finite dirichlet mixture models and applications. IEEE Trans. Neural Netw. Learn. Syst. **23**(5), 762–774 (2012)
8. Liu, Y., Du, F., Sun, J., Jiang, Y.: ILDA: an interactive latent dirichlet allocation model to improve topic quality. J. Inf. Sci. **46**(1), 23–40 (2020)
9. Mcauliffe, J., Blei, D.: Supervised topic models. In: 20th Proceedings of the Conference on Advances in Neural Information Processing Systems (2007)
10. Mimno, D., Wallach, H.M., Talley, E., Leenders, M., McCallum, A.: Optimizing semantic coherence in topic models. In: Proceedings of the Conference on Empirical Methods in Natural Language Processing, pp. 262–272. EMNLP '11, Association for Computational Linguistics, USA (2011)
11. Opper, M., Saad, D.: Mean-field theory of learning: from dynamics to statics (2001)
12. Saravia, E., Liu, H.C.T., Huang, Y.H., Wu, J., Chen, Y.S.: CARER: contextualized affect representations for emotion recognition. In: Proceedings of the 2018 Conference on Empirical Methods in Natural Language Processing, pp. 3687–3697. Association for Computational Linguistics, Brussels, Belgium, October–November 2018

Classifying Me Softly: A Novel Graph Neural Network Based on Features Soft-Alignment

Alessandro Bicciato[1]📀, Luca Cosmo[1]📀, Giorgia Minello[1,2]📀,
Luca Rossi[3(✉)]📀, and Andrea Torsello[1]📀

[1] Ca' Foscari University of Venice, Venice, Italy
[2] IESE Business School, Barcelona, Spain
[3] Queen Mary University of London, London, UK
luca.rossi@qmul.ac.uk

Abstract. Graph neural networks are increasingly becoming the framework of choice for graph-based machine learning. In this paper we propose a new graph neural network architecture based on the soft-alignment of the graph node features against sets of learned points. In each layer of the network the input node features are transformed by computing their similarity with respect to a set of learned features. The similarity information is then propagated to other nodes in the network, effectively creating a message passing-like mechanism where each node of the graph individually learns what is the optimal message to pass to its neighbours. We perform an ablation study to evaluate the performance of the network under different choices of its hyper-parameters. Finally, we test our model on standard graph-classification benchmarks and we find that it outperforms widely used alternative approaches, including both graph kernels and graph neural networks.

Keywords: Graph neural network · Deep learning

1 Introduction

Graphs have long been used as a powerful yet natural abstraction for data where structure plays a key role, ranging from biological data [6] to collaboration networks [24] and 3D shapes [10,11]. With respect to standard vectorial data, their rich information content however introduces further difficulties when one is faced with the task of learning on this type of data.

For several years, a standard approach to tackle these difficulties has been to either directly or indirectly embed graphs into vectorial spaces where a kernel measure had been defined, allowing to frame learning on graphs in the context of kernel machines [3–5,23,26,30,32,33]. In recent years, however, graph neural networks (GNNs) have quickly emerged as efficient and effective models to learn on graph data [1,2,8,9,17,20,22,31,35], in many cases outperforming traditional approaches based on graph kernels. One of the key advantages of GNNs is the

A. Krzyzak et al. (Eds.): S+SSPR 2022, LNCS 13813, pp. 43–53, 2022.
https://doi.org/10.1007/978-3-031-23028-8_5

ability to provide an end-to-end learning framework where input features do not need to be handcrafted anymore but can be learned directly by the model itself given the task at hand. Recent works [27], however, have shown that a certain degree of caution is required when designing neural architectures for graphs as simple message-passing mechanisms [17,37], on which many GNNs rely, constrain the resulting network to be at most as powerful as standard graph kernels based on the one-dimensional Weisfeiler-Lehman isomorphism test [32].

In this paper we propose a novel GNN based on the soft-alignment of the node features (FSA-GNN). Each FSA layer transforms the input node features as follows. Given a node n, we construct the egonet of radius r around it. For each node in the egonet, we then compute the dot product of the corresponding node feature with a set of features, which we allow our model to learn. Each learned feature correspond to a different message, and the one that maximizes the similarity is then propagated back to the center of the egonet n, where it is used to update the corresponding feature. The result is a reminiscent of a message-passing network [17] where, however, we do not have a simple linear message generating function, but rather the correct message is selected from a mixture of different modes, thus allowing each node of the graph to individually learn the message to pass to its neighbours.

We perform an extensive ablation study to evaluate the impact of the choice of the network parameters on its performance. We find that, on most datasets considered, our model is generally robust to the choice of its hyper-parameters. Finally we test FSA-GNN on both graph classification and regression tasks against widely used alternative methods, both neural and traditional, and we find that the performance of our model is superior or comparable to that of competing ones.

The remainder of this paper is organised as follows. In Sect. 2 we review the related work and in Sect. 3 we present the proposed neural architecture. In Sect. 4 we perform an extensive experimental analysis to validate our model and compare it against widely used alternatives on graph classification benchmarks, while Sect. 5 concludes the paper.

2   Related Work

In recent years, with the advent of deep learning and the renewed interest in neural architectures, the focus of graph-based machine learning researchers has quickly moved to extending deep learning approaches to deal with graph data. The fundamental idea underpinning most GNNs is that of exploiting the structure of the input graph by iteratively propagating node information between groups of nodes. The features obtained after several rounds (i.e., layers) of propagation are then used, in conjunction with suitably defined pooling operations and fully connected layers, for tasks such as graph classification, node classification, and link prediction. One of the first papers to propose this idea is that of [31], where an information diffusion mechanism is used to learn the nodes latent representations by exchanging neighbourhood information.

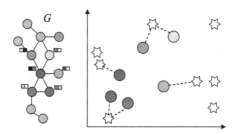

Fig. 1. The features soft-alignment procedure at the core of the FSA layer. The input graph G with the nodes of the egonet (shaded area) highlighted in colour (left) and the corresponding soft-aligned node features, with the trained features shown as stars (right).

Depending on the form this diffusion takes, we can distinguish between convolutional [1,22], attentional [35], or message passing [17] GNNs, with the latter being the most general and formally equivalent to the Weisfeiler-Lehman graph isomorphism test under some conditions [27,37]. More recently, [14] proposed a reinterpretation of graph convolution in terms of partial differential equations on graphs, noting that different problems can benefit from different networks dynamics, hence suggesting the need to look beyond diffusion. For a comprehensive survey of GNNs we refer the reader to [36].

3   Features Soft-Alignment Graph Neural Networks

The core component of our model is the FSA layer, where the input node features are transformed through a soft-alignment procedure against a set of learned features. Figure 1 illustrates the alignment procedure, while Fig. 2 shows this in the wider context of the FSA layer.

Given an input graph G, we update the node features as follows. Consider the node v of the graph and the egonet of radius r centered on it, i.e., the egonet is the set of nodes with shortest-path distance at most r from v. We start by computing the similarity between the feature of each node u in the egonet and a set of trained features. Let f_u^l denote the feature associated to the egonet node u and let m_i be the i-th trained feature, where l indicates the number of the current FSA layer. Then we compute the matrix S_v containing the pairwise similarity information (dot product) between egonet node features and learned features, i.e.,

$$S_v^l(u, i) = f_u^{l\top} m_i^l. \tag{1}$$

For each node u in the egonet, we select the trained feature i with the highest similarity and we aggregate the corresponding similarity information as the new feature f_v of v, i.e.,

$$f_v^{l+1} = \sum_u \max_i (S_v^l(u, i)). \tag{2}$$

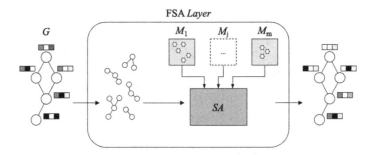

Fig. 2. Each FSA layer receives in input a graph and a set of node features. On these, the soft alignment procedure shown in Fig. 1 is performed, producing a new set of node features as the output of the layer.

To ensure that the computation of the new features is differentiable, we replace the max operation in Eq. 2 with the softmax and we obtain

$$f_v^{l+1} = \sum_{u,i} \frac{exp(\beta S_v^l(u,i))}{\sum_k exp(\beta S_v^l(u,k))} S_v^l(u,i). \tag{3}$$

Figure 3 shows the FSA layer in the context of the proposed neural architecture. Each FSA layer is equipped with m sets of trained features M_j (see Fig. 2), so the dimensionality of f_v is m, with exception of the first layer where the dimensionality of the node features depends on the dataset. We apply sum pooling to the output of each FSA layer and we concatenate the results into a single vector, effectively creating skip connections from the FSA layers to the final multi-layer perceptron module.

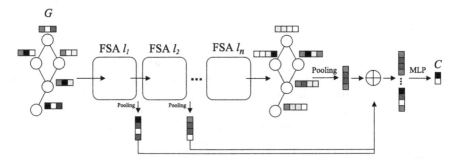

Fig. 3. The FSA-GNN architecture. After l_n FSA layers, we obtain a graph-level feature vector through sum pooling on the nodes features, which is then fed to an MLP to output the final classification label. Note the presence of skip connections which effectively mean we apply sum pooling to the output of each FSA layer and concatenate the corresponding results into a single vector.

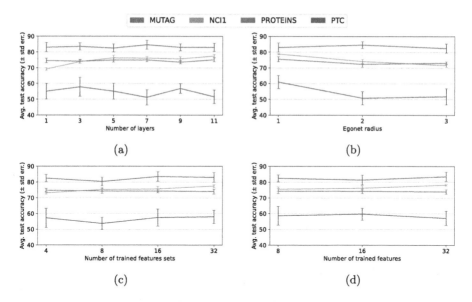

Fig. 4. Average classification accuracy (± standard error) on the four bio/chemo-informatics datasets as we vary (a) n, the number of FSA layers; (b) r, the radius of the egonets; (c) $|M_j|$, the number of trained features sets for each layer; (d) m, the number of learned features in each set.

4 Experiments

We run an extensive set of experiments to evaluate the performance of the proposed model on both graph classification and graph regression tasks. We also perform a thorough ablation study to assess the sensitivity of our model to the choice of its hyper-parameters.

4.1 Experimental Setup

We compare our model against 6 state-of-the-art GNNs (DiffPool [38], GIN [37], DGCNN [39], ECC [34], GraphSAGE [18], and (s)GIN [12]), a baseline method for molecular graphs as suggested in [15] (Molecular Fingerprint, [25,29]), the WL kernel [32] together with a C-SVM [7] classifier, and two recent GNNs employing a differentiable graph kernel (RWNN [28] and KerGNN [16]).

We use publicly available graph classification datasets frequently employed in the graph learning literature [21]. In particular, we use 4 bio/chemo-informatics datasets collecting molecular graphs (MUTAG, NCI1, PROTEINS, and PTC). All the datasets we consider have categorical node features. For the graph regression task, we use the drug constrained solubility prediction dataset ZINC [19]. Specifically, we use a subset (12K) of the ZINC molecular graphs (250K), as in [13], to regress the desired molecular property, i.e., the constrained solubility. For each molecular graph, the node features identify the types of heavy atoms.

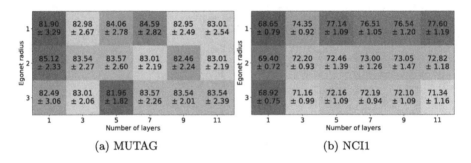

Fig. 5. Average classification accuracy on (a) MUTAG and (b) NCI1 dataset as we vary both the egonet radius and the number of FSA layers.

Although the dataset also considers the types of bonds between nodes, encoded as edge features, we do not utilise them in our experiments.

We ensure a fair comparison by following the same experimental protocol for each of the competing methods. For the graph classification task, we perform 10-fold cross validation where in each fold the training set is further subdivided in training and validation with a ratio of 9:1. The validation set is used for both early stopping and to select the best model within each fold. Importantly, folds and train/validation/test splits are consistent among all the methods. For the graph regression task, we evaluate the performance of our method and competing ones using the mean absolute error (MAE) between the predicted and the ground-truth constrained solubility for each molecular graph. We use the training, validation and test splits provided with the dataset.

For all the methods we perform grid search to optimize the hyper-parameters. In particular, for the WL method we optimize the value of C and the number of WL iterations $h \in \{4, 5, 6, 7\}$. For the RWGNN we investigate the hyper-parameter ranges used by the authors [28], while for all the other GNNs we follow [15] and we perform a full search over the hyper-parameters grid. For our model, we explore the following hyper-parameters: number of structural masks in $\{4, 8, 16, 32\}$, maximum number of nodes of a structural mask in $\{8, 16, 32\}$, subgraph radius in $\{1, 2, 3\}$, and number of layers in $\{1, 3, 5, 7, 9, 11\}$.

In all our experiments, we train our model for 2000 epochs, using the Adam optimizer with learning rate of 0.001 and a batch size of 32. The MLP takes as input the sum pooling of the node features and is composed by two layers of output dimension m and c (# classes).

4.2   Ablation Study

We commence by performing an ablation study to investigate how the various components and hyper-parameters of our architecture affect the performance of our model in a graph classification task. We perform a full search over the hyper-parameters grid where we calculate the average classification accuracy over 10 folds while fixing the hyper-parameter being investigated. Figure 4 shows how the

Table 1. Classification results (mean accuracy ± standard error) on the selected chemo/bio-informatics datasets. The best performance (per dataset) is highlighted in bold, the second best is underlined.

	MUTAG	PTC	NCI1	PROTEINS
Baseline	78.57 ± 4.00	58.34 ± 2.02	68.50 ± 0.87	73.05 ± 0.90
WL	82.67 ± 2.22	55.39 ± 1.27	**79.32 ± 1.48**	74.16 ± 0.38
DiffPool	81.35 ± 1.86	55.87 ± 2.73	75.72 ± 0.79	73.13 ± 1.49
GIN	78.13 ± 2.88	56.72 ± 2.66	<u>78.63 ± 0.82</u>	70.98 ± 1.61
DGCNN	**85.06 ± 2.50**	53.50 ± 2.71	76.56 ± 0.93	<u>74.31 ± 1.03</u>
ECC	79.68 ± 3.78	54.43 ± 2.74	73.48 ± 0.71	73.76 ± 1.60
GraphSAGE	77.57 ± 4.22	<u>59.87 ± 1.91</u>	75.89 ± 0.96	73.11 ± 1.27
sGIN	<u>84.09 ± 1.72</u>	56.37 ± 2.28	77.54 ± 1.00	73.59 ± 1.47
KerGNN	82.43 ± 2.73	53.15 ± 1.83	74.16 ± 3.36	72.96 ± 1.46
RWGNN	82.51 ± 2.47	55.47 ± 2.70	72.94 ± 1.16	73.95 ± 1.32
FSA-GNN	84.04 ± 2.48	**63.18 ± 3.63**	78.51 ± 0.40	**75.13 ± 1.53**

model classification accuracy changes as we vary: (a) n, the number of FSA layers; (b) r, the radius of the egonets; (c) $|M_j|$, the number of trained features sets for each layer; (d) m, the number of learned features in each set. The results show that our model is largely robust with respect to the choice of hyper-parameters, with the exception of NCI1, where the value of all four hyper-parameters has a statistically significant impact on the model performance. Moreover, on all the datasets except for MUTAG, the best performance is achieved for low values of the egonet radius r. This in turn may indicate that in these datasets the local feature information captured for low values of r is highly discriminative.

In Fig. 5 we show the result of varying both the egonet radius and the number of FSA layers on the MUTAG (a) and NCI1 (b) datasets. While one would intuitively expect there to be a trade-off between these two hyper-parameters, we do not observe it in practice. Indeed, this is likely a consequence of the skip connections introduced in Sect. 3, which, by concatenating the output of each FSA layer, make deep networks with low egonet radius able to capture information at varying levels of scale, as opposed to a shallow network with large egonet radius. Indeed, on NCI1 we clearly see that the optimal performance peaks for low values of the egonet radius r and high number of layers. This suggests again that local feature information captured for low levels of r is highly discriminative, yet more global information captured in deeper layers is also important.

4.3 Graph Classification Results

We now evaluate the accuracy of our model on the graph classification task. The results for each method and dataset are shown in Table 1. For the bio/chemo-informatics datasets PROTEINS and PTC, our approach is better than the

Table 2. Graph regression performance measured as MAE between the predicted and the ground-truth constrained solubility on the ZINC dataset [19].

	Baseline	WL	DiffPool	GIN	DGCNN	ECC	GraphSAGE	RWGNN	**FSA-GNN**
ZINC	0.66	1.24	0.49	0.38	0.60	0.73	0.46	0.55	**0.37**

competitors, being better in accuracy from one to nearly three percent compared to the best of the others. On the remaining datasets, MUTAG and NCI1, the algorithm performs on par with the second best competitor. More generally, our approach is satisfactory with all the datasets where the node features have a consistent impact for the classification of the graphs, resulting in a high classification accuracy with small set of node features - MUTAG (7 node features), PROTEINS (3 node features) - as well with big ones - NCI1 (37 node features), PTC (18 node features).

4.4   Graph Regression Results

Finally, we show the results of the graph regression task in Table 2. Note that as the performance of the various methods is computed in terms of MAE, lower values correspond to better performance. FSA-GNN clearly outperforms all competing methods, with only GIN achieving a comparable, yet lower, performance. It should be stressed that in our experimental evaluation we did not make use of the edge features, both for our method and for the competing ones. Finally, note that the methods that outperformed us in the graph classification task perform very poorly in the context of regression, with the MAE for DGCNN and WL being respectively two and three times that of FSA-GNN.

5   Conclusion

In this paper we proposed FSA-GNN, a novel neural architecture for graphs based on the soft-alignment of the graph node features. Our model is inspired by message-passing architectures but it goes beyond these by allowing the nodes to learn which message to propagate to the neighbouring nodes in order to best solve the optimisation problem at hand. We showed through an extensive ablation study that our model is relatively robust to the choice of its hyperparameters and that there seems to be a trade-off between the radius of the egonet over which the features propagate and the number of layers on which this propagation happens. Finally, a thorough experimental evaluation on widely-used graph classification datasets showed that our model is able to outperform both standard machine learning approaches and neural ones.

References

1. Atwood, J., Towsley, D.: Diffusion-convolutional neural networks. In: Advances in Neural Information Processing Systems, pp. 1993–2001 (2016)
2. Bai, L., Cui, L., Jiao, Y., Rossi, L., Hancock, E.: Learning backtrackless aligned-spatial graph convolutional networks for graph classification. IEEE Trans. Pattern Anal. Mach. Intell. **44**, 783–798 (2020)
3. Bai, L., Hancock, E.R.: Graph kernels from the Jensen-Shannon divergence. J. Math. Imaging Vis. **47**(1), 60–69 (2013)
4. Bai, L., Rossi, L., Torsello, A., Hancock, E.R.: A quantum Jensen-Shannon graph kernel for unattributed graphs. Pattern Recogn. **48**(2), 344–355 (2015)
5. Borgwardt, K.M., Kriegel, H.P.: Shortest-path kernels on graphs. In: Fifth IEEE International Conference on Data Mining (ICDM 2005), pp. 8-pp. IEEE (2005)
6. Borgwardt, K.M., Ong, C.S., Schönauer, S., Vishwanathan, S., Smola, A.J., Kriegel, H.P.: Protein function prediction via graph kernels. Bioinformatics **21**(suppl_1), i47–i56 (2005)
7. Chang, C.C., Lin, C.J.: LIBSVM: a library for support vector machines. ACM Trans. Intell. Syst.Ttechnol. **2**(3), 1–27 (2011)
8. Cosmo, L., Kazi, A., Ahmadi, S.-A., Navab, N., Bronstein, M.: Latent-graph learning for disease prediction. In: Martel, A., et al. (eds.) MICCAI 2020. LNCS, vol. 12262, pp. 643–653. Springer, Cham (2020). https://doi.org/10.1007/978-3-030-59713-9_62
9. Cosmo, L., Minello, G., Bronstein, M., Rodolà, E., Rossi, L., Torsello, A.: Graph kernel neural networks. arXiv preprint arXiv:2112.07436 (2021)
10. Cosmo, L., Minello, G., Bronstein, M., Rodolà, E., Rossi, L., Torsello, A.: 3D shape analysis through a quantum lens: the average mixing kernel signature. Int. J. Comput. Vis. **130**, 1–20 (2022)
11. Cosmo, L., Minello, G., Bronstein, M., Rossi, L., Torsello, A.: The average mixing kernel signature. In: Vedaldi, A., Bischof, H., Brox, T., Frahm, J.-M. (eds.) ECCV 2020. LNCS, vol. 12365, pp. 1–17. Springer, Cham (2020). https://doi.org/10.1007/978-3-030-58565-5_1
12. Di, X., Yu, P., Bu, R., Sun, M.: Mutual information maximization in graph neural networks. In: 2020 International Joint Conference on Neural Networks (IJCNN), pp. 1–7. IEEE (2020)
13. Dwivedi, V.P., Joshi, C.K., Laurent, T., Bengio, Y., Bresson, X.: Benchmarking graph neural networks. arXiv preprint arXiv:2003.00982 (2020)
14. Eliasof, M., Haber, E., Treister, E.: PDE-GCN: novel architectures for graph neural networks motivated by partial differential equations. In: 34th Proceedings of the Conference on Advances in Neural Information Processing Systems (2021)
15. Errica, F., Podda, M., Bacciu, D., Micheli, A.: A fair comparison of graph neural networks for graph classification. In: Proceedings of the 8th International Conference on Learning Representations (ICLR) (2020)
16. Feng, A., You, C., Wang, S., Tassiulas, L.: KerGNNs : interpretable graph neural networks with graph kernels. arXiv preprint arXiv:2201.00491 (2022)
17. Gilmer, J., Schoenholz, S.S., Riley, P.F., Vinyals, O., Dahl, G.E.: Neural message passing for quantum chemistry. In: International Conference on Machine Learning, pp. 1263–1272. PMLR (2017)
18. Hamilton, W., Ying, Z., Leskovec, J.: Inductive representation learning on large graphs. In: 30th Proceedings of Conference on Advances in Neural Information Processing Systems (2017)

19. Irwin, J.J., Sterling, T., Mysinger, M.M., Bolstad, E.S., Coleman, R.G.: Zinc: a free tool to discover chemistry for biology. J. Chem. Inf. Model. **52**(7), 1757–1768 (2012)
20. Kazi, A., Cosmo, L., Ahmadi, S.A., Navab, N., Bronstein, M.: Differentiable graph module (DGM) for graph convolutional networks. IEEE Trans. Pattern Anal. Mach. Intell. Early Access (2022)
21. Kersting, K., Kriege, N.M., Morris, C., Mutzel, P., Neumann, M.: Benchmark data sets for graph kernels (2016). https://ls11-www.cs.tu-dortmund.de/staff/morris/graphkerneldatasets
22. Kipf, T.N., Welling, M.: Semi-supervised classification with graph convolutional networks. In: Proceedings of the 5th International Conference on Learning Representations. ICLR 2017 (2017)
23. Kriege, N., Mutzel, P.: Subgraph matching kernels for attributed graphs. arXiv preprint arXiv:1206.6483 (2012)
24. Lima, A., Rossi, L., Musolesi, M.: Coding together at scale: Github as a collaborative social network. In: Eighth International AAAI Conference on Weblogs and Social Media (2014)
25. Luzhnica, E., Day, B., Liò, P.: On graph classification networks, datasets and baselines. arXiv preprint arXiv:1905.04682 (2019)
26. Minello, G., Rossi, L., Torsello, A.: Can a quantum walk tell which is which? a study of quantum walk-based graph similarity. Entropy **21**(3), 328 (2019)
27. Morris, C., et al.: Weisfeiler and leman go neural: Higher-order graph neural networks. In: Proceedings of the AAAI Conference on Artificial Intelligence, pp. 4602–4609 (2019)
28. Nikolentzos, G., Vazirgiannis, M.: Random walk graph neural networks. Adv. Neural. Inf. Process. Syst. **33**, 16211–16222 (2020)
29. Ralaivola, L., Swamidass, S.J., Saigo, H., Baldi, P.: Graph kernels for chemical informatics. Neural Netw. **18**(8), 1093–1110 (2005)
30. Rossi, L., Torsello, A., Hancock, E.R.: Measuring graph similarity through continuous-time quantum walks and the quantum Jensen-Shannon divergence. Phys. Rev. E **91**(2), 022815 (2015)
31. Scarselli, F., Gori, M., Tsoi, A.C., Hagenbuchner, M., Monfardini, G.: The graph neural network model. IEEE Trans. Neural Netw. **20**(1), 61–80 (2008)
32. Shervashidze, N., Schweitzer, P., Van Leeuwen, E.J., Mehlhorn, K., Borgwardt, K.M.: Weisfeiler-Lehman graph kernels. J. Mach. Learn. Res. **12**(9) (2011)
33. Shervashidze, N., Vishwanathan, S., Petri, T., Mehlhorn, K., Borgwardt, K.: Efficient graphlet kernels for large graph comparison. In: Proceedings of the 12th International Conference on Artificial Intelligence and Statistics, pp. 488–495. PMLR (2009)
34. Simonovsky, M., Komodakis, N.: Dynamic edge-conditioned filters in convolutional neural networks on graphs. In: Proceedings of the IEEE Conference on Computer Vision and Pattern Recognition, pp. 3693–3702 (2017)
35. Veličković, P., Cucurull, G., Casanova, A., Romero, A., Liò, P., Bengio, Y.: Graph attention networks .In: International Conference on Learning Representations (2018)
36. Wu, Z., Pan, S., Chen, F., Long, G., Zhang, C., Philip, S.Y.: A comprehensive survey on graph neural networks. IEEE Trans. Neural Netw. Learn. Syst. **32**(1), 4–24 (2020)
37. Xu, K., Hu, W., Leskovec, J., Jegelka, S.: How powerful are graph neural networks? arXiv preprint arXiv:1810.00826 (2018)

38. Ying, R., You, J., Morris, C., Ren, X., Hamilton, W.L., Leskovec, J.: Hierarchical graph representation learning with differentiable pooling. arXiv preprint arXiv:1806.08804 (2018)
39. Zhang, M., Cui, Z., Neumann, M., Chen, Y.: An end-to-end deep learning architecture for graph classification. In: Thirty-Second AAAI Conference on Artificial Intelligence (2018)

Review of Handwriting Analysis for Predicting Personality Traits

Yan Xu[1,2](✉) [iD], Yufang Tang[2](✉) [iD], and Ching Y. Suen[1] [iD]

[1] Concordia University, Montreal, QC H3G 1M8, Canada
[2] Shandong Normal University, Jinan, Shandong 250014, China
yan.soe1011@gmail.com, tangyufang@outlook.com

Abstract. Most research on document analysis is carried out through human intervention, such as manually counting the parameters of document analysis as the input of neural networks. Document analysis via computer vision methods is a relatively less explored area of research.

In this paper, we investigated the literature on document analysis in recent years, and summarized its development process and the commonly used research methods, discussed their advantages and disadvantages. Meanwhile, we put forward the general research ideas and steps for document analysis, highlight the limitations of existing processes and the challenges typically faced when designing such systems, provide potential, feasible solutions, and point out the direction of further research, which give guidance to novice researchers and have reference value for subsequent researchers. Our experiments verify that our method is feasible and can be used as a substitute for specific application scenarios without professional handwriting experts.

Keywords: Handwriting analysis · Personality trait · Feature extraction · Prediction model · Computer vision

1 Introduction

Graphology (or Handwriting Analysis) is an emerging interdisciplinary study evaluating the relationship between graphological feature and personality trait (temperament, character, ability, etc.), gender, age, occupation, appearance, experience, disease, etc. It involves psychology, sociology, physiology, pathology and linguistics, which belongs to a branch of applied psychology.

Psychology has the inseparable relationship with graphology. Psychology studies general phenomenon, whereas graphology focuses on very special and individual phenomena. Thus, psychology can play a useful guiding role in solving and explaining the problems graphology is facing.

Handwriting is more important in describing personality because the human brain is more engaged when writing than when speaking or listening. Like unique fingerprint, handwriting is carrying the richest information related to the physical, mental and emotional state of the writer.

© The Author(s), under exclusive license to Springer Nature Switzerland AG 2022
A. Krzyzak et al. (Eds.): S+SSPR 2022, LNCS 13813, pp. 54–63, 2022.
https://doi.org/10.1007/978-3-031-23028-8_6

Handwriting is essentially a kind type of human language, which is the form of a text, not the content of the text itself. Handwriting is usually not the mirror of thought, but the reflection of consciousness and subconsciousness.

1.1   History

In 1622 [1], Camillo Baldi, an Italian doctor of medicine and philosophy, a professor at the University of Bologna, and the inventor of graphology, wrote the first graphological essay to carry out the systematic observation on the manner of handwriting. In 1871, Abbé Jean-Hippolyte Michon, a French priest, an archaeologist and the founder of graphology, who formulated almost all the signs and rules, coined the term 'graphology' derived from two Greek words 'graphein' (writing or drawing) and 'logos' (theory or doctrine). He published two books named 'Système de Graphologie' in 1875, introducing Michon's system of handwriting signs, and 'Méthode pratique de Graphologie' in 1878, explaining the principles of graphological analysis. In 1895, Wilhelm Preyer, a German professor of physiology at the University of Jena, made a highly significant contribution to handwriting analysis when he discovered the similarity of writing carried out by the left hand, right hand, toes and even the teeth. Alfred Binet, a lawyer in Paris and the father of intelligence testing in the 20th century, scientifically controlled tests with the help of experts in the field, and searched for a correlation between specific character traits and specific handwriting characteristics, achieving affirmative results with respect to the graphic indices of honesty and intelligence.

In 1932, Dr. Ludwig Klages, a German philosopher, psychologist and the father of modern graphology, wrote 'Handschrift und Charakter' (Writing and Personality), creating a complete and systematic theory of graphology with Gestalt theory. Max Pulver, a Swiss professor and psychologist at the University of Zurich, wrote the book 'Symbolism of Handwriting' in 1940, discussing the interpretation of handwriting and its relationship to various unconscious mythological or ancient symbols, and developing a theory of the space symbolism which emphasized the importance of interpreting upper, lower and middle zones of writing. Werner Wolff, a German experimental psychologist, was recognized for his contribution to the graphological examination of rhythm and subconscious personality tendencies. In his book 'Diagrams of the Unconscious' in 1948, he stated that man in his handwriting or artistic expression communicated not only his conscious thoughts but also his underlying thoughts of which he was unaware. In 1983, British Institute of Graphologists was founded in London by Frank Hilliger, and the first edition of 'The Graphologist' was produced later in the same year.

In China, Xu [2] wrote the masterpiece of modern Chinese graphology in 1943. Yang [3] wrote the first empirical research paper about Chinese handwriting in 1964. Weng [4] made a similar study after ten years. Yuan [5], a well-known handwriting expert, proposed a theory with guiding significance for further study of handwriting in 1993. Mrs. Menas, a famous European graphologist, established the first 'International Institute of Chinese Characters Graphology Science' in

Belgium in 1990. She once visited China and discussed with the Chinese psychology community on the issue of graphology, open a door to the world for Chinese graphology. In 1994, the first Chinese graphology seminar was held in Department of Psychology of Beijing Normal University, and 'China Graphology Research Association' was established, which was a milestone in the history of Chinese graphology development. In 1999, the national Ministry of personnel in China completed 'Research on the Application of Handwriting Analysis in Talent Recruitment'.

1.2   Applications

Graphology can be used in a vast range of applications [6], such as healthcare to predict certain disease, recruitment to evaluate personality, identification to analyze criminal investigation cases, education to improve the development of children, and so on.

1.3   Requirements

The handwriting samples should be: 1) written under normal circumstances; 2) from skilled writers; 3) with the number of words above 30, best in different periods; and 4) including age, gender, and occupational information of the writers.

Writing instruments can include: bamboo pen, brush pen, fountain pen (ideal), ballpoint pen (small friction, more stretched stroke, rounder angle, with the most obvious pen pressure), pencil (not clear enough), signature pen, etc. No matter what to use, it must be comfortable. Otherwise, the resulting handwriting is a fake one and should not be analyzed. Measuring tools used for handwriting analysis include: protractor, ruler, magnifier, pen, etc.

The Chinese fonts include: Oracle, Jinwen, Dazhuan, Xiaozhuan, Lishu, Cursive, Kaishu, Xingshu, Imitation Song and various artistic fonts. Cursive and Xingshu are used for normal handwriting. The handwriting from a skilled writer is considered as a normal handwriting, while if written by an unskilled writer, it is a fake handwriting. Any handwriting is deliberately disguised, even if from a skilled writer, cannot be treated as a normal handwriting.

2   Research Progress

There are many papers on graphology in recent years, we selected some of them with different foci on handwriting analysis.

2.1   Advantages

Mihai and Nicolae [7] provide a general idea and details on how to design and realize the structure of three-layer neural networks. The experiments are organized and designed well, considering the handwriting samples from the same and different writers with controlled and random databases, finding the links between handwriting features and personality traits by counting, and comparing state-of-the-art methods on their own database.

Chen and Lin [8] use online handwriting samples to predict the QZPS (Chinese Personality Scale) model for Chinese people. The handwriting features are divided into three categories: dynamic, text-dependent, and text-independent features. Four kinds of classification methods are chosen: SVM, kNN, AdaBoost, and ANN with common parameter settings. As a kind of binary classification task, the performance measures are accuracy, sensitivity, and specificity. The experiment shows the writing speed can reflect on the personality traits, but the minimum spacing between two consecutive letters do not affect that much. Furthermore, the class distribution should be considered in building a database.

Afnan et al [9] summarize the recent development of computational graphology with more details in the form of table, and conduct the experiment with the handwriting samples from four groups of people to calculate the statistical data (average value, variance), and from the same person across languages (English, French, Chinese) over time with the new sight.

In recent years, some researchers have started the direct method with computer vision to let computer extract the features from a whole handwriting image or segmentation, with deep neural networks, such as CNN [10], Resnet50 [11] and lightweight deep convolutional neural network (LWDCNN) [12]. Those methods can fundamentally solve the problem of manual selection of handwriting features and achieve truly intelligent document analysis.

2.2 Disadvantages

The public database is rare, most databases are private and self-built, and the number of handwriting samples may about two hundred, only for a special research topic, which increases the difficulty to repeat the experiments and validates the results.

Most automatic computational graphology systems output a naive analysis result by lots of manual intervention to predict certain personality type based on rule base, not really intelligent in the true sense.

Most handwriting features are statistical features, simply simulating the manual feature extraction of graphologists. Moreover, the procedure of feature extraction is actually manually chosen, not automatically extracted by neural networks. Different feature extraction methods can result in different accuracies of final prediction.

There exist many personality classification standards, making handwriting features different. Those features are served for the classification task, but the related work summaries rarely mention the personality traits used. The prediction accuracy is influenced by database employed, extracted features, and personality traits, if those factors can be fixed, the algorithm comparison will be more convincing.

3 Research Steps

From some perspectives, handwriting analysis can be considered as a kind of character recognition (CR) problem. Thus, several established CR methods in the steps of pre-processing, segmentation, feature extraction and classification can be used for reference. Likewise, the problem of document analysis requires many steps to consider.

3.1 Database

Few public handwriting database can be found, such as English database IAM [13], Chinese databases IAAS-4M [14], HCL2000 [15], HIT-MW, HIT-OR3C, SCUT-COUCH2009, CASIA-AHCDB [16]. There is no standard how to build a database(including size, feature, label, etc.). Most studies are based on the knowledge and experience of graphologists, to build their own database for specific applications and label manually, taking a lot of time to collect and label samples, which might be one of reasons why the public database is hard to find and share.

3.2 Pre-processing

Pre-processing is an indispensable step before analyzing the handwriting. The purpose is to separate foreground text from background as much as possible, removing possible noises to eliminate the impact on the text. There is no unified process, some are simple, others are complex. However, some essential and optional operations can be learned from pre-processing of CR method according to different applications.

Among all the operations of pre-processing, binarization is very important one and worth further studying for the following processing. It not only cleans the background with ink spots, bleed through, but also enhances the foreground to recover strokes.

3.3 Feature Extraction

For the same feature, there are various feature extraction methods, which directly affect the accuracy of extracted features. The process of extraction is to simulate and repeat the manual measurement process, which can be considered in a semi-automatic manner. Fully automated realization will rely on the deep learning technology, the input is handwriting samples instead of handwriting features.

In fact, more and more traits and symbols in handwriting have been explored. There are over 100 features, even 800 features, defined by graphologists and psychologists. Do all these features make sense and have the scientific meanings? Why are they extracted? How to select suitable features for a specific task? However, only a few of them are used in the personality traits. That is because the majority of human heart is stable, and the remaining part is changed, but only a small part. The handwriting feature that represents psychological stability is the main focus, which is stable and less variable.

Ahmed and Mathkour [17] provide three kinds of measurements: 1) General measurement, which indicates an overall impression from global perspective (e.g., stroke feature), 2) Fundamental measurement, which is local feature (e.g., baseline, slant, pressure, size, spacing, speed, margin, shape features), and 3) Accessories measurement, which is a graphic symbol (e.g., zone, connecting strokes, connectedness, ending stroke feature, features of English letter 'd'/'i'/'t'/'y', personal pronoun 'I', capital English letters, and small English letters).

However, those features are all statistical features. Non-statistical features [18] should also be considered, such as high-order local autocorrelation (HLAC, insensitive to shift) and generalized discriminant analysis (GDA, for feature combination and increased resolution).

Referring to CR method, feature extraction method can be categorized into three major groups [19]: 1) Global transformation and series expansion (Fourier transform, Gabor transform, Wavelets, moments, and Karhunen-Loeve expansion), 2) Statistical representation (zoning, crossings and distances, and projections), and 3) Geometrical and topological representation (extracting and counting topological structures, measuring and approximating the geometrical properties, coding, graph and tree). Another problem is after selecting features, how to choose proper feature extraction method. There are some literature can be studied and summarized [20, 21].

3.4 Personality Trait

We have discussed the input of a computerized graphological system, the output should be personality traits. The same question will come up as feature extraction above, how to decide the personality traits, or the classification type. The answer might be kept on the side of psychologist.

Psychology is a mature discipline, but nobody can confidently say that he is a person with certain a character, not a person with a mixture personality. Meanwhile, there exist many personality classification standards [6] (e.g., Myers-Briggs type indicator (MBTI), sixteen personality factor questionnaire (16PF), and FFM), making it hard to compare the accuracy of different papers.

3.5 Prediction Model

It is worth noting that in most automatic handwriting analysis software, feature extraction and feature analysis are always employed separately, combining these two steps into a deep learning model is a feasible approach to reduce manual intervention. Computer vision method is a good option.

Furthermore, if the training model can ignore the form of different languages, we can design a universal model to apply to a wider application scenario. It is possible because the graphology is to analyze the handwriting strokes and movements, not the handwriting content.

Numerous techniques for CR can be investigated in five general approaches [19]: 1) Template matching (direct matching, deformable templates and elastic matching, and relaxation matching), 2) Statistical technique (non-parametric / parametric recognition, clustering analysis, hidden Markov modeling (HMM), and fuzzy set reasoning), 3) Structural technique (grammatical method, and graphical method), 4) Neural networks (NNs), and 5) Combined technique (serial, parallel, and hybrid architecture).

3.6 Performance Measurement

Although the reliability of graphical measurement is satisfactory, its validity needs further research. Validity is effectiveness, which refers to the extent to measure things accurately by tools or means. Its general test criterion is in accordance with: psychological test, evaluation from experts and managers, and factor analysis technique.

The validity of the criteria may also be a problem. According to the statistics of the British Psychological Association, the average validity of personality test is only 0.15. It is recognized by the psychology community that the result of psychological measurement is not very effective and reliable. However, there is no better way to test and compare different approaches, so objective psychological measurement has become a method or a reference for graphology.

4   Experiment and Future Work

From the respective of computer vision, we use deep learning method to predict personality traits from handwriting with the help of psychology questionnaire.

For our deep convolutional neural network(CNN), the input are the original handwriting images after preprocessing (e.g., grayscale, binarization and dilation), and the output are two dimensions (Extroversion, Conscientiousness) of the Big Five dimensions [22]: Extroversion, Neuroticism, Agreeableness, Conscientiousness and Openness. The features are extracted with convolutional kernel size 3×3, and filtered by 'relu' activation function. The classifier is 'softmax', the loss function is 'categorical_crossentropy', the optimizer is 'Adadelta', the metrics is 'accuracy'.

4.1 Experiment

Our dataset was collected by CENPARMI at Concordia University in 2019. It includes 234 handwriting samples with the resolution 600 dpi. Its label of two dimensions is from two parts, one is from psychology questionnaire, the other is from graphologist, and we discard samples with large differences from two parts. The labels/scores are from 1 to 5(very low, low, average, high, very high), see Fig. 1 . However, there are 43 missing questionnaire. It also has 211 Wartegg test(a psychometric personality test, drawings organized in 8 quadrants, designed by Erik Wartegg in the 1930 s) samples, which are not considered in our research domain. We carried out two different experiments as follows:

(a) score=1   (b) score=2   (c) score=3   (d) score=4   (e) score=5

(f) score=1   (g) score=2   (h) score=3   (i) score=4   (j) score=5

Fig. 1. Two dimensions(Extroversion(1st row), Conscientiousness(2nd row)) with scores from 1(very low) to 5(very high).

Two Classes Without Deleting Samples. We only use the score of 4 and 5 as the dimension, if there is conflict with the questionnaire and graphologist, we just ignore them. Therefore, there are 173 conscientiousness samples, 70 extroversion samples, and 32 unused samples. The experiment trains on 121 samples, validates on 122 samples with epoch $= 100$. And the loss and accuracy of the validation data is respectively 2.0716 and 0.6148.

Two Classes with Deleting Samples. The similar experiment is done as before, but with deleting samples. Therefore, there are 107 conscientiousness samples, 42 extroversion samples. The experiment trains on 74 samples, validates on 75 samples with epoch $= 100$. And the loss and accuracy of the validation data is respectively 1.8104 and 0.6800.

Conclusion. We could see from Fig. 2, after training without and with deleting samples, the second experiment has better and smooth results than the first one, with a higher accuracy and a lower loss. The reason maybe the deletion operation may remove the distracting samples, making models more effective.

4.2 Future Work

In the future work, we will do more research to summarize many kinds of feature extraction methods for the same graphological feature, provide more information

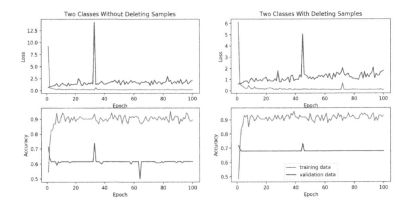

Fig. 2. Experimental results with loss and accuracy.

about databases and personality traits to analyze different computerized graphology systems.

We will do more experiments with different neural networks, and classify into different classes (such as 10 classes, because in each dimension there are scores from 1 to 5). Currently, our experiment is for a whole image of handwriting, later we will choose different granularity, like couples of paragraphs, lines and words. This will reduce the influence of our small-size of datasets.

We will research on a deep learning model to combine the steps of feature extraction and feature analysis for an English public database, then apply to a Chinese public database.

References

1. Kedar, S., Nair, V., Kulkarni, S.: Personality identification through handwriting analysis: a review. Int. J. Adv. Res. Comput. Sci. Softw. Eng., **5**(1), 548–556 (2015)
2. Shengxi, X.: Graphology. Fudan University Law School, Shanghai (1943)
3. Yang, G., Lin, B.: Chinese Characters and Personality: an Exploratory Study, PhD dissertation of Department of Psychology, National Taiwan University (1964)
4. Weng, S.: Study on Chinese Handwriting and Personality. Bulletin of the Department of Chinese Literature National Chengchi Univesity, (2), pp. 146–160 (1981)
5. Yuan, Z.: Research and application on handwriting. Beijing Mass press (1993)
6. Survey on handwriting-based personality trait identification: K. Chaudhari and A. Thakkar. Expert Syst. Appl. **124**, 282–308 (2019)
7. Gavrilescu, M., Vizireanu, N.: Predicting the Big Five personality traits from handwriting. EURASIP J. Image Video Process.**57**(1), 1–17 (2018)
8. Chen, Z., Lin, T.: Automatic personality identification using writing behaviours: an exploratory study. Behaviour Info. Technol. **36**(8), 839–845 (2017)
9. Garoot, A., Safar, M., Nobile, N.: Computational graphology applied to handwriting images. In: Proceedings of the International Conference on Pattern Recognition and Artificial Intelligence. Montreal, Canada, pp. 677–682 (2018)

10. Chaubey, G., Arjaria, S.K.: Personality Prediction Through Handwriting Analysis Using Convolutional Neural Networks. Tiwari, R., Mishra, A., Yadav, N., Pavone, M. (eds) In: Proceedings of International Conference on Computational Intelligence. Algorithms for Intelligent Systems. Springer, Singapore (2022). https://doi.org/10.1007/978-981-16-3802-2_5

11. Nair, G.H., Rekha, V., Soumya Krishnan, M.: Handwriting Analysis Using Deep Learning Approach for the Detection of Personality Traits. In: Karuppusamy, P., Perikos, I., García Márquez, F.P. (eds) Ubiquitous Intelligent Systems. Smart Innovation, Systems and Technologies, vol 243. Springer, Singapore (2022). https://doi.org/10.1007/978-981-16-3675-2_40

12. Anari, M.S., Rezaee, K., Ahmadi, A.: TraitLWNet: a novel predictor of personality trait by analyzing Persian handwriting based on lightweight deep convolutional neural network. Multimedia Tools Appl. **81**(8), 10673–10693 (2022). https://doi.org/10.1007/s11042-022-12295-3

13. Sudholt, S., Fink, G.A.: PHOCNet: A deep convolutional neural network for word spotting in handwritten documents. In: 15th International Conference on Frontiers in Handwriting Recognition (ICFHR) Shenzhen, China, pp. 277–282 (2016)

14. Yin, F., Wang, Q.F. Zhang, X.Y. et al.: Chinese handwriting recognition competition (ICDAR)Washington D.C., USA, pp. 1464–1469 (2013)

15. Liu, C.-L., Yin, F., Wang, D.-H., et al.: Chinese Handwriting Database Building and Benchmarking. Adv. Chin. Doc. Text Proc. **2**, 31–55 (2017)

16. Xu, Y., Yin, F., Wang, D.-H., et al.: CASIA-AHCDB: a Large-scale Chinese Ancient Handwritten Characters Database. International Journal on Document Analysis and Recognition (IJDAR),Sydney. Australia, pp. 793–798 (2019)

17. Ahmed, P., Mathkour, H.: On The Development of an Automated Graphology System. In: Proceedings of the 2008 International Conference on Artificial Intelligence (IC-AI) Las Vegas, USA, pp. 897–901 (2008)

18. Fallah, B., Khotanlou, H.: Identify human personality parameters based on handwriting using neural network. In Artificial Intelligence and Robotics (IRANOPEN). Qazvin, Iran, pp. 120–126 (2016)

19. Chaudhuri, A., Mandaviya, K., Badelia, P.: Optical character recognition systems. In: Optical Character Recognition Systems for Different Languages with Soft Computing, pp. 9–41. Springer, Cham (2017). https://doi.org/10.1007/978-3-319-50252-6_2

20. Bal, A., Saha, R.: An improved method for handwritten document analysis using segmentation, baseline recognition and writing pressure detection. In: 6th International Conference on Advances in Computing & Communications (ICACC) Cochin, India, pp. 403–415 (2016)

21. Hashemi, S., Vaseghi, B., Torgheh, F.: Graphology for Farsi handwriting using image processing techniques. IOSR J. Electron. Commun. Eng. (IOSR-JECE), **10**(3), 1–7 (2015)

22. Norman, W. T.: Toward an adequate taxonomy of personality attributes: replicated factor structure in peer nomination personality ratings. J. Abnorm. Soc. Psychol. **66**(6), 574–583(1963)

Graph Reduction Neural Networks for Structural Pattern Recognition

Anthony Gillioz[1](✉)(iD) and Kaspar Riesen[1,2](iD)

[1] Institute of Computer Science, University of Bern, Bern, Switzerland
[2] Institute for Informations Systems, University of Applied Science Northwestern Switzerland, Olten, Switzerland
{anthony.gillioz,kaspar.riesen}@inf.unibe.ch

Abstract. In industry, business, and science large amounts of data are produced and collected. In some of these data applications, the underlying entities are inherently complex, making graphs the representation formalism of choice. Actually, graphs endow us with both representational power and flexibility. On the other hand, methods for graph-based data typically have high algorithmic complexity hampering their application in domains that comprise large entities. This motivates to research graph-based algorithms that are both efficient and reliable. In the present paper, we introduce a novel pattern recognition framework that consists of two basic parts. First, we substantially reduce the size of the original graphs by means of a graph neural network. Eventually, we use the reduced graphs in conjunction with graph edit distance and a distance-based classifier. On five datasets we empirically confirm the benefit of the novel reduction scheme regarding both classification performance and computation time.

Keywords: Structural pattern recognition · Graph matching · Graph reduction · Graph neural network

1 Introduction and Related Work

Due to their representational power and flexibility, graphs are regarded as universal data representation that is widely used in pattern recognition and related fields (e.g. in [1–3]). A crucial task in graph-based pattern recognition is *graph matching* [4,5]. The problem of graph matching consists of finding common, or similar, (sub-)structures in pairs of graphs and define a similarity or dissimilarity score based on the found matching. Due to the extensive use of graphs in diverse applications, the development and research of efficient methods for graph matching is an active field of research [6].

A standard method to solve the task of graph matching is *Graph Edit Distance* (GED) [7]. GED defines a graph dissimilarity based upon the minimum amount of distortion necessary to convert a source graph into a target graph. Unfortunately, finding an exact solution for GED is known to be an $\mathcal{NP}$-complete

© The Author(s), under exclusive license to Springer Nature Switzerland AG 2022
A. Krzyzak et al. (Eds.): S+SSPR 2022, LNCS 13813, pp. 64–73, 2022.
https://doi.org/10.1007/978-3-031-23028-8_7

problem for general graphs. To cope better with the computational complexity of GED, various approximation algorithms have been proposed in the literature(e.g., [8,9]).

A somehow complementary idea to better handle large graphs, is to work with size-reduced versions of the original graphs [10,11]. *Graph reduction methods* produce a graph $G_R = (V_R, E_R)$ from the original graph $G = (V, E)$, with reduced node and/or edge sets $|V_R|$ and/or $|E_R|$, respectively. The aim of such a reduction is to obtain a smaller graph G_R that maintains the main topology and properties of the original graph G. Graph reduction methods have been successfully applied in diverse pattern recognition applications. In [12], for instance, a multiple classifier system is presented that combines the GED on differently reduced graphs, while in [13], various levels of reduced graphs are used to perform classification in a coarse-to-fine fashion.

In the present paper, we introduce a novel graph reduction method that learns the relevant features of the graph topology by means of *Graph Neural Networks* (GNN). Research on GNNs is a rapidly emerging field in structural pattern recognition [14,15]. The general idea of GNNs is to learn vector representations for nodes and/or (sub-)graphs such that given criteria are optimized. From a broader perspective, GNNs can be seen as a mapping from a graph domain into a vector space. Once an embedding is computed for a node and/or a (sub-)graph, the vectorial representation can be used to solve downstream tasks such as node or graph classification.

The contribution of the present paper is twofold. First, we show how a learned GNN model can be used to select the nodes of graph to be removed (instead of applying handcrafted graph coarsening methods as proposed in [12], for instance). Once the GNN model is trained, we are able to readily reduce arbitrary graphs from various datasets. Second, we use the learned graph reductions in a graph matching scenario. In particular, we approximate the GED on the reduced graphs in order to solve an underlying graph classification task. To the best of our knowledge, such an approach – that combines GNNs for graph reduction with fast approximate GED – has not been proposed before.

The remainder of the present paper is organized as follows. In the next Section we describe in detail the reduction algorithm based on GNNs as well as the complete graph matching framework. In Sect. 3, we present a thorough experimental evaluation on five different graph data sets. We first evaluate and analyze the structure and characteristics of the reduced graphs, and eventually, we conduct a large scale classification experiment (including an ablation study). Finally, in Sect. 4 we draw conclusions and make suggestions for future research activities.

2   Graph Matching on GNN Reduced Graphs

Major goal of the present paper is to make graph-based pattern recognition more efficient. To this end, we combine two complementary research directions, viz. Graph Neural Networks (GNNs) (for graph reduction) and approximate

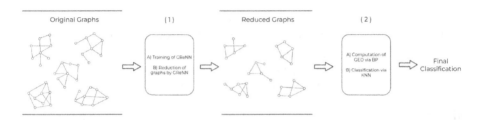

Fig. 1. The proposed graph matching framework consists of two basic parts. (1) Training of a GNN and reduction of the graphs with the optimized GNN. (2) Graph classification with a KNN using approximated GED that is computed on the reduced graphs.

Graph Edit Distance (GED) (for graph matching). Hence, the proposed framework can be split into two main parts as illustrated in Fig. 1.

- In part (1), we aim at producing reduced graphs via GNNs. This part includes both training of the network model and the actual reduction of the graphs. Goal of the GNN based reduction is to obtain strongly reduced graphs, that are still representative enough such that they can be used for pattern recognition tasks. In Subsect. 2.1 we give more details on this part.
- In part (2) of our framework, we use the reduced graphs in a classification scenario by means of approximate graph matching. It is our main hypothesis that the power and flexibility of GED in conjunction with the strong generalization and learning power of GNNs lead to a fast and accurate graph recognition framework. In Subsect. 2.2 we give more details on this part.

2.1   Graph Reduction Neural Network (GReNN)

In contrast with previously proposed reduction methods (e.g., graph coarsening [10] or graph sparsification [11]), we aim for a reduction method that learns the nodes to be deleted. Actually, a learned model has the advantage that once it is trained it can readily be applied on unseen data. In particular, we propose to use a GNN to learn the graph reduction on a training set and eventually apply the model to the complete dataset.

In Table 1 we present an overview of the architecture of the proposed *Graph Reduction Neural Network* (GReNN). We use a mix of graph convolution layers (*GCNConv*) [15] to learn the node representation, and *gPool layers* [16] as node sampling method.

The gPool layer is based upon the projection of the feature vector $\boldsymbol{x}_u$ (attached to node u) on a trainable vector $\boldsymbol{p}$ (i.e., $y_u = \boldsymbol{x}_u \boldsymbol{p}/\|\boldsymbol{p}\|$). The scalar value y_u quantifies the retained information when projecting the feature vector of node $u \in V$ onto the direction of vector $\boldsymbol{p}$. The node sampling is done according to the largest scalar projection value in order to preserve the maximum of information.

Table 1. Overview of the architecture of our graph reduction scheme GReNN for both the 50% and 25% setting. Reduced graphs are obtained after the last graph convolution layer (marked with an asterisk *)

GReNN-50	GReNN-25
GCNConv($\mathbb{R}^n$, 64)	
gPool(0.5)	gPool(0.5)
GCNConv(64, 64)	GCNConv(64, 64)
–	gPool(0.5)
–	GCNConv(64, 64)
GCNConv(64, 64)*	
POOLING	
Linear(64, #classes)	

The gPool layer also incorporates a graph connectivity augmentation that turns out to be particularly advantageous in our framework. This augmentation method is originally employed to improve the information flow in subsequent layers. In our case it prevents the creation of reduced graphs that are sparsely connected or even completely edge-free.

We propose two versions of the reduction model, termed GReNN-50 and GReNN-25 (where the number indicates the percentage of remaining nodes). For GReNN-50 we apply a pooling ratio of 0.5 (means that we are deleting 50% of the nodes). For GReNN-25 we evaluate two ways to reduce the graphs. First, we train the network from scratch with a pooling ratio of 0.25. Second, we train the model with a reduction level of 50%, freeze the already trained layers, and add a new layer (again with pooling ratio 0.5) to obtain graphs with 25% of the nodes. The newly updated network is then trained with half of the original epochs. Preliminary experiments show that for the architecture of GReNN-25 the two step optimization process actually achieves better results than training from scratch. Hence, we apply this specific architecture of an extra pooling layer in our framework.

The last two layers of our model are used for training only. That is, they are used to produce a graph embedding which in turn is used to back-propagate the classification error during the training of the reduction model. The reduced graphs – actually used for classification in part (2) of our framework – are obtained after the last graph convolution layer (marked with an asterisk in Table 1).

2.2 Classification of GReNN Reduced Graphs

For classification, we employ a *K-Nearest Neighbor* classifier (KNN) that operates on the GReNN reduced graphs. In particular, we compute the GED on GNN reduced graphs using the fast approximation algorithm *BP* [9]. The algorithm BP[1] basically consists of a special formatting of the underlying graphs as

---

[1] BP stands for BiPartite graph matching.

sets of local substructures of equal size. Eventually, the entities of these sets are optimally assigned to each other in cubic time by means of a bipartite matching algorithm [17]. Based on the found assignment on can instantly derive an approximation of the GED.

Rather than a KNN any other dissimilarity based classification algorithm could be employed as well in our framework. We feel, however, that the KNN is particularly well suited because of its direct use of the dissimilarity information without any additional training. The same accounts somehow for the GED computation via algorithm BP. That is, any other approximation algorithm could be used for this task (e.g., suboptimal algorithms surveyed in [4]).

3 Empirical Evaluations

3.1 Datasets and Experimental Setup

We use five standard datasets, viz. one from the IAM graph repository [18][2] (Mutagenicity) and four datasets from the TUDataset Repository [19][3] (NCI1, Proteins, Enzymes, DD). Two datasets represent molecules from real-world applications (Mutagenicity and NCI1) and three datasets represent protein structures from bioinformatic applications (Enzymes, Proteins, and DD).

The datasets used in the present paper are split into three disjoint sets for training, validation, and testing[4]. The splitting is carried out once at the beginning of our evaluation, and – for the sake of consistency – we maintain this splitting throughout each run. Yet, it is known that the performance of neural networks depend on the initialization of its weights and the actual split of the dataset [20]. To disentangle this random factor from our method and be sure that our reduction scheme works properly, each experiment is repeated five times with different weight initialization and different dataset splits.

The training and optimization of the hyperparameters for the GReNNs are exclusively achieved on the training and validation sets, respectively. The same accounts for the optimization of the KNN classification and GED computation via BP in the second part of our framework.

For the training of the GReNNs we use the hyperparameters originally proposed in [16]. Yet, we reduce the number of epochs from 200 to 20 as preliminary experiments show limited improvements when using more than 20 epochs. For optimizing the KNN we evaluate the number of nearest neighbors $k \in \{1, 3, 5\}$ and for the computation of GED we optimize a factor α that weights the relative importance between node and edge edit costs in the range $]0, 1]$ with a step size of 0.05.

The following empirical evaluation consists of three major parts, viz. an analysis of the characteristic and structure of the GReNN reduced graphs (see

---

[2] www.iam.unibe.ch/fki/databases/iam-graph-database.

[3] http://www.graphlearning.io/.

[4] During the splitting of the datasets we sample the graphs such that the class balance is preserved in each subset.

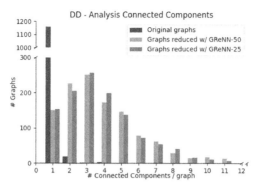

Fig. 2. Histogram that shows the number of graphs (on the y-axis) that have a given number of connected components per graph (on the x-axis) on the dataset DD (the frequencies of graphs with more than 11 connected components are clipped for the sake of conciseness).

Sect. 3.2), a classification experiment that compares our novel framework with a reference system (see Sect. 3.3), and an ablation study that investigates the usefulness of the two major parts of our framework (see Sect. 3.4).

3.2 Analysis of the Structure of the Reduced Graphs

When reducing a graph by removing nodes and their incident edges one might obtain edge-free, or at least sparsely connected graphs. To confirm that the proposed reduction does not produce graphs that consists of sets of unconnected nodes only, we start our evaluation by thoroughly analyzing the reduced graphs. The following evaluations and visualizations are related to one dataset only, viz. DD. However, on the other datasets we obtain similar results and make similar observations.

In Fig. 2 we visualize the number of graphs on the y-axis that have a certain number of connected components (x-axis) in a histogram. We compare the original graphs with the reduced graphs obtained with GReNN-50 and GReNN-25. We observe that the vast majority of original graphs consist of one connected component. On the other hand, the reduced graphs tend to have more than only one connected component. However, we see that the number of connected components is smaller than, or equal to, five for about 80% of the graphs (for both reduction levels).

We also analyze the number of isolated nodes per graph obtained after reduction on all datasets. We can report that on three datasets (Mutagenicity, Enzymes, and Proteins) there are less than three isolated nodes per graph for both reductions. On NCI1 and DD we observe a maximum of seven isolated nodes per graph when reducing the graphs with GReNN-25. This implies that the connected components of the reduced graphs do not mainly consist of single nodes but rather of connected subgraphs. Hence, the reduced graphs maintain their connectivity in general, which is mainly due to the connectivity augmentation method of the gPool layer (as discussed in Sect. 2).

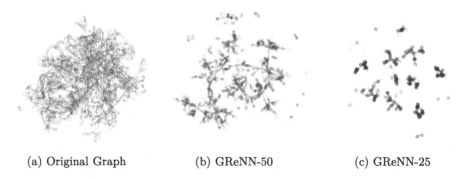

(a) Original Graph   (b) GReNN-50   (c) GReNN-25

Fig. 3. Example of an original graph and the corresponding reduced graphs via GReNN-50 and GReNN-25.

In Fig. 3 we illustrate the effects of our graph reduction on a sample graph. Reductions from the original graph to 50% and 25% of the available nodes clearly lead to an increased number of connected components. However, we also observe the effects of the built-in edge augmentation. That is, the remaining nodes stay - in general - highly connected via dense and local edge structures.

3.3   Classification Results

Next, we conduct classification experiments in order to evaluate our complete framework (using both parts (1) and (2)). The first experiment compares our framework (GReNN-50 and GReNN-25) with a KNN classifier that operates on the original graphs (termed reference system from now on).

In Table 2 we show the classification accuracies of all systems and the speed up of the matching times achieved with our framework (compared to the reference system). With GreNN-50 we achieve quite similar classification accuracies as the reference system on four out of five datasets (the only statistically significant deterioration is observed on the Enzymes dataset). As expected, we also observe a clear speed up on all datasets (we observe speed ups by factors of about three).

The improvement regarding computation time is even more pronounced when using graphs reduced to 25% of their original size. That is, with GReNN-25 we reduce the runtime by factors of about ten on four out of five datasets (on Enzymes the speed up is about a factor of five).

Using such a strong reduction of the graphs, however, leads to more substantial deteriorations w.r.t the classification accuracy. That is, with GreNN-25 we observe three statistically significant deteriorations. On the other two datasets, however, no statistically significant difference is visible between our system and

Table 2. Classification accuracies obtained on test sets with the reference system and our novel framework (GReNN-50, GReNN-25). We also show the relative speed up of the matching times. (•: indicates a statistically deterioration compared to the reference system using a t-test with p-value = 0.05).

Dataset	Ref. System	GReNN-50	Speed Up	GreNN-25	Speed Up
Mutagenicity	74.3 ± 1.8	72.6 ± 2.6	3.6	68.1 ± 2.3 •	8.5
NCI1	72.0 ± 1.1	71.6 ± 1.7	3.7	63.8 ± 1.6 •	9.9
Enzymes	50.3 ± 4.4	43.5 ± 2.9 •	2.1	33.2 ± 5.1 •	4.6
Proteins	71.8 ± 4.2	70.5 ± 1.6	3.4	69.3 ± 1.6	8.8
DD	73.4 ± 1.7	74.3 ± 2.2	2.2	70.1 ± 2.0	12.6

the reference system. In general, the results obtained are in a fairly similar range to the original results – this is quite astonishing, considering that we are using 25% of the nodes only.

3.4 Ablation Study

Finally, we conduct an ablation study in order to investigate the usefulness of the two separate parts of our framework (for the sake of conciseness we use GReNN-50 only). We compare the complete framework with the following systems:

- *Without-1*: This refers to a system in which we replace the first part of our framework (1) with a random reduction of the graphs to 50% of the available nodes. This system helps us to verify whether or not our framework actually benefits from the elaborated GNN reduction.
- *Without-2*: This system refers to the proposed GNN architecture of the GReNN that is directly employed for graph classification without taking the detour of GED computation (that is, we omit part (2) of our framework). This system helps us to verify whether the proposed framework actually benefits from the combination of GNN reductions and GED computations.

In Table 3 we observe that our novel framework GReNN-50 outperforms the system Without-1 on all five data sets (three of the five improvements are statistically significant). On four out of five data sets GReNN-50 also outperforms the second reference system Without-2 (three out of five improvements are statistically significant).

The main finding of this comparison is twofold. First, it clearly shows that our novel learning-based reduction scheme outperforms a naïve graph reduction, and second, the proposed graph reduction based on GNNs in conjunction with GED computations is clearly beneficial as it outperforms the isolated GNN in general.

Table 3. Ablation study where we compare the classification accuracy obtained with and without the use of the two parts (1) and (2) of our framework. Symbols ①/② indicate a statistically significant improvement with and without the use of the two parts (1) and (2), respectively (using a t-test with p-value = 0.05).

Dataset	Mutagenicity	NCI1	Enzymes	Proteins	DD
Without-1	64.4 ± 1.3	59.5 ± 0.9	23.2 ± 3.3	69.7 ± 1.5	72.5 ± 1.7
Without-2	73.3 ± 4.4	67.0 ± 2.3	26.1 ± 3.7	70.2 ± 2.4	67.9 ± 3.1
Ours	72.6 ± 2.6 ①/-	71.6 ± 1.7 ①/②	43.5 ± 2.9 ①/②	70.5 ± 1.6 -/-	74.3 ± 2.2 -/②

4   Conclusions and Future Work

In the present paper we introduce a novel graph-based pattern recognition framework. In particular, we propose to use recent GNN layers and adopt them for the specific task of graph reduction. Eventually, we classify the learned graph reductions with well-known pattern recognition techniques based on GED. To the best of our knowledge, such an architecture has not been proposed before.

By means of an experimental evaluation on diverse datasets we empirically confirm that our graph reduction process is useful for downstream graph classification tasks. That is, we show that our framework maintains satisfactory classification accuracy when deleting 50%, and on some datasets even 25%, of the nodes. Simultaneously, we show that the runtime can – as expected – be substantially reduced by means of our method. The conducted experiments also clearly reveal that the novel approach for graph reduction performs significantly better than randomized graph reductions. Last but not least, we observe that combining graph matching with our specific GNN for graph reduction achieves better results than an isolated GNN model (that employs the same architecture as we use for reduction).

Future research objectives comprise to further improve the reduction so that the GReNN-25 reduced graphs also achieve classification accuracies similar to the ones obtained with GReNN-50. Our long term objective is to propose a graph reduction framework that produces extremely small graphs that are still useful for pattern recognition (e.g., graphs reduced to 10% of their original size). Moreover, we aim at researching whether we are able to learn a network model for graph reductions such that inter-class distances remain large (or even increase) while intra-class distances remain small (or even decrease).

References

1. Bai, L., Cui, L., Jiao, Y., Rossi, L., Hancock, E.R.: Learning backtrackless aligned-spatial graph convolutional networks for graph classification. IEEE Trans. Pattern Anal. Mach. Intell. **44**(2), 783–798 (2022)
2. Blumenthal, D.B., Boria, N., Bougleux, S., Brun, L., Gamper, J., Gaüzère, B.: Scalable generalized median graph estimation and its manifold use in bioinformatics, clustering, classification, and indexing. Inf. Syst. **100**, 101766 (2021)

3. Bahonar, H., Mirzaei, A., Sadri, S., Wilson, R.C.: Graph embedding using frequency filtering. IEEE Trans. Pattern Anal. Mach. Intell. **43**(2), 473–484 (2021)
4. Conte, D., Foggia, P., Sansone, C., Vento, M.: Thirty years of graph matching in pattern recognition. Int. J. Pattern Recognit Artif Intell. **18**(3), 265–298 (2004)
5. Foggia, P., Percannella, G., VentoM, M.: Graph matching and learning in pattern recognition in the last 10 years. Int. J. Pattern Recognit. Artif. Intell., **28**(1), 1450001 (2014)
6. Riesen, K.: Structural Pattern Recognition with Graph Edit Distance. ACVPR, Springer, Cham (2015). https://doi.org/10.1007/978-3-319-27252-8
7. Bunke, H., Allermann, G.: Inexact graph matching for structural pattern recognition. Pattern Recognit. Lett., **1**(4), 245–253 (1983)
8. Fischer, A., Riesen, K., Bunke, H.: Improved quadratic time approximation of graph edit distance by combining hausdorff matching and greedy assignment. Pattern Recognit. Lett. **87**, 55–62 (2017)
9. Riesen, K., Bunke, H.: Approximate graph edit distance computation by means of bipartite graph matching. Image Vis. Comput. **27**(7), 950–959 (2009)
10. Chen, J., Saad, Y., Zhang, Z.: Graph coarsening: from scientific computing to machine learning. CoRR, abs/2106.11863 (2021)
11. Spielman, D.A., Teng, S.-H.: Spectral sparsification of graphs. SIAM J. Comput. **40**(4), 981–1025 (2011)
12. Gillioz, A., Riesen, K.: Improving graph classification by means of linear combinations of reduced graphs. In: Proceedings of the 11th International Conference on Pattern Recognition Applications and Methods, ICPRAM 2022, Online Streaming, February 3–5, 2022, pp. 17–23 SCITEPRESS, 2022
13. Riba, P., Lladós, J., Fornés, A.: Hierarchical graphs for coarse-to-fine error tolerant matching. Pattern Recognit. Lett. **134**, 116–124 (2020)
14. Hamilton, W.L., Ying, R., Leskovec, J.: Representation learning on graphs: Methods and applications. IEEE Data Eng. Bull. **40**(3), 52–74 (2017)
15. Kipf, T.N., Welling, M.: Semi-supervised classification with graph convolutional networks. In: 5th International Conference on Learning Representations, ICLR 2017, Toulon, France, 24–26 April 2017, Conference Track Proceedings. OpenReview.net (2017)
16. Gao, H., Ji, S.: Graph u-nets. In: Proceedings of the 36th International Conference on Machine Learning, ICML 2019, 9–15 June 2019, Long Beach, California, USA, Proceedings of Machine Learning Research. PMLR **97**, 2083–2092 (2019)
17. Rainer, E.: Burkard, Mauro Dell'Amico, and Silvano Martello. SIAM, Assignment Problems (2009)
18. Riesen, K., Bunke, H.: IAM Graph Database Repository for Graph Based Pattern Recognition and Machine Learning. In: da Vitoria Lobo, N., et al. (eds.) SSPR /SPR 2008. LNCS, vol. 5342, pp. 287–297. Springer, Heidelberg (2008). https://doi.org/10.1007/978-3-540-89689-0_33
19. Morris, C., Kriege, N.M., Bause, F., Kersting, K., Mutzel, P.: and Marion Neumann. A collection of benchmark datasets for learning with graphs. CoRR, TUDataset (2020)
20. Fellicious, C., Weissgerber, T., Granitzer, M.: Effects of Random Seeds on the Accuracy of Convolutional Neural Networks. In: Nicosia, G., et al. (eds.) LOD 2020. LNCS, vol. 12566, pp. 93–102. Springer, Cham (2020). https://doi.org/10.1007/978-3-030-64580-9_8

Sentiment Analysis from User Reviews Using a Hybrid Generative-Discriminative HMM-SVM Approach

Rim Nasfi$^{(\boxtimes)}$ and Nizar Bouguila

Concordia Institute for Information Systems Engineering, 1455 De Maisonneuve Blvd. West, Montreal, QC, Canada
r_nasfi@encs.concordia.ca, nizar.bouguila@concordia.com

Abstract. Sentiment analysis aims to empower automated methods with the capacity to recognize sentiments, opinions and emotions in text. This recognition capacity is now highly demanded to process and extract proper knowledge from the exponentially-growing volume of user-generated data. Applications such as analyzing online products reviews on e-commerce marketplaces, opinion mining from social networks and support chat-bots optimization are putting into practice various methods to perform this complex natural language processing task. In this paper, we apply a hybrid generative-discriminative approach using Fisher kernels with generalized inverted Dirichlet-based hidden Markov models to improve the recognition performance in the context of textual analysis. We propose a method that combines HMMs as a generative approach, with the discriminative approach of Support Vector Machine. This strategy allows us to deal with sequential information of the text, and at the same time use the special focus on the classification task that SVM could provide us. Experiments on two challenging user reviews datasets i.e. Amazon for products reviews and IMDb movies reviews, demonstrate an effective improvement of the recognition performance compared to the standard generative and Gaussian-based HMM approaches.

Keywords: Sentiment analysis · Opinion mining · Hybrid Generative-Discriminative · Hidden Markov models · Support Vector Machines

1 Introduction

Opinion mining received massive interest in recent years, particularly with the important role that reviews and shared experiences over e-commerce and marketplaces platforms play in shaping purchase intentions. User-generated data is increasing drastically, especially when it comes to reviews and feedback shared over the internet [1]. This huge volume of data calls for automated methods to process and extract proper knowledge from it [2]. Analyzing users' opinions from different perspectives can considerably help not only the customer to buy or adopt the best product available in the market, but also the merchant, to better understand what are the good or bad features related to their products and determine their effect on the buyers' opinion and feeling regarding the

A. Krzyzak et al. (Eds.): S+SSPR 2022, LNCS 13813, pp. 74–83, 2022.
https://doi.org/10.1007/978-3-031-23028-8_8

product [3]. These reviews are for the most part available in a text format in an unstructured way and naturally need to be modeled appropriately in order to provide useful insights to both customer and seller. Therefore, recognizing sentiments and attitudes in textual data can provide a better understanding of trends and tendencies related to products [4].

Sentiment analysis, also known as opinion mining analyzes people's opinions as well as their emotions towards a product, an event or an organization [5]. It has been widely investigated in different research works and approached through different methodologies such as lexicon-based approaches as well as hybrid approaches [6]. Nevertheless, there has been rarely a solid explainability or knowledge behind decisions resulting from these methods, and the latter were oftentimes handled as black-box methods. Challenges in sentiment analysis as a natural language processing application are numerous. In fact, analyzing text reviews implies dealing with text sequences that are usually limited in length, have many misspellings and shortened forms of words [7]. As a result, we have an immense vocabulary size and vectors representing each review are highly sparse. Two main approaches are used in machine learning to perform recognition tasks: generative techniques that model the underlying distributions of classes, and discriminative techniques that give a sole focus on learning the class boundaries [8]. Both techniques have been widely used in sentiment analysis to effectively recognize divergent users' attitudes [9].

Hidden Markov Models (HMMs) represent a powerful tool to properly model sequential information within textual data. Their generative aspects constitute a quite potent way to handle sentiment recognition and they tend to require less training data than discriminative models. In the case where the task to be performed is classification, Support Vector Machines (SVM) can clearly distinguish the differences between categories and can thus outperform generative models especially if a large number of training examples are available. SVM is extensively used due to its great capacity to generalize, often resulting in better performance than traditional classification techniques [10]. As a discriminative approach, the main functioning of SVM is to find surfaces that better separate the different data classes using a kernel that allows efficient discrimination in non-linearly separable input feature spaces. Hence the importance of adopting the convenient kernel function which needs to be suitable for the classified data and the objective task. Standard kernels include linear polynomial and radial basis function kernels [11]. Adopting these kernels is not always possible, especially when it comes to classifying objects represented by sequences of different lengths [12]. Consequently, the mentioned kernels may not be a good fit to model our text data. Therefore, a hybrid generative-discriminative approach is adopted to permit the conversion of data into fixed-length and hence provide additional performance to the model.

In this work, we introduce a novel implementation of a hybrid generative-discriminative model and examine its performance on real-life benchmark datasets. We propose the use of Fisher Kernels (FK) generated with Generalized inverted Dirichlet-based HMMs (GIDHMM) to model textual data. Moreover, the use of GID (Generalized Inverted Dirichlet) to model emission probabilities is backed by the several interesting mathematical properties that this distribution has to offer. These properties allow for a representation of GID samples in a transformed space where features are independent and follow inverted Beta distributions. Adopting this distribution allows us to take advantage of conditional independence among features. This interesting strength is used in this paper to develop a statistical model that essentially handles positive vectors.

In light of the existing methods in sentiment analysis, the main contributions of this paper are the following: First, we apply for the first time a non-Gaussian HMM, i.e. generalized inverted Dirichlet-based HMM on the challenging product reviews benchmark by Amazon and the IMDb movie reviews dataset. Second, we derive a hybrid generative-discriminative approach of our HMM-based framework with FK for SVM-based modeling of positive vectors. This novel approach is also tested on the aforementioned datasets, as an unprecedented attempt of using Fisher Vectors-based hybrid generative-discriminative models to handle textual data analysis. The remainder of the paper is organized as follows, Sect. 2 presents background topics on sentiment analysis and examines related works. Section 3 discusses the proposed model. Section 4 presents the performed experiments and obtained results. We finally conclude the paper in Sect. 5.

2   Related Work

Sentiment analysis has been the focus of numerous research works, where it has been approached in different levels namely document, sentence and aspect level [13]. While the document level focuses on classifying the whole opinion document into either a positive or negative sentiment, sentence-level looks at determining whether the sentence expresses the nature of opinion (negative, positive, neutral). On the other hand, the aspect-level analysis provides a detail-oriented approach to handle the broad aspect. Thus, it focuses on determining whether or not a part of the text is opinion-oriented towards a certain aspect. It can present a positive polarity towards one aspect and a negative polarity towards another. Classifying the text as positive or negative depends on the chosen aspect and applied knowledge [14]. It is noteworthy to mention that expressions associated with sentiment are mainly the words or features that express the sentiment of the text, such as adjectives or adverbs. Furthermore, when tackling sentiment analysis, there are mainly three types of machine learning approaches, i.e. supervised, unsupervised and semi-supervised learning and they are respectively used in cases where data is labelled, unlabeled and partially labelled [15].

In HMM-based sentiment analysis, models analyze the input textual data and formulate clusters. After that HMMs are utilized to perform the categorization by considering the clusters as hidden states. Every model analyzes a given instance in order to specify its sentimental polarity. In comparison to related works in the literature, HMM-based methods possess higher interpretability and can model the changing aspects of sentiment information. Multiple sentiment analysis applications adopt HMMs as the main model. In [16] Rabiner proposes a method of predicting sentiments from voice. Also, in [17] authors use HMMs to detect sentiments by considering the label information as positive, negative or neutral. Knowledge about the words position and hidden states is available and injected into the model. While this approach has shown effective results, it clearly assumes knowledge of the labels and thus requires a significant human effort.

In our work, we do not require knowledge about the states labels and we propose another alternative where we estimate the similarity between the pattern of input text and that of sentences expressing either a positive or negative sentiment, plus we make use of SVM to increase the model's performance when it comes to the classification accuracy. We detail our method in the next section.

3 Hybrid Generative-Discriminative Approach with Fisher Kernels

When it comes to our adopted approach, a single HMM is trained for every class in the data depending on the context aspect. The resulting likelihoods will be further classified by the SVM classifier to identify the sentiment. In this section, we present the proposed approach. To illustrate our model, we are first listing various HMM notations and enumerating the upcoming used work script. We then recall the main process behind the forward-backward algorithm. Lastly, we perform a complete derivation of the FK-based model.

3.1 Hidden Markov Models

A Hidden Markov Model is a statistical model that can be used to describe real-world processes with observable output signals. HMMs are defined as an underlying stochastic process formed by a Markov chain that is not observable (hidden). For each hidden state, a stochastic model creates observable output signals or observations, based on which hidden states can be estimated [18]. We consider a HMM with continuous emissions and K hidden states. We put a set of hidden states $H = \{h_1, ..., h_T\}; h_j \in [1, K]$. The transition probabilities matrix: $B = \{b_{ij} = P(h_t = j | h_{t-1} = j')\}$ and the emission probabilities matrix: $C = \{c_{ij} = P(m_t = i | h_t = j)\}; i \in [1, M]$ where M is the number of mixture components associated with state j. We define the initial probability: π_j which is the probability to start the observation sequence from the state j. We denote an HMM as: $\Delta = \{B, C, \varphi, \pi\}$ where φ is the set of mixture parameters depending on the chosen type of mixture. In this work, we focus on the generalized inverted Dirichlet distribution. Let $\overrightarrow{X}$ a D-dimensional positive vector following a GID distribution. The joint density function is given by Lingappaiah [19] as:

$$p(\overrightarrow{X} | \overrightarrow{\alpha}, \overrightarrow{\beta}) = \prod_{d=1}^{D} \frac{\Gamma(\alpha_d + \beta_d)}{\Gamma(\alpha_d)\Gamma(\beta_d)} \frac{X_d^{\alpha_d - 1}}{\left(1 + \sum_{l=1}^{d} X_d\right)^{\eta_d}} \tag{1}$$

where $\overrightarrow{\alpha} = [\alpha_1, ..., \alpha_D]$, $\overrightarrow{\beta} = [\beta_1, ..., \beta_D]$. η is defined such that $\eta_d = \alpha_d + \beta_d - \beta_{d+1}$ for $d = 0, ..., D$ with $\beta_{D+1} = 0$.

3.2   Inference on Hidden States: Forward-Backward Algorithm

The forward algorithm computes the probability of being in state h_j up to time t for the partial observation sequence produced by the model Δ. We consider a forward variable $\gamma_t(i) = P(X_1, X_2, \ldots, X_t, , i_t = h_i | \Delta)$. There is a recursive relationship that is used to compute the former probability. We can resolve for $\gamma_t(i)$ recursively as follows:

1. Initialization:

$$\gamma_t(i) = \pi_i \varphi_i(X_1) \quad 1 \leq i \leq K \tag{2}$$

2. Recursion:

$$\gamma_{t+1}(j) = \left[\sum_{i=1}^{K} \gamma_t(i) b_{ij} \right] \varphi_j(X_{t+1}) \quad for \ \ 1 \leq t \leq T-1, \ 1 \leq j \leq K \tag{3}$$

3. Termination:

$$P(X|\Delta) = \sum_{i=1}^{K} \gamma_T(i) \tag{4}$$

The backward variable, which is the probability of the partial observation sequence $X_{t+1}, X_{t+2}, \ldots, X_T$ given the current state is denoted by $\delta_t(i)$ and can similarly be determined as follows:

1. Initialization:

$$\delta_t(i) = 1, \ 1 \leq i \leq K \tag{5}$$

2. Recursion:

$$\delta_t(i) = \sum_{j=1}^{K} b_{ij} \varphi_j(X_{t+1} \delta_{t+1}(j) \quad for \ \ t = T-1, T-2, \ldots, 1 \ \ 1 \leq i \leq K \tag{6}$$

3. Termination:

$$P(X|\Delta) = \sum_{i=1}^{K} \gamma_i(T) = \sum_{i=1}^{K} \pi_i b_i(X_1) \varphi_i(1) = \sum_{i=1}^{K} \sum_{j=1}^{K} \gamma_t(i) b_{ij} \varphi_j(X_{t+1}) \delta_{t+1}(j) \tag{7}$$

3.3   Fisher Kernels

Non-linear SVM serves our discrimination needs in the context of realistic recognition tasks. The strategy is to use a Kernel method to avoid calculation cost and memory consumption problems that might arise from performing inner product calculation of high-dimensional feature vectors. It will allow us to implicitly project objects to high-dimensional space by using a kernel function $\kappa(x_\zeta, x_\eta) = \langle \phi(x_\zeta), \phi(x_\eta) \rangle$ and solving the problem with observations x_ζ and x_η represented as Bag of Features (BoF) or Bag of Words (BoW) [20] in general with ϕ being a projection function and $\langle \cdot, \cdot \rangle$ meaning the inner product. Here we choose Fisher Kernel as the kernel function. This choice is motivated by FK being a general way of fusing generative and discriminative approaches for classification. FK is formulated as

$$FK(X_\zeta, X_\eta) = \langle FS(X_\zeta, \Delta), FS(X_\eta, \Delta) \rangle \tag{8}$$

where X_ζ and X_η are two observations, Δ is the parameters set of a generative model defined by $P(X|\Delta)$ and $FS(X_\zeta,\Delta)$ is the Fisher score.

$$FS(X,\Delta) = \nabla_\Delta \log P(X|\Delta) \qquad (9)$$

Given a particular HMM:

$$L(X|\Delta) = \log P(X|\Delta) = \log \sum_{i=1}^{K} \gamma_T(i) = \log \sum_{i=1}^{K} \pi_i \varphi_i(X_1)\delta_1(i) \qquad (10)$$

The derivatives for GID-based HMM can be defined as follows

$$\nabla_\Delta L(X|\Delta) = \left[\frac{\partial L(X|\Delta)}{\partial \pi_i}, \frac{\partial L(X|\Delta)}{\partial b_{ij}}, \frac{\partial L(X|\Delta)}{\partial \alpha_{id}}, \frac{\partial L(X|\Delta)}{\partial \beta_{id}} \right] \qquad (11)$$

$$\frac{\partial L(X|\Delta)}{\partial \pi_i} = \frac{\varphi_i(X_1)\delta_1(i)}{\sum_{i=1}^{K} \pi_i \varphi_i(X_1)\delta_1(i)} \qquad (12)$$

$$\frac{\partial L(X|\Delta)}{\partial b_{ij}} = \frac{1}{P(X|\Delta)} \sum_{k=1}^{K} \frac{\partial \gamma_T(k)}{\partial b_{ij}}$$
$$= \frac{1}{P(X|\Delta)} \sum_{k=1}^{K} \sum_{l=1}^{K} \frac{\partial \gamma_{T-1}(l)}{\partial b_{ij}} b_{lk}\varphi_k(X_T) + \partial \gamma_{T-1}(i)\varphi_{ij}(X_T) \qquad (13)$$

$$\frac{\partial L(X|\Delta)}{\partial \alpha_{id}} = \frac{1}{P(X|\Delta)} \left(\sum_{j=1}^{K} \sum_{k=1}^{K} \frac{\partial \gamma_{T-1}(k)}{\partial \alpha_{id}} b_{kj}\varphi_j(X_T) + \sum_{k=1}^{K} \partial \gamma_{T-1}(k) b_{ki} \frac{\partial \varphi_i(X_T)}{\partial \alpha_{id}} \right) \qquad (14)$$

$$\frac{\partial L(X|\Delta)}{\partial \beta_{id}} = \frac{1}{P(X|\Delta)} \left(\sum_{j=1}^{K} \sum_{k=1}^{K} \frac{\partial \gamma_{T-1}(k)}{\partial \beta_{id}} b_{kj}\varphi_j(X_T) + \sum_{k=1}^{K} \partial \gamma_{T-1}(k) b_{ki} \frac{\partial \varphi_i(X_T)}{\partial \beta_{id}} \right) \qquad (15)$$

$$\frac{\partial \varphi_i(X_t)}{\partial \alpha_{id}} = \Psi(\alpha_{id}+\beta_{id}) - \Psi(\alpha_{id}) + \log\left(\frac{X_d}{1+X_d}\right) \qquad (16)$$

$$\frac{\partial \varphi_i(X_t)}{\partial \beta_{id}} = \Psi(\alpha_{id}+\beta_{id}) - \Psi(\beta_{id}) + \log\left(\frac{1}{1+X_d}\right) \qquad (17)$$

4   Experiments

4.1   Problem Modeling

The main motive behind the use of HMMs in sentiment analysis is the strong analogy behind the process of understanding a sentiment from a text in real-life and the predictive aspect of HMMs. In fact, an opinion consists of a number of words that together represent what the person is trying to express. To understand it, a person would first proceed by reading the words sequentially from left to right, knowing that each word would normally be related to the previous one in a certain way to create a meaningful sentence. Accordingly, words forming an emotion are modeled as observations in a HMM, while the emotion is the hidden state, which needs to be unveiled. In this work, we propose to tackle this problem of sentiment analysis in a hybrid way; where the HMM states can be modeled approximately by considering a hidden variable given by patterns that are independent of the class of text. An abstraction of this process is illustrated in Fig. 1. We first perform word clustering to indicate a certain word pattern, in a way that all negative and positive connotations are clustered separately. After that, we make use of our SVM to further classify the output into negative and positive classes. This treatment requires a dictionary constructed for use by sentiment analysis models.

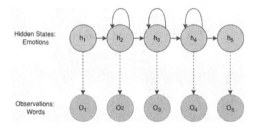

Fig. 1. Problem modeling through hidden state-observation HMM

4.2 Datasets

We choose to experiment on the Amazon[1] reviews dataset from the Stanford Network Analysis Project (SNAP), which spans 18 years period of product reviews [21]. Amazon dataset is a popular corpus of product reviews collected from the Amazon marketplace, which includes ratings and plain text reviews for a multitude of product niches. In this work, we choose to work on the Electronics niche and randomly pick a mix of 30,000 sample reviews from training and testing sets with a vocabulary size of 72.208 unique words. As a word segmentation approach, we use the Part-of-Speech tagger designed by the Stanford NLP group applying default settings. We have also removed numerals, auxiliary words and verbs, punctuation and stop words as a part of the pre-processing. An output vector is then generated corresponding to the input word. Each review is modeled by a text vector that is obtained after adding up all the word vectors and dividing by the number of words. We also test our work on the IMDb dataset [22], which is mainly developed for the task of binary sentiment classification of movie reviews. It consists of an equal number of positive and negative reviews. The dataset is evenly divided between training and test sets with 25,000 reviews each. We choose to work on both subsets uniformly and we hence deal with 50,000 samples from each group with 76,340 unique words in total. A similar pre-processing to the one adopted with the Amazon dataset is then applied.

---

[1] Publicly available at: https://snap.stanford.edu/data/web-Amazon.html.

Experiments are carried out in a total of 10 independent runs, each time starting with a different initial observation. The resulting log-likelihoods and accuracies are averaged on these 10 independent runs. It is worth mentioning that through the training process our aim is to maximize the log-likelihood of the data and we target by performing these experiments a higher recognition accuracy each time. Therefore we will not focus on assessing the time complexity in this work. We also use the totality of the dataset for training and testing and do not opt for splitting the data into train and test segments for both datasets.

4.3 Results

A HMM is trained for each class, in this case, positive and negative. A test text is sent to each model and the probabilities of occurrence are computed. Each HMM returns a probability and the model with the highest probability of occurrence will indicate the class of this text. We choose the number of hidden states to be $K = 2$ for both experiments. In our discussion, we focus on the effect of using a hybrid model as opposed to a HMM-based model. We also shed some light on the usefulness of using the GID distribution as emission probabilities. Results on both datasets are presented in Fig. 2. We notice that, on both datasets, Hybrid HMM-SVM models perform remarkably better than Generative-only models, be it GID-based or Gaussian-based. The hybrid GID-based HMM/SVM model achieved average accuracies of 86.40% and 88.94% on the Amazon and IMDb datasets respectively, while we only yielded 79.72% and 82.09% using the GID-HMM Generative-only model. Most importantly, we notice that GID-based models achieved the highest accuracy on both hybrid and generative approaches compared to the Gaussian-based models. This increasing recognition capacity was expected and is once more validated when it comes to positive vector modeling. Using GID as an emission probability distribution, clearly improved the modeling accuracy. Conclusively, results achieved by applying the hybrid generative-discriminative approach demonstrate the striking increase in terms of the modeling accuracy and further validate the improved performance that SVM provided to the generative technique.

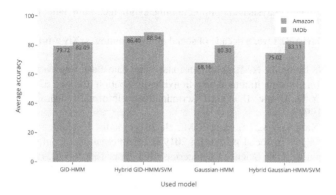

Fig. 2. Average accuracies for sentiment recognition on the Amazon and IMDb datasets with each of the tested models

5   Conclusion

In this work, we presented a hybrid generative-discriminative approach to automatically identify sentiments expressed in user reviews online, using a combination of HMMs as a generative approach, along with the discriminative SVM. The main motivation behind this choice is to be able to enhance the model's capacity by taking advantage of the powerful classification role that SVM plays without neglecting the sequential aspect of text data. We also gave a special focus on modeling positive vectors by using non-Gaussian Generalized Inverted Dirichlet distributions as emission probabilities for our HMM. The interest in adopting the GID for modeling our data arose from the limitations encountered when inverted Dirichlet was adopted, in particular its restraining strictly positive covariance. We carried out what we believe to be the first attempt of applying GIDHMM both in generative and hybrid modes on the challenging Amazon product reviews and IMDb reviews. A comprehensive solution for sentiment detection was introduced, where we allowed the automatic recognition of positive and negative emotions, in textual data. According to the results obtained from the conducted experiments, we proved that the proposed approach obtained highly accurate recognition rates compared to both generative GIDHMM and Gaussian-based HMM. Future works are intended to be done in the near future extending this work to different Natural Language Process applications, such as product recommendation and understanding user intent.

References

1. Chakraborty, K., Bhattacharyya, S., Bag, R.: A survey of sentiment analysis from social media data. IEEE Trans. Comput. Soc. Syst. **7**(2), 450–464 (2020)
2. Liu, B.: Sentiment analysis: mining opinions, sentiments, and emotions. Cambridge University Press, (2020)
3. Wang, Y., Pal, A.: Detecting emotions in social media: a constrained optimization approach, In: Proceedings of the 24th International Conference on Artificial Intelligence, ser. IJCAI'15. AAAI Press, pp. 996–1002 (2015)
4. Liang, R., Wang, J.-Q.: A linguistic intuitionistic cloud decision support model with sentiment analysis for product selection in e-commerce. Int. J. Fuzzy Syst. **21**(3), 963–977 (2019)
5. Liu, B.: Sentiment analysis and opinion mining. synth. lect. hum. lang. technol. **5**(1), 1–167 (2012)
6. Cortis, K., Davis, B.: Over a decade of social opinion mining, arXiv e-prints, pp. arXiv-2012, (2020)
7. Zamzami, N., Bouguila, N.: High-dimensional count data clustering based on an exponential approximation to the multinomial beta-liouville distribution. Inf. Sci. **524**, 116–135 (2020)
8. Rubinstein, Y. D., Hastie, T., et al.: Discriminative vs informative learning. in KDD, vol. 5, pp. 49–53 (1997)
9. Perikos, I., Kardakis, S., Paraskevas, M., Hatzilygeroudis, I.: Hidden markov models for sentiment analysis in social media, In: 2019 IEEE International Conference on Big Data, Cloud Computing, Data Science Engineering (BCD), pp. 130–135 (2019)
10. Nasfi, R.,Bouguila , N.: Online learning of inverted beta-Liouville HMMs for anomaly detection in crowd scenes, In: Hidden Markov Models and Applications. Springer, 2022, pp. 177–198 https://doi.org/10.1007/978-3-030-99142-5_7
11. Bdiri, T., Bouguila, N.: Bayesian learning of inverted dirichlet mixtures for svm kernels generation. Neural Comput. Appl. **23**(5), 1443–1458 (2013)

12. Wang, C., Zhao, X., Wu, Z., Liu, Y.: Motion pattern analysis in crowded scenes based on hybrid generative-discriminative feature maps, In 2013 IEEE International Conference on Image Processing. IEEE, pp. 2837–2841 (2013)

13. Feldman, R.: Techniques and applications for sentiment analysis. Commun. ACM **56**(4), 82–89 (2013)

14. Liu, N., Shen, B.: Rememnn: a novel memory neural network for powerful interaction in aspect-based sentiment analysis. Neurocomputing **395**, 66–77 (2020)

15. G. Gautam, G., Yadav, D.: Sentiment analysis of twitter data using machine learning approaches and semantic analysis, In: 2014 Seventh International Conference on Contemporary Computing (IC3), pp. 437–442 (2014)

16. Rabiner, L.: A tutorial on hidden markov models and selected applications in speech recognition. Proc. IEEE **77**(2), 257–286 (1989)

17. Jin, W., Ho, H. H., Srihari, R. K.: Opinionminer: a novel machine learning system for web opinion mining and extraction, In: Proceedings of the 15th ACM SIGKDD International Conference on Knowledge Discovery and Data Mining, ser. KDD '09. New York, NY, USA: Association for Computing Machinery, pp. 1195–1204. (2009) https://doi.org/10.1145/1557019.1557148

18. Rabiner, L., Juang, B.: An introduction to hidden markov models, IEEE assp magazine, 3(1), pp. 4–16 (1986)

19. Lingappaiah, G.: On the generalised inverted dirichlet distribution. Demostratio Math. **9**(3), 423–433 (1976)

20. Harris, Z.S.: Distributional structure. Word **10**(2–3), 146–162 (1954)

21. McAuley, J., Leskovec, J.: Hidden factors and hidden topics: understanding rating dimensions with review text, In: Proceedings of the 7th ACM Conference on Recommender Systems, ser. RecSys '13. Association for Computing Machinery, pp. 165–172 (2013)

22. Maas, A.L., Daly, R.E., Pham, P.T., Huang, D., Ng, A.Y., Potts, C.: Learning word vectors for sentiment analysis, In: Proceedings of the 49th Annual Meeting of the Association for Computational Linguistics: Human Language Technologies Portland, Oregon, USA: Association for Computational Linguistics, pp. 142–150 June 2011 http://www.aclweb.org/anthology/P11-1015

Spatio-Temporal United Memory for Video Anomaly Detection

Yunlong Wang(ID), Mingyi Chen(ID), Jiaxin Li(ID), and Hongjun Li(✉)(ID)

School of Information Science and Technology, Nantong University, Nantong 226019, China
lihongjun@ntu.edu.cn

Abstract. Video anomaly detection aims to identify the anomalous that do not conform to normal behavior. The abnormal events tend to relate to appearance and motion, in which there are considerable difference in each other. From the perspective of philosophy, "act according to circumstances", we propose a dual-flow network to dissociate appearance information and motion information, processing these information in two individual branches. In addition, we employ a spatio-temporal united memory module to bridge the relationship between appearance and motion, since there is another saying in philosophy that "The things are universal and interact with each other", thus the hidden relationship in them also ought to be utilized as a useful clue for anomaly detection. To the best of our knowledge, this is the first work to detect anomalies with a spatio-temporal united memory. The model is able to achieve AUC 96.92%, 87.43%, and 75.42% on UCSD Ped2, Avenue, and ShanghaiTech, respectively. Extensive experiments on three publicly available datasets demonstrate the excellent generalization and high effectiveness of the proposed method.

Keywords: Video anomaly detection · Information dissociation · Spatio-temporal united memory

1 Introduction

Video anomaly detection is an open and challenging task, since the occurrence probability of abnormal events is usually much less than that of normal events, and in real scenarios, the type of abnormal events is unpredictable due to the boundlessness and rarity of itself [1, 2]. It is usually a unsupervised learning problem, in which there are priori unknown abnormal samples, it is assumed that most training datasets are only composed of normal data. Therefore, under the premise of unsupervised, anomaly detection intends to acquire the hidden feature representation behind the video sequences of the testing set.

Most existing deep learning models employ an autoencoder as the infrastructure, which attempts to induce the network to extract the significant features of some datasets through using a low dimensional intermediate representation (or some sparse regularization) as a bottleneck [1, 3–5]. Although autoencoders have occupied a significant place in anomaly detection, some deficiency remain in most approaches. Concretely, some

A. Krzyzak et al. (Eds.): S+SSPR 2022, LNCS 13813, pp. 84–93, 2022.
https://doi.org/10.1007/978-3-031-23028-8_9

approaches utilize a single-flow network which tends to specializes in learning rules of single type information, yet ignoring some crucial factor, namely abnormal appearance or motion cues [1, 3]. Some methods comprehensively consider motion and appearance cues, and use dual-flow approaches for anomaly detection tasks, but most of these methods employ an optical flow to capture motion features, in which the optical flow has a shortcoming of massive running-time [4, 5].

We propose a dual-flow architecture to separate the appearance information and the motion information because of the inherent difference in them. Rather than the previous approaches which takes an optical flow as the input of motion branch, we utilize the RGB difference of the consecutive video frames. Further, a memory module is affixed in the bottleneck layer of the dual-flow network to unite the spatio-temporal information, matching the motion features to the corresponding appearance features and acquiring the hidden relationship in them, since this hidden relationship between appearance and motion plays a significant role in anomaly detection.

2 Related Work

2.1 Dual-Flow Structure Based on Autoencoder

Deep learning models for anomaly detection mainly utilize autoencoders as feature extractors to learn the normality behind video sequences [6–8], thereby utilizing the hidden data features for reconstruction or prediction tasks. Some methods [9, 10] combine the scene features with the complex motion features to perform a modeling task, proving the effectiveness of dual-flow architecture in modeling complex dynamic scenes. To fully utilize the spatio-temporal information in video frames, Stewart et al. [4] create a precedent for a dual-flow architecture including a RGB flow and an optical flow. Nguyen et al. [5] employ a reconstruction network with an image translation model. Specially, the former sub-network determines the most significant structures that appear in video sequences and the latter one attempts to associate motion templates to such structures. Similarly, Chang et al. [11] introduce two streams to separate and process the spatio-temporal information, the spatial part models the normality on the appearance feature space through a reconstruction task, while the temporal part learns the regularity of motion with a prediction task. Apart from the methods which tend to directly take video frames and optical flow as input, Morais et al. [12] break new ground, modeling the normal motion mode from 2D human skeleton tracks, and learning the regular spatio-temporal pattern of skeleton features through two interacting branches.

2.2 Memory

In recent years, memory module has attracted more and more attention for its powerful ability to enhance the neural network performance, which occupies a firm foothold in many fields, such as multi-mode data generation, one-time learning, and natural language processing. Quan et al. [13] propose a new external memory architecture which has two distinct advantages over existing neural external memory architectures, namely the division of the external memory into two parts-long-term memory and working memory. In order to consider the diversity of normal patterns explicitly, Gong et al. [14] design a memory module with a new update scheme where items in the memory record prototypical patterns of normal data. Inspired by this, we utilize a memory module to automatically memorize the appearance features and the corresponding motion features in video clips, bridging relationship between features so as to detect the mismatched objects or/and motion patterns.

3   Methodology

In our proposed method, we dissociate the appearance information and the motion information with a dual-flow architecture, and utilize a memory module as the auxiliary task for bridging relationship between different information, as shown in Fig. 1. Specifically, our model takes the first four consecutive frames (empirical results indicate that too many or too few frames cannot capture reasonable motion features) as the input of the motion flow encoder to acquire motion features representation, the first frame as the input of the static flow encoder to acquire appearance features representation. Then, we concatenate these representations in the feature channels at the same spatial location to fuse the two flows, memorizing and reconstructing the concatenated representation in the memory module. The static decoder and the motion decoder take the corresponding reconstructed representations to generate a prediction frame.

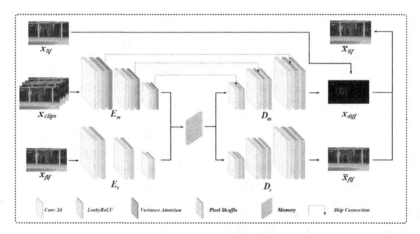

Fig. 1. Overview of our spatio-temporal dual-flow network for video anomaly detection.

Encoding and Decoding. Since abnormal events tend to differ from the normal in objects and motion patterns, both the appearance information and the motion information can be significant clues for anomaly detection. To obtain the useful intermediate features of video frame, the two encoders generate a appearance representation z_a and a motion representation z_m, respectively. For the reconstructed representation $\bar{z}_a$ and $\bar{z}_m$, the corresponding decoder decode them to generate $\bar{x}_{diff}$ and $\bar{x}_{lif}$.

Variance Attention. Attention mechanism based on variance can highlight the fast moving region precisely and suppress the irrelevant static region at the same time. Therefore, with the pretreatment of attention mechanism, motion representation acquires the moving object features and motion pattern features.

Memory Module. The memory module is designed as a matrix $M \in R^{N \times 2C}$ which mainly refers to two operations: reading and writing. For the intermediate representation $z_a \in R^{H \times W \times C}$ and $z_m \in R^{H \times W \times C}$, it firstly concatenate them as a feature map r, and generates one real value vector with size of $1 \times 1 \times 2C$ along the dimension of the feature map, which contains partial features of motion and appearance. The memory module bridge relationship between this two kinds features, reconstructing the concatenated representation r as $\bar{r}$ through the historical information stored in memory module.

For the query item q_r^t generated from r and the memory item m_i, we calculate the matching probabilities α_i and β_t in reading and writing, respectively.

$$\alpha_i = \frac{\exp(d(q_r^t, m_i))}{\sum\limits_{i'=1}^{N} \exp(d(q_r^t, m_{i'}))}, \quad \sum_{i=1}^{N} \alpha_i = 1 \tag{1}$$

$$\beta_t = \frac{\exp(d(q_r^t, m_i))}{\sum\limits_{t'=1}^{T} \exp(d(q_r^{t'}, m_i))}, \quad \sum_{t=1}^{T} \beta_t = 1 \tag{2}$$

where N denotes the number of the memory items number, $T = H \times W$ denotes the number of query items, and $d(\cdot)$ denotes cosine similarity function.

Rebuilding in memory with a finite number of normal patterns contributes to generate small reconstruction errors in normal cases. Nevertheless, some anomalies may still have a chance of being well reconstructed through a complex combination of tiny memory items. In order to alleviate this problem, we adopt a hard shrinkage operation to improve the sparsity of α.

$$\hat{\alpha}_i = \frac{\max(\alpha_i - \varepsilon, 0) \cdot \alpha_i}{|\alpha_i - \varepsilon| + \delta} \bigg/ \left\| \frac{\max(\alpha_i - \varepsilon, 0) \cdot \alpha_i}{|\alpha_i - \varepsilon| + \delta} \right\|_1 \tag{3}$$

where ε is a sufficiently small tunable hyperparameter. Futher, the query item and the memory item can be expressed as follows.

$$\hat{q}_r^t = \hat{\alpha} M = \sum_{i=1}^{N} \hat{\alpha}_i m_i \tag{4}$$

$$\hat{m}_i = \sum_{t=1}^{T} f(\beta_t q_r^t + m_i) \tag{5}$$

where $f(\cdot)$ denotes L2 normalization function, and $T = H \times W$.

Anomaly Score. We quantify the extent of normality or abnormality in a video frame at the testing phase. Assuming that the queries obtained from a normal video frame are similar to the memory items, as they record prototypical patterns of normal data. We compute the L2 distance between each query and the nearest item.

$$D_{mem} = D(q_r^t, m_m) = \frac{1}{K} \sum_{k=1}^{K} \left\| q_r^k - m_m \right\|_2 \qquad (6)$$

where $D(\cdot)$ denotes the L2 distance function. Next, we combine the output $\bar{x}_{fif}$ with $\bar{x}_{diff}$ to generate the prediction error of the dual-flow network.

$$D_{pre} = D(x_{fif}, \bar{x}_{fif}) + D(x_{diff}, \bar{x}_{diff}) + \sum_{d \in \{x,y\}} \left\| \left| g_d(x_{diff}) \right| - \left| g_d(\bar{x}_{diff}) \right| \right\|_1 \qquad (7)$$

where g_d denotes the image gradient of the video frame along the x and y axes of space. In order to make a more reasonable score for a testing frame, we comprehensively consider both the D_{mem} and the D_{pre} to generate a normal score, normalizing it to obtain $Score(t)$ within the range of [0, 1] for each video frame.

$$Score(t) = \frac{1/D_{mem}D_{pre} - \min_t(1/D_{mem}D_{pre})}{\max_t(1/D_{mem}D_{pre}) - \min_t(1/D_{mem}D_{pre})} \qquad (8)$$

4   Experiments

4.1   Datasets and Evaluation Metrics

We evaluate our model on three publicly available datasets. UCSD Pedestrian [10]: We select the UCSD Ped2 subset containing 16 training videos and 12 test videos, and the testing videos including 12 abnormal events with a resolution of 240×360. Avenue [15]: Avenue dataset contains 16 training videos and 21 test videos with a resolution of 360×640. ShanghaiTech [3]: Compared with other datasets, the ShanghaiTech dataset contains 13 different scenes with different lighting conditions and camera angles. We resize all input video frames to 256×256 and trained our network on a single NVIDIA GeForce 1080 Ti using Adam optimizer [16], and the learning rate is set to 2e−5. We evaluate the performance of our method by measuring the area under the ROC curve (AUC), the higher the AUC value is, the more accurate the anomaly detection result is.

4.2   Comparison with Existing Methods

To prove the effectiveness of our method, We compare it with the state of the art which can be specific divided into reconstruction based methods [1, 5, 17, 18], and prediction based methods [3, 11, 14, 19], on the publicly available datasets.

Table 1. AUC of Different Methods on UCSD Ped2, CUHK Avenue and ShanghaiTech.

Algorithm		AUC (%)		
		Ped2	Avenue	ShanghaiTech
Reconstruction based methods	Conv2DAE [1]	85.00	80.00	60.90
	Nguyen [5]	96.20	86.90	–
	StackRNN [17]	92.20	81.70	68.00
	Abati [18]	95.41	–	72.50
Prediction based methods	Liu [3]	95.40	84.90	72.80
	Cluster-AE [11]	96.70	87.10	73.70
	MemAE [14]	94.10	83.30	71.20
	AnoPCN [19]	96.80	86.20	73.60
Ours		**96. 92**	**87.43**	**75. 42**

From the AUC results of our method and other methods in Table 1, we observe three things: (1) Our method gives the best results on all three datasets, achieving the average AUC of 96.92%, 87.43%, and 75.42%, respectively. Especially on ShanghaiTech, our method appears at least 1.72 (75.42% *vs* 73.70%) points higher than other methods. This demonstrates the effectiveness of our approach to exploit a memory module for anomaly detection. (2) Similar to our method, although MemAE [14] propose to augment the autoencoder with a memory module and develop an improved autoencoder called memory-augmented autoencoder, the performance of our method is vividly better than that of MemAE (96.92% *vs* 94.10% & 87.43% *vs* 83.30% & 75.42% *vs* 71.20%). The reason for this can be attributed to that our method unite the spatial and temporal information, obtaining the hidden relationship in different features which is ignored in MemAE. (3) Cluster-AE [11] employs a K-means clustering strategy to acquire a more compact intermediate representation, the performance of which is slightly inferior to ours on USCD Ped2 and Avenue (96.70% vs 96.92% & 87.10% vs 87.43%). However, the clustering strategy it used is kind of double-edged sword to some extent, concretely, on the one hand a manual selection is needed in the choice of the clustering numbers, on the other hand it occupies more running-time while our method is superior to it (36 FPS *vs* 32 FPS).

To qualitatively analyze the anomaly detection performance of our model, we visualize the anomaly detection examples on three datasets, as shown in Fig. 2. The blue shaded region in the figure indicates anomalies in Ground Truth while the red curve represents the normal score of the frames in a specific video, a higher score denotes a higher normality degree. As we can see that, when a abnormal event appears (disappears), the anomaly score will increase (decrease) rapidly, which verifies that our method does detect the anomalies in a high speed.

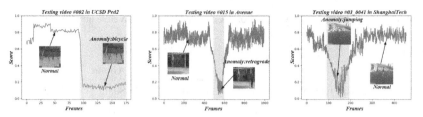

Fig. 2. Some examples of video anomaly detection.

4.3 Ablation Experiments

To evaluate the effectiveness of the memory module we used, several ablation studies on three datasets are conducted, the corresponding AUC results and ROC curves are shown in Table 2 and Fig. 3, respectively.

Table 2. Ablation study results on three datasets.

	Ped2		Avenue		ShanghaiTech	
Basic	✓	✓	✓	✓	✓	✓
Memory	–	✓	–	✓	–	✓
AUC (%)	95.13	**96.92**	85.37	**87.43**	72.31	**75. 42**

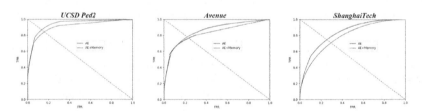

Fig. 3. ROC curve with or without memory module on three datasets.

As we can find in Table 2, the AUC of combining memory module with Basic model are 1.79%, 2.06%, and 3.11% improvement on UCSD Ped2, Avenue, and ShanghaiTech, respectively. Which indicates that the hidden relationship between appearance and motion plays a significant role in detecting abnormal events, our method can acquire this kind of relationship and thus tend to be in a better state. In addtion, it is interesting that the boosting effect on ShanghaiTech is much higher than that of the other two datasets, which can be considered as the memory is superior in handling complex data, while the ShanghaiTech dataset contains more complex scenes and objects. This further illustrates that the combination of dual-flow network and memory module is more suitable for practical and complex scenarios.

Figure 3 shows the comparison results of ROC curves for Basic model and Basic model with a memory on the three publicly available datasets. The network with memory module tends to have a better ROC curve compared to the no memory case, which demonstrates that our spatio-temporal united memory network has significant advantages in identifying abnormal events in videos.

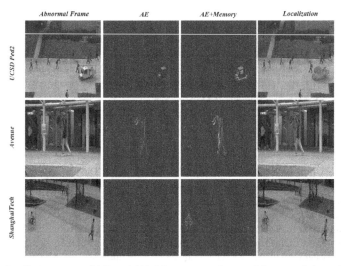

Fig. 4. Prediction error of AE and AE+Memory on abnormal frames of three datasets.

Figure 4 show the qualitative results of our model for anomaly localization on all three datasets. It shows input frames, prediction error without memory, prediction error with memory, and abnormal regions overlaid to the frame. In comparison to the no memory case, the network with a memory module can realize a more vivid localization for the abnormal events. Concretely, we can see that normal regions are predicted well, while abnormal regions are not, and abnormal events, such as the appearance of car, throwing, and bicycle, are highlighted. It can be concluded that our method has an extremely accurate effect on the localization of abnormal events.

4.4 Running Time

With an NVIDIA GeForce 1080 Ti, our current implementation takes on average 0.027 s to estimate abnormality for an image of size 256×256 on Avenue, namely, we achieve 36 FPS for anomaly detection.

5 Conclusion

In this paper, we have proposed a dual-flow network with a spatio-temporal memory module for video anomaly detection. We have shown that processing the appearance representation and the motion representation separately, enabling our model make reasonable response to different information. We have also presented spatio-temporal united memory, which matches the motion information with the corresponding appearance information. Extensive evaluations on standard benchmarks show that our method outperforms existing methods by a large margin, which proves the effectiveness of our method in anomaly detection. In our further work, we will further study how to combine dual-current network and memory module to achieve more ideal results under the premise of lightweight.

Acknowledgment. This work is supported in part by National Natural Science Foundation of China under Grant 61871241, Grant 61971245 and Grant 61976120, in part by Jiangsu Industry University Research Cooperation Project BY2021349, in part by Nantong Science and Technology Program JC2021131 and in part by Postgraduate Research and Practice Innovation Program of Jiangsu Province KYCX21_3084 and KYCX22_3340.

References

1. Hasan, M., Choi, J., Neumann, J., Roy-Chowdhury, A.K., Davis, L.S.: Learning temporal regularity in video sequences. In: Proceedings of the IEEE Conference on Computer Vision and Pattern Recognition (CVPR), Las Vegas, NV, USA, pp. 733–742, June 2016
2. Luo, W.X., et al.: Video anomaly detection with sparse coding inspired deep neural networks. IEEE Trans. Pattern Anal. Mach. Intell. **43**(3), 1070–1084 (2021)
3. Liu, W., Luo, W., Lian, D., Gao, S.: Future frame prediction for anomaly detection-a new baseline. In: Proceedings of the IEEE Conference on Computer Vision and Pattern Recognition (CVPR), Salt Lake City, UT, USA, pp. 6536–6545, June 2018
4. Xu, D., Yan, Y., Ricci, E., Sebe, N.: Detecting anomalous events in videos by learning deep representations of appearance and motion. Comput. Vis. Image Underst. **156**, 117–127 (2017)
5. Mahadevan, V., Li, W., Bhalodia, V., Vasconcelos, N.: Anomaly detection in crowded scenes. In: Proceedings of the IEEE Conference Computer Vision and Pattern Recognition (CVPR), San Francisco, CA, USA, pp. 1975–1981, June 2010
6. Chong, Y.S., Tay, Y.H.: Abnormal event detection in videos using spatiotemporal autoencoder. In: Cong, F., Leung, A., Wei, Q. (eds.) ISNN 2017. LNCS, vol. 10262, pp. 189–196. Springer, Cham (2017). https://doi.org/10.1007/978-3-319-59081-3_23
7. Stewart, R., Ermon, S.: Label-free supervision of neural networks with physics and domain knowledge. In: Proceedings of the 31st AAAI Conference on Artificial Intelligence (AAAI), San Francisco, USA, pp. 2576–2582, February 2017
8. Nguyen, T.-N., Meunier, J.: Anomaly detection in video sequence with appearance-motion correspondence. In: Proceedings of IEEE International Conference on Computer Vision (ICCV), Seoul, Korea (South), pp. 1273–1283, October 2019
9. Kang, M., Lee, K., Lee, Y.H., Suh, C.: Autoencoder-based graph construction for semi-supervised learning. In: Vedaldi, A., Bischof, H., Brox, T., Frahm, J.-M. (eds.) ECCV 2020. LNCS, vol. 12369, pp. 500–517. Springer, Cham (2020). https://doi.org/10.1007/978-3-030-58586-0_30
10. Zhao, Y., Deng, B., Shen, C., Liu, Y., Lu, H., Hua, X.-S.: Spatiotemporal AutoEncoder for video anomaly detection. In: Proceedings of the 25th ACM Multimedia Conference (MM), New York, NY, USA, pp. 1933–1941, October 2017
11. Chang, Y.P., Tu, Z.G., Xie, W., Luo, B., Zhang, S.F., Sui, H.G.: Video anomaly detection with spatio-temporal dissociation. Pattern Recogn. **122**, 1–12 (2022)
12. Morais, R., Le, V., Tran, T., Saha, B., Mansour, M., Venkatesh, S.: Learning regularity in skeleton trajectories for anomaly detection in videos. In: Proceedings of the IEEE Computer Society Conference on Computer Vision and Pattern Recognition (CVPR), Long Beach, CA, USA, pp. 11988–11996, June 2019

13. Quan, Z., Zeng, W., Li, X., Liu, Y., Yu, Y., Yang, W.: Recurrent neural networks with external addressable long-term and working memory for learning long-term dependences. IEEE Trans. Neural Netw. Learn. Syst. (TNNLS) **31**(3), 813–826 (2020)
14. Gong, D., et al.: Memorizing normality to detect anomaly: memory-augmented deep autoencoder for unsupervised anomaly detection. In: Proceedings of the IEEE International Conference on Computer Vision (ICCV), Seoul, Korea (South), pp. 1705–1714, October 2019
15. Lu, C., Shi, J., Jia, J.: Abnormal event detection at 150 FPS in MATLAB. In: Proceedings of the IEEE International Conference on Computer Vision (ICCV), Sydney, NSW, Australia, pp. 2720–2727, December 2013
16. Kingma, D., Ba, J.: Adam: a method for stochastic optimization. In: Proceedings of the International Conference on Learning Representations (ICLR), San Diego, California, USA, May 2015. https://doi.org/10.48550/arXiv.1412.6980
17. Luo, W., Liu, W., Gao, S.: A revisit of sparse coding based anomaly detection in stacked RNN framework. In: Proceedings of the IEEE International Conference on Computer Vision (ICCV), Venice, Italy, pp. 341–349, October 2017
18. Abati, D., Porrello, A., Calderara, S., Cucchiara, R.: Latent space autoregression for novelty detection. In: Proceedings of the IEEE Conference on Computer Vision and Pattern Recognition (CVPR), Long Beach, CA, USA, pp. 481–490, June 2019
19. Ye, M., Peng, X., Gan, W., Wu, W., Qiao, Y.: AnoPCN: video anomaly detection via deep predictive coding network. In: Proceedings of the 27th ACM International Conference on Multimedia (MM), Nice, France, pp. 1805–1813, October 2019

A New Preprocessing Method for Measuring Image Visual Quality Robust to Rotation and Spatial Shifts

Guang Yi Chen[1], Adam Krzyzak[1(✉)], and Ventzeslav Valev[2]

[1] Department of Computer Science and Software Engineering, Concordia University, Montreal, QC H3G 1M8, Canada
{guang_c,krzyzak}@cse.concordia.ca
[2] Institute of Mathematics and Informatics, Bulgarian Academy of Sciences, 1113 Sofia, Bulgaria
valev@math.bas.bg

Abstract. Measuring the visual quality of an image is an extremely important task in computer vision. In this paper, we perform 2D fast Fourier transform (FFT) to both test and reference images and take the logarithm of their spectra. We convert both log spectra images to polar coordinate system from cartesian coordinate system and use FFT to extract features that are invariant to translation and rotation. We apply the existing structural similarity (SSIM) index to the two invariant feature images, where no extra inverse transform is needed. Experimental results show that our proposed preprocessing method, when combined with the mean SSIM (MSSIM), performs better than the standard MSSIM significantly in terms of visual quality scores even when no distortions are introduced to the images in the LIVE Image Quality Assessment Database Release 2. In addition, when images are distorted by small spatial shifts and rotations, our new preprocessing step combined with MSSIM still performs better than the standard MSSIM.

Keywords: Translation invariant · Rotation invariant · Image visual quality · Quality metrics

1 Introduction

Measuring image visual quality is extremely important in image processing and computer vision. Its aim is to develop metrics that can automatically assess the visual quality of images in a consistent way. There exist three types of metrics for measuring image visual quality: (a) Full reference (FR) metrics assess visual quality by comparing the test image with the reference image; (b) Reduced reference (RR) metrics measure the quality of the test image by extracting features from both the test and the reference images; (c) No reference (NR) metrics estimate the quality of the test image without comparing it with the reference image.

We briefly review several metrics for measuring image visual quality here. Wang et al. [1] developed a structural similarity (SSIM) index by comparing local correlations

© The Author(s), under exclusive license to Springer Nature Switzerland AG 2022
A. Krzyzak et al. (Eds.): S+SSPR 2022, LNCS 13813, pp. 94–102, 2022.
https://doi.org/10.1007/978-3-031-23028-8_10

in luminance, contrast, and structure between the reference and distorted images. This SSIM metric is defined as:

$$SSIM(x, y) = \frac{2\mu_x\mu_y + C_1}{\mu_x^2 + \mu_y^2 + C_1} \times \frac{2\sigma_x\sigma_y + C_2}{\sigma_x^2 + \sigma_y^2 + C_2} \times \frac{\sigma_{xy} + C_3}{\sigma_x\sigma_y + C_3}$$

where μ_x and μ_y are sample means; σ_x^2 and σ_y^2 are sample variances, and σ_{xy} is the sample cross-covariance between images x and y. The constants C_1, C_2 and C_3 are useful when the means and variances are small. The mean SSIM (MSSIM) on the whole image is the final quality score for the image. Rezazadeh and Coulombe [2, 3] proposed two wavelet-based methods for FR image quality assessment. They automatically computed the suitable scales for wavelet decomposition before measuring visual quality. In addition, their proposed methods defined a multi-level edge map for each image. Qian and Chen [4] proposed four RR metrics for measuring the visual quality of hyperspectral images after spatial resolution enhancement. These metrics can measure the visual quality of HSI whose high spatial resolution reference image is not available whereas the low spatial resolution reference image is available. Keller et al. [5] proposed a pseudo polar-based method to compute the large translation, rotation, and scaling in images. Standard metrics such as MSSIM can be applied after aligning both images by compensating the translation, rotation, and scaling that are present in the original images. Chen and Coulombe [6] developed a novel preprocessing step for MSSIM by taking the 2D Fourier transform (FFT) and better results were obtained. This method is more accurate than the standard MSSIM in measuring image visual quality, and more importantly it needs less computational time than the standard MSSIM.

In this paper, we propose a new preprocessing method for MSSIM, which is invariant to small spatial shifts and rotations of the images. We take a forward 2D FFT to both the reference and test images and compute the logarithm of both spectra images. We then apply a polarization transform to both log spectra images such that we can convert rotation to spatial shift. Another 2D FFT is applied to both polarized images so that they are invariant to small rotations and spatial shifts. We apply the MSSIM to both extracted feature images to generate a visual quality score. Experimental results show that our preprocessing method combined with MSSIM improves the quality scores significantly for the LIVE image quality assessment database release 2. More importantly, it generates better results even when no additional distortions are introduced to the testing images.

The major difference between this paper and our previous work [6] is that this paper is invariant to both rotation and spatial shifts whereas our previous work [6] is invariant to only spatial shifts. As a result, our new method should perform better than our previous work [6] in measuring image visual quality.

The organization of the rest of this paper is as follows. Section 2 proposes a novel preprocessing method so that the extracted features are invariant to spatial shifts and rotation. Section 3 conducts experiment to verify if the proposed preprocessing method works or not. Finally, Sect. 4 concludes the paper and proposes future research directions.

2 Proposed Preprocessing Method

Measuring image visual quality is an extremely important task in image processing and computer vision. Existing metric such as MSSIM performs poorly when a small deformation is introduced to the images. For example, small rotation, small spatial shifts, or small scaling factors can cause significantly score difference for standard MSSIM. This problem can be eased by preprocessing the images to be measured so that these deformations can be ignored. It is preferable to combine a preprocessing technique with existing metric such as MSSIM to improve image visual scores. Several image deformation factors need to be considered: (a) small rotations; (b) small spatial Shifts; and (c) small scaling factors. We do not consider small scaling factors in this paper because it introduces big visual difference and as a result causes big difference in image visual quality scores, which is undesirable in practical image processing.

In this paper, we propose a novel preprocessing method for measuring image visual quality that is invariant to translation and rotation of the input images. We utilize the property that the Fourier spectra are invariant to spatial shifts, and the polarization transform converts rotation into spatial shifts. As a result, taking another forward Fourier transform and obtaining their Fourier spectra will be invariant to translation and rotation of the input images. We use both invariant feature images as input to the MSSIM, and we obtain improved visual quality scores.

For polarizing the input images, we use Matlab function [7]:

$$imP = ImToPolar \, (imR, \, rMin, \, rMax, \, M, \, N)$$

which converts rectangular image imR to polar form imP. The output image is an M × N image with M points along the r axis and N points along the θ axis. The origin of the image is assumed to be at the center of the given image. The image is assumed to be grayscale. Bilinear interpolation is used to interpolate between points not exactly in the image. rMin and rMax should be between 0 and 1 and rMin < rMax. r = 0 is the center of the image and r = 1 is half the width or height of the image.

The proposed preprocessing method can be summarized as follows:

a) Take the 2D FFT of both the reference image x and the test image y and obtain the magnitude of the Fourier coefficients F_1 and F_2, which are of size M × N.
b) Take the logarithm of both spectra images: $LF_1 = \log(1 + F_1)$ and $LF_2 = \log(1 + F_2)$,
c) Convert the two generated images LF_1 and LF_2 from cartesian coordinate system to polar coordinate system, and we obtain $G_1(\xi, \theta)$ and $G_2(\xi, \theta)$.
d) Take another forward 2D FFT to both polarized images G_1 and G_2, and compute their Fourier spectra, denoted as S_1 and S_2.
e) Extract the center regions:

$$Im_1 = S_1(M/4 : 3M/4, \, N/4 : 3N/4)$$
$$Im_2 = S_2(M/4 : 3M/4, \, N/4 : 3N/4)$$

f) Use MSSIM to measure the visual quality of the images Im_1 and Im_2.

Figure 1 shows the flow diagram of the proposed preprocessing method in this paper: (a) The original image; (b) The magnitude of the Fourier coefficients; (c) The polarized image; and (d) The Fourier spectra of the polarized image.

The major advantages of our preprocessing method are that we do not need to estimate the parameters for the translation and rotation factors in the input images. This can significantly reduce the computational time and avoid errors generated in parameter estimation. The second advantage is that it does not align the spatial origins of the input images, whereas the alignment is extremely important to the polarization technique. Because the magnitude of the Fourier coefficients is invariant to spatial shifts, the origin of step a) in our preprocessing method is always at the center of the Fourier spectra images. This makes our preprocessing method combined with MSSIM more accurate than standard metric MSSIM. Experimental results demonstrate that our preprocessing method combined with MSSIM is better than the standard MSSIM even when there are no distortions to the images in the LIVE image database. In addition, when images are distorted by small spatial shifts and rotations, our new preprocessing step combined with MSSIM still performs better than the standard MSSIM. Our new preprocessing method is simple and easy to implement, so it should be particularly fast in term of computational time.

3 Experimental Results

We perform a couple of experiments by using the LIVE Image Quality Assessment Database Release 2 [8], which has 779 deformed images from 29 original images with five types of deformations. These deformations are: (a) JPEG compression; (b) JPEG2000 compression; (c) Gaussian white noise (GWN); (d) Gaussian blurring (GBlur); and (e) the Rayleigh fast fading (FF) channel model. We investigate three performance measures here: (a) The correlation coefficients (CC) between the difference mean opinion score (DMOS) and the objective model outputs after nonlinear regression; (b) The root mean square error (RMSE); and (c) The Spearman rank order correlation coefficient (SROCC), which is a nonparametric rank-based correlation metric and it is not related to any monotonic nonlinear mapping between subjective and objective scores.

Table 1 presents the MSSIM scores for the LIVE image database without adding extra deformation. Our preprocessing method combined with MSSIM performs better than the standard MSSIM in this experiment, which is particularly important in computer vision. This demonstrates that our preprocessing step is extremely useful in measuring image visual quality. Table 2 presents experimental results when every image in the LIVE database is deformed by rotation angle $0.02 \times 180/\pi$ degrees, spatial shift 2 pixels horizontally, and spatial shift 2 pixels vertically. Table 3 shows experimental results when every image in the LIVE database is deformed by rotation angle $0.02 \times 180/\pi$ degrees only. Table 4 presents experimental results when every image in the LIVE database is deformed by spatial shift 2 pixels horizontally, and spatial shift 2 pixels vertically. The standard metric MSSIM generates very low-quality scores, whereas our preprocessing method combined with MSSIM yields higher visual quality scores. We assume that we have the same DMOS scores for the original images in the LIVE image database and their distorted images.

Figure 2 depicts the scatter plots of DMOS versus MSSIM for all images in the LIVE image database without adding extra deformations, whereas Fig. 3 shows the scatter plots of DMOS versus MSSIM for the LIVE image database by introducing a combination of spatial shifts and rotations, where each image is deformed by rotation angle 0.02 × 180/π degrees, spatial shift 2 pixels horizontally, and spatial shift 2 pixels vertically. Figure 4 illustrates the scatter plots of DMOS versus MSSIM for the LIVE image database by introducing rotation only, where each image is deformed by rotation angle 0.02 × 180/π degrees. Figure 5 shows the scatter plots of DMOS versus MSSIM for the LIVE image database by introducing spatial shifts only, where each image is deformed by spatial shift 2 pixels horizontally, and spatial shift 2 pixels vertically. Our preprocessing method combined with MSSIM is very robust against both spatial shifts and rotations, whereas the standard MSSIM performs poorly in this case.

Table 1. A comparison of image visual quality assessment for images in the LIVE image database without adding any extra deformations. The best method is highlighted in bold font.

MSSIM	Method	CC	SROCC	RMSE
	Original	0.9040	0.9104	11.6828
	Proposed	**0.9146**	**0.9125**	**11.0466**

Table 2. A comparison of image visual quality assessment by deforming all images in the LIVE image database, where each image in this database is deformed by rotation angle 0.02 × 180/π degrees, spatial shift 2 pixels horizontally, and spatial shift 2 pixels vertically. The best method is highlighted in bold font.

MSSIM	Method	CC	SROCC	RMSE
	Original	0.3949	0.2058	25.1017
	Proposed	**0.7987**	**0.7958**	**16.4400**

Table 3. A comparison of image visual quality assessment by deforming all images in the LIVE image database, where each image in this database is deformed by rotation angle 0.02 × 180/π degrees. The best method is highlighted in bold font.

MSSIM	Method	CC	SROCC	RMSE
	Original	0.3940	0.2079	25.1124
	Proposed	**0.7997**	**0.7969**	**16.4055**

Table 4. A comparison of image visual quality assessment by deforming all images in the LIVE image database, where each image in this database is deformed by spatial shift 2 pixels horizontally, and spatial shift 2 pixels vertically. The best method is highlighted in bold font.

MSSIM	Method	CC	SROCC	RMSE
	Original	0.3894	0.1382	25.1653
	Proposed	**0.9161**	**0.9163**	**10.9529**

4 Conclusions

Image visual quality can be assessed by two approaches: (a) subjective and (b) objective. Subjective approaches rely on the perceptual assessment of a person about the attributes of an image, while objective approaches are based on computational models, which predict perceptual image visual quality. Both objective and subjective approaches are not necessarily agreeing with each other. A person can detect small differences in visual quality in an image whereas a computer program cannot. Furthermore, image visual quality assessment needs to develop methods for objective assessment, which are consistent with subjective assessment.

In this paper, we have proposed a novel preprocessing method for measuring image visual quality that is invariant to the translation and rotation of the images. We apply the forward 2D FFT to both the reference and test images, take the logarithm of both spectra images, perform polarization transform to both log spectra images, and then apply the forward 2D FFT to the polarized images. In this way, the extracted coefficients are invariant to both spatial shifts and rotations. We select the MSSIM to measure the visual quality of the two generated coefficient images. Experimental results demonstrate that the proposed preprocessing step combined with MSSIM metric outperforms the standard metric MSSIM significantly under both kinds of distortions. More importantly, our method yields better results even when no additional distortions are introduced to the testing images.

Future research will be conducted in the following ways. We would like to develop a novel metric for measuring image visual quality by means of Radon transform. In addition, we would like to develop a new metric that is invariant to affine transform (translation, rotation, and scaling). Also, we would like to develop a new metric that is based on dual-tree complex wavelet transform (DTCWT [9]). This transform is approximate shift invariant, which is especially important in image processing and computer vision. We would also apply our newly developed preprocessing step combined with MSSIM metric to hyperspectral imagery denoising [10], illumination invariant face recognition [11, 12], and hyperspectral face recognition [13, 14].

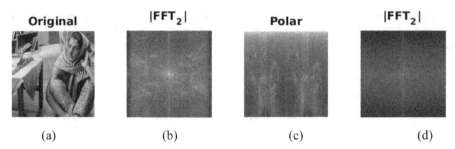

Fig. 1. The flow diagram of the proposed method in this paper: (a) The original image; (b) the Fourier spectra; (c) the polarized image; and (d) the Fourier spectra of the polarized image.

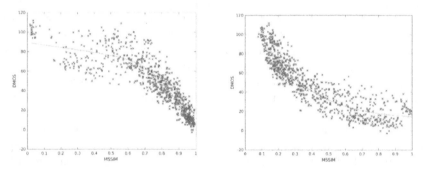

Fig. 2. Scatter plots of DMOS versus the original MSSIM (left) and the proposed method (right) for all images in the LIVE image database without introducing extra deformations.

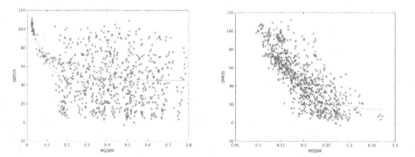

Fig. 3. Scatter plots of DMOS versus the original MSSIM (left) and the proposed method (right) for all images in the LIVE image database, where each image in the LIVE database is deformed by rotation angle $0.02 \times 180/\pi$ degrees, spatial shift 2 pixels horizontally, and spatial shift 2 pixels vertically.

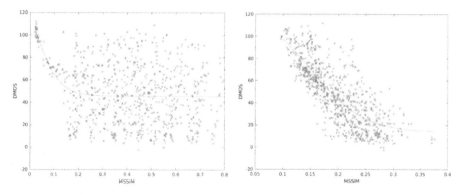

Fig. 4. Scatter plots of DMOS versus the original MSSIM (left) and the proposed method (right) for all images in the LIVE image database, where each image in the LIVE database is deformed by rotation angle $0.02 \times 180/\pi$ degrees.

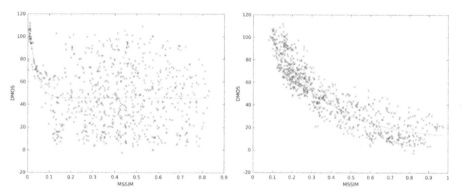

Fig. 5. Scatter plots of DMOS versus the original MSSIM (left) and the proposed method (right) for all images in the LIVE image database, where each image in the LIVE database is deformed by spatial shift 2 pixels horizontally, and spatial shift 2 pixels vertically.

Data Availability Statement. The datasets analysed during the current study are available in [8]. These datasets were derived from the following public domain resources: [LIVE image quality assessment database release 2, http://live.ece.utexas.edu/research/quality].

Conflict of Interests. The authors of this article declare that there are not conflict of interests in this article.

References

1. Wang, Z., Bovik, A.C., Sheikh, H.R., Simoncelli, E.P.: Image quality assessment: from error visibility to structural similarity. IEEE Trans. Image Process. **13**(4), 600–612 (2004)
2. Rezazadeh, S., Coulombe, S.: Novel discrete wavelet transform framework for full reference image quality assessment. Signal Image Video Process. **7**, 559–573 (2013)

3. Rezazadeh, S., Coulombe, S.: A novel discrete wavelet domain error-based image quality metric with enhanced perceptual performance. Int. J. Comput. Electr. Eng. **4**(3), 390–395 (2012)

4. Qian, S.E., Chen, G.Y.: Four reduced-reference metrics for measuring hyperspectral images after spatial resolution enhancement. In: ISPRS International Archives of the Photogrammetry, Remote Sensing and Spatial Information Sciences, Vienna, Austria, 5–7 July 2010, pp. 204–208 (2010)

5. Keller, K., Averbuch, A., Israeli, M.: Pseudo polar-based estimation of large translations, rotations, and scalings in images. IEEE Trans. Image Process. **14**(1), 12–22 (2005)

6. Chen, G.Y., Coulombe, S.: An FFT-based visual quality metric robust to spatial shift. In: The 11th International Conference on Information Science, Signal Processing, and their Applications (ISSPA), Montreal, Quebec, Canada, 2–5 July 2012, pp. 372–376 (2012)

7. Manandhar, P.: Polar To/From Rectangular Transform of Images (2021). MATLAB Central File Exchange. https://www.mathworks.com/matlabcentral/fileexchange/17933-polar-to-from-rectangular-transform-of-images. Accessed 8 Jan 2021

8. Sheikh, H.R., Wang, Z., Cormack, L., Bovik, A.C.: LIVE image quality assessment database release 2. http://live.ece.utexas.edu/research/quality

9. Kingsbury, N.G.: Complex wavelets for shift invariant analysis and filtering of signals. J. Appl. Comput. Harmon. Anal. **10**(3), 234–253 (2001)

10. Chen, G.Y., Qian, S.E.: Denoising of hyperspectral imagery using principal component analysis and wavelet shrinkage. IEEE Trans. Geosci. Remote Sens. **49**(3), 973–980 (2011)

11. Chen, G.Y., Bui, T.D., Krzyzak, A.: Filter-based face recognition under varying illumination. IET Biom. **7**(6), 628–635 (2018)

12. Chen, G.Y., Bui, T.D., Krzyzak, A.: Illumination invariant face recognition using dual-tree complex wavelet transform in logarithm domain. J. Electr. Eng. **70**(2), 113–121 (2019)

13. Chen, G.Y., Sun, W., Xie, W.F.: Hyperspectral face recognition using log-polar Fourier features and collaborative representation-based voting classifiers. IET Biom. **6**(1), 36–42 (2017)

14. Chen, G.Y., Li, C.J., Sun, W.: Hyperspectral face recognition via feature extraction and CRC-based classifier. IET Image Proc. **11**(4), 266–272 (2017)

Learning Distances Between Graph Nodes and Edges

Elena Rica⬤, Susana Álvarez⬤, and Francesc Serratosa⁽⊠⁾⬤

Universitat Rovira i Virgili, Tarragona, Spain
francesc.serratosa@urv.cat

Abstract. Several applications can be developed when graphs represent objects composed of local parts and their relations. For instance, chemical compounds are characterised by nodes that represent chemical elements and edges that represent bonds between them. Given this representation, applications such as drug discovery (graph generation), toxicity prediction (graph regression) or drug analysis (graph classification) can be developed. In all of these applications, it is crucial to properly define how similar are the local parts and how important are them in the application at hand. We present a method that learns these similarities of local parts of the objects and also how important are when objects are represented by attributed graphs and attributes on the graphs are categorical values. Although the method is independent of the application, we have empirically tested on drug classification obtaining competitive results.

Keywords: Graph edit distance · Graph matching · Learning edit costs

1 Introduction

In fields like cheminformatics, bioinformatics, character recognition, computer vision and many others, graphs are commonly used to represent objects when it is needed to capture local features of the objects (represented by nodes on the graphs) and the relation between them (represented by edges on the graphs). These objects can range from chemical compounds to hand written characters. One of the key points in these applications is properly characterising the objects and the similarity among their local parts such that some properties are hold.

In this paper, we present a method to learn the distance between nodes and between edges and also how important are they to characterise the object. We have concretised the problem to the applications where nodes and edges have a categorical attribute. For instance, while representing chemical compounds, nodes are chemical elements and the attribute is the type of chemical element (for instance, "Hydrogen" or "Oxygen"). Moreover, edges represent chemical bonds and the attributes are the types of bonds, for instance, "Ionic" or "Covalent". Thus, the aim of our method is to learn how important is a type of chemical

A. Krzyzak et al. (Eds.): S+SSPR 2022, LNCS 13813, pp. 103–112, 2022.
https://doi.org/10.1007/978-3-031-23028-8_11

element in a specific position of the compound or how important is having each type of bond instead of another.

The rest of the document is organised as follows: in Sect. 2, we define the graphs and describe the GED. Moreover, we summary the learning algorithms applied to learn the GED costs and the K-NN. In Sect. 3, we explain our learning method. In Sect. 4, we show the experimental results. Finally, in Sect. 5, we conclude the paper.

2   Related Work

We define a graph $G = (V, E)$ as a set of nodes V and a set of edges E. G_i is the ith node in V and $G_{i,j}$ is the edge in E between the ith node and the jth node in V. Moreover, γ_i is the attribute of node G_i and $\beta_{i,j}$ is the attribute of edge $G_{i,j}$.

2.1   Graph Edit Distance

The Graph Edit Distance (GED) [21] between two attributed graphs is defined as the transformation from one graph into another, through the edit operations, that obtains the minimum cost. These edit operations are: Substitution, deletion and insertion of nodes and also edges.

Having a pair of graphs, G and G', a correspondence f between these graphs is a bijective function that assigns one node of G to only one node of G'. We suppose that both graphs have the same number of nodes since they have been expanded with new nodes that have a specific attribute. We call these new nodes as *Null*. Note that the mapping between edges is imposed by the mapping of the nodes whose edges are connected.

We also define the mapping $f(i) = a$ from G_i to G'_a and we define the GED as follows:

$$GED(G, G') =$$

$$\min_{\forall f} \left\{ \sum_{\forall G_i} C^v(i, f(i)) + \sum_{\forall G_{i,j}} C^e(i, j, f(i), f(j)) \right\} \tag{1}$$

where, functions $C^v(i, f(i))$ and $C^e(i, j, f(i), f(j))$ represent the cost of mapping a pair of nodes (G_i and $G'_{f(i)}$) and a pair of edges ($G_{i,j}$ and $G'_{f(i),f(j)}$).

2.2   Learning the Edit Costs

We categorise the algorithms of learning costs in two classes, depending on the nature of the attributes on nodes and edges. In the first ones, attributes on nodes and edges (C_S^v, C_D^v, C_I^v, C_S^e, C_D^e and C_I^e) are vectors of numbers and, in the second ones, the attributes are categorical values.

In the first case, methods such as [13, 14, 19] or [2, 10, 11] have been presented using optimisation functions such as Davies-Bouldin, Dunn, C, Goodman-Krusk,

Calinski-Haraba, Rand index, Jaccard, or Fowlkes-Mallo. The methods in [1,15] solve the same costs configuration learning the weights of the weighted Euclidean distance to define the costs. Other methods [3,4] learn the substitution functions on nodes and edges through neural networks.

Contrarily, the algorithms that tackle with categorical attributes define the six node and edge edit costs as tables of constants that represent the degree of similarity between categories and also how important they are. Three papers have been published related to this case [5,6,8]. The first one [8] does not present a learning method but the value of the costs given the chemical knowledge of the authors are called *Harper costs*. The second paper [5] applied exactly the same edit costs but used the *GED* to compute the distance between compounds. The third paper [6] presents a method to learn exactly the same combination of categories of nodes and edges than in [6,8] but in an automatised way.

3   Method

We present an iterative algorithm that in each iteration one graph that has been incorrectly classified becomes to be correctly classified. The aim is to keep iterating and modifying the edit costs until all the graphs are properly classified or there are no more modifications. The classification is performed through the K-Nearest Neighbours classifier. Unfortunately, some other graphs that were properly classified in the previous iteration may become incorrectly classified. The algorithm begins given some initial edit cost and in each iteration they are smoothly updated. We want to generate the minimum modification on the edit costs with the aim of properly classifying the selected graph but reducing, to the most, the impact of the new edit costs to the other properly classified graphs. To do so, the selected graph is the one that it is easier to be moved from the incorrectly classified ones to the correctly classified ones. In the next section, our learning algorithm is explained in detail.

3.1   The Learning Method

In our approach, attributes on the nodes and edges are defined as categorical, although several nodes and edges can have the same category.

Given the K-NN classifier and the current costs $C_1, ..., C_n$, let G_j be a graph in the learning set that has been incorrectly classified. Moreover, we define D_j as the minimal GED between G_j and all the graphs that have a different class than G_j:

$$D_j = \min_q GED(G_j, G_q, C_1, ..., C_n)$$

$$\text{where class}(G_j) \neq \text{class}(G_q). \tag{2}$$

Contrarily, we define D'_j as the minimal GED between G_j and all the graphs that have the same class than G_j:

$$D'_j = \min_p GED(G_j, G_p, C_1, ..., C_n)$$

$$\text{where class}(G_j) = \text{class}(G_p)$$

(3)

Note that we can confirm that $D'_j > D_j$ since G_j is incorrectly classified.

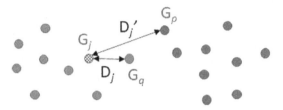

Fig. 1. Classification of G_j. True classes in solid colours. G_j is classified in the wrong class (blue). The distance between G_j and G_q is lower than the distance between G_j and G_p. (Color figure online)

Given the 1-NN, a representation of an umproperly classified graph is shown in Fig. 1. In this example, G_j and G_q belong to different classes although the distance between them is smaller than the distance between G_j and its closest graph, which has the same class than G_j, which is G_p.

The basis of our method is to permute D'_j and D_j modifying the edit costs the minimum possible. Having swapped these distances, we achieve the distance between G_j and the graph of its same class (G_p) to be lower than the distance between G_j and the graph with different class (G_q). Thus, G_j becomes to be properly classified. Nevertheless, adapting these distances affects to all the graphs classifications in the learning set. For this reason, we select a graph G_i among the incorrectly-classified ones, $\{G_j | D'_j > D_j, \forall G_j\}$ which satisfies that the difference of the distances $D'_j - D_j$ is the minimum one, as shown in Eq. 4. We highlight that in Eq. 4, all the values of $D'_j - D_j$ are always positive because $D'_j > D_j$.

$$G_i = arg \min_{\{G_j | D'_j > D_j\}} (D'_j - D_j)$$

(4)

Figure 2 shows the process of properly classifying the selected graph. We consider that it is imperative to understand that this modification is performed in the distances since the graph representations are not modified. And this process is carried out by modifying the edit costs. Thus, the aim is to define the new edit costs such that D'_i becomes D_i and vice-versa. We explain how to modify the edit costs in the rest of this section.

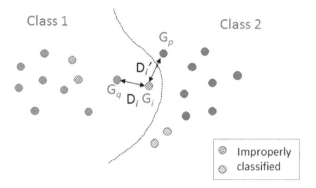

Fig. 2. Stripped graphs are improperly classified using K-NN. G_i is the one that minimises $D'_j - D_j$ being $D'_j > D_j$.

The GED is composed of edit costs $C_1, ..., C_n$ given Eq. 1, where $C_1, ..., C_n$ represent any cost on nodes and edges. Moreover, we also have the number of times the specific edit operations have been taken $N_1, ..., N_n$. Our algorithm updates the edit costs without modifying the number of operations $N_1, .., N_n$.

Considering this point of view of the GED, we define D_i and D'_i as follows:

$$D_i = \frac{C_1 N_1 + ... + C_n N_n}{L}$$
$$D'_i = \frac{C_1 N'_1 + ... + C_n N'_n}{L'} \tag{5}$$

Then, we swap the GEDs D_i and D'_i and update the edits costs adding new terms:

$$D_i = \frac{(C_1 + \alpha'_1)N'_1 + ... + (C_n + \alpha'_n)N'_n}{L'}$$
$$D'_i = \frac{(C_1 + \alpha_1)N_1 + ... + (C_n + \alpha_n)N_n}{L} \tag{6}$$

It is important to realise that these new terms $\alpha'_1, ..., \alpha'_n$ and also $\alpha_1, ..., \alpha_n$ are defined such that the new GED of D_i is D'_i instead of D_i and vice-versa. Besides, the edit costs $C_1, ..., C_n$ are the same in both expressions. We detail below how to compute the terms $\alpha'_1, ..., \alpha'_n$ and also $\alpha_1, ..., \alpha_n$.

From Eq. 6, we achieve the following normalised expressions:

$$1 = \frac{\alpha'_1 N'_1}{(D_i - D'_i)L'} + ... + \frac{\alpha'_n N'_n}{(D_i - D'_i)L'}$$
$$1 = \frac{\alpha_1 N_1}{(D'_i - D_i)L} + ... + \frac{\alpha_n N_n}{(D'_i - D_i)L} \tag{7}$$

It is important to recall that some terms in Eq. 7 are zero because there are some edit operations that are not used to transform a graph into another. These edit operations are the ones with $N_t = 0$ or $N'_t = 0$. At this point, we call m and m' as the number of edit operations that have been used, that is, the ones with $N_t \neq 0$ or $N'_t \neq 0$, respectively. If we impose that each non-zero term in Eq. 7 equals $1/m'$ or $1/m$, respectively, we obtain α_t and $\alpha'_{t'}$ that allow the modification from D_i to D'_i and from D'_i to D_i, respectively.

$$
\begin{aligned}
\alpha'_t &= \frac{(D_i - D'_i)L'}{m'N'_{t'}}, N'_t > 0 \\
\alpha_t &= \frac{(D'_i - D_i)L}{mN_t}, N_t > 0
\end{aligned}
\tag{8}
$$

Note that given Eq. 5, Eq. 6 and Eq. 8, we have, from one side that, the new costs $\overline{C}_t = C_t + \alpha_t$ and from the other side that $\overline{C}_t = C_t + \alpha'_t$. Since it may happen that $\alpha_t \neq \alpha'_t$, we assume the mean option is the best choice when both weights are computed,

$$
\overline{C}_t = \begin{cases}
C_t + \frac{\alpha_t + \alpha'_t}{2}, & \text{if } N_t > 0 \text{ and } N'_t > 0 \\
C_t + \alpha_t, & \text{if } N_t > 0 \text{ and } N'_t = 0 \\
C_t + \alpha'_t, & \text{if } N_t = 0 \text{ and } N'_t > 0 \\
C_t, & \text{if } N_t = 0 \text{ and } N'_t = 0
\end{cases}
\tag{9}
$$

3.2   The Algorithm

Our method consists in an iterative algorithm in which, in each iteration, the edit costs are updated to correct the classification of one selected graph. The classification is carried out through the K-NN algorithm. The updated costs are used in the next iteration to classify again all the graphs, select a graph and modify the costs again.

In *line* 2, the initial costs are imposed. In the experimental section two different initial costs are tested. In *line* 3, graphs are classified, given the current costs. Note that we are using the K-NN although other classifiers could be used. In *line* 4 and *line* 5, one graph is selected that has been improperly classified. In *line* 6, *line* 7 and *line* 8, costs are updated.

The computational cost of our algorithm is linear with respect to the number of iterations. Besides, the cost of each iteration is quadratic with respect the number of graphs. Clearly, the cost of computing the GED depends on the used algorithm. Usually, an approximate distance is returned and the algorithms range from cubic [20] to linear [18] with respect to the number of nodes of the graphs.

We stress that the convergence is not demonstrated, although in our experiments, a stationary value has always been achieved in few iterations. This is the reason why the stopping criterion is not an optimisation equation but the number of iterations.

Algorithm
Input (Learning Set, Initial edit costs, Max_Iter)
Output (Learnt edit costs)

1. $iter = 1$.
2. $C_1,..., C_n =$ Initial edit costs.
  While $iter \leq Max\_Iter$:
3.   **Classify** graphs by 1-KNN, Eq. 1 and $C_1, ..., C_n$.
4.   **Compute** D_j **and** D'_j: (Eq. 2 and 3).
5.   **Deduce** G_i (Eq. 4).
6.   **Compute** α_t **and** α'_t: (Eq. 8).
7.   **Compute** $\overline{C}_1,..., \overline{C}_n$ (Eq. 9).
8.   **Update costs:** $C_t = \overline{C}_t, t = 1, ..., n$.
9. $iter = iter + 1$.
  End While

  End Algorithm

4   Practical Experiments

Table 1. Accuracy obtained in each data set. Mean accuracy in the last column. OM = Our method. DB1 = CAPST, DB2 = DUD-E, DB3 = GLL & GDD, DB4 = MUV, DB5 = NRLiSt_BDB, DB6 = ULS-UDS, Me = Mean.

	DB1	DB2	DB3	DB4	DB5	DB6	Me
Harper	93,8	95,9	85,7	**92,8**	93,2	**96,1**	92,9
1s	92,9	91,3	93,0	56,0	94,8	92,9	86,8
C1	89,3	92,7	82,5	86,0	88,6	89,7	88,1
C2	89,8	91,1	82,5	87,4	88,2	91,7	88,4
C3	91,3	91,3	83,3	86,7	87,8	92,3	88,8
C4	89,5	90,9	82,4	86,0	89,9	92,6	88,6
OM: Harper init.	**95,9**	**96,4**	**93,7**	88,7	**95,9**	94,0	**94,1**
OM: 1s init.	88,2	93,5	93,3	61,8	95,0	95,3	87,8

Our method has been compared to the learning algorithm in [6] and the manually imposed configuration in [8] in a classification framework.

We have used six available public datasets, which are composed of several databases: ULS-UDS [23] (4 databases), GLL&GDD [7] (69 databases), CAPST [17] (4 databases), DUD-E [12] (9 databases), NRLiSt-BDB [9] (24 databases) and MUV [16] (17 databases). All these datasets had been formatted and standardized by the LBVS benchmarking platform developed by Skoda and Hoksza [22]. The databases are composed of active and inactive molecules (two classes) arranged according to the purpose of a target. Each group is split in two halves, the test (100 graphs) and train sets (100 graphs). In total there are 127 databases and 200 graphs per database.

Table 1 shows the classification ratios obtained in each data set using different edit cost configurations, algorithms and initialisation. The first row corresponds to the accuracies obtained by the costs proposed by Harper [8], the second row corresponds to the accuracies deduced by setting all the costs to 1 (no learning algorithm). The next four rows correspond to the accuracies obtained using the costs deduced in García et al. [6] in their four experiments (C1, C2, C3, and C4). Finally, the two last rows present the accuracies obtained by our method: the first row by initialising the algorithm by Harper costs and the second one by initialising all the costs to 1.

The databases and the four tables of the edit costs (node substitutions, edge substitution, node insertion/deletion and edge insertion/deletion) deduced by [6,8] and our algorithm are available at [1]. Note that [6] not only presented a table per experiment (C1, C2, C3, and C4) but also per database. Usually, being much more specific in the training set makes the recognition ratio returned by the pattern classification method to be higher.

We realise that in all the data sets, except in MUV and ULS-UDS, our costs with Harper initialisation obtained the highest classification ratios. In these two data sets, the best accuracy is obtained by Harper costs. Note that our method initialised by all-ones costs returns lower accuracies than our method initialised by Harper costs except for the ULS-UDS dataset. This behaviour makes us think that the initialisation point is very important in this type of algorithms.

Another highlight is that we have achieved better accuracies than the four experiments presented by García et al. [6] in all the tests. In ULS-UDS dataset, our method returns close accuracy to the Harper costs. Nevertheless, it is not the case of MUV dataset. As previously commented, the experiments in [6] use a different set of four tables per database and experiment (C1, C2, C3 and C4), which could be an advantage while obtaining the accuracy.

5   Conclusions

We have presented a method that automatically learns the edit costs of the graph edit distance when attributes are categorical values. When nodes and edges represent different categories, the edit cost can be assimilated to the distance between these local elements and how important are in the application at hand. For this reason, these learned relations can be used in a range of different applications, for instance, in toxicology prediction, drug discovery or protein classification. The experimental section shows that our method achieves higher accuracies than another algorithm and also than the edit costs manually imposed by a specialist.

Acknowledgements. This project has received funding from Martí-Franquès Research Fellowship Programme of Universitat Rovira i Virgili.

---

[1] https://deim.urv.cat/francesc.serratosa/databases/.

References

1. Algabli, S., Serratosa, F.: Embedding the node-to-node mappings to learn the graph edit distance parameters. Pattern Recogn. Lett. **112**, 353–360 (2018)
2. Caetano, T.S., McAuley, J.J., Cheng, L., Le, Q.V., Smola, A.J.: Learning graph matching. IEEE Trans. Pattern Anal. Mach. Intell. **31**(6), 1048–1058 (2009)
3. Cortés, X., Conte, D., Cardot, H.: Learning edit cost estimation models for graph edit distance. Pattern Recogn. Lett. **125**, 256–263 (2019)
4. Cortés, X., Conte, D., Cardot, H., Serratosa, F.: A deep neural network architecture to estimate node assignment costs for the graph edit distance. In: Proceedings of Structural, Syntactic, and Statistical Pattern Recognition - Joint IAPR International Workshop, S+SSPR 2018, Beijing, China, 17–19 August 2018, Vol. 4, pp. 326–336 (2018)
5. Garcia-Hernandez, C., Fernández, A., Serratosa, F.: Ligand-based virtual screening using graph edit distance as molecular similarity measure. J. Chem. Inf.. Model. **59** (2019)
6. Garcia-Hernandez, C., Fernández, A., Serratosa, F.: Learning the edit costs of graph edit distance applied to ligand-based virtual screening. Curr. Top. Med. Chem. **20**(18), 1582–1592 (2020)
7. Gatica, E.A., Cavasotto, C.N.: Ligand and decoy sets for docking to g protein-coupled receptors. J. Chem. Inf. Model. **52**(1), 1–6 (2011)
8. Harper, G., Bravi, G.S., Pickett, S.D., Hussain, J., Green, D.V.S.: The reduced graph descriptor in virtual screening and data-driven clustering of high-throughput screening data. J. Chem. Inf. Comput. Sci. **44**(6), 2145–2156 (2004). https://doi.org/10.1021/ci049860f
9. Lagarde, N., et al.: NRLIST BDB, the manually curated nuclear receptors ligands and structures benchmarking database. J. Med. Chem. **57**(7), 3117–3125 (2014)
10. Leordeanu, M., Sukthankar, R., Hebert, M.: Unsupervised learning for graph matching. Int. J. Comput. Vision **96**(1), 28–45 (2012)
11. Martineau, M., Raveaux, R., Conte, D., Venturini, G.: Learning error-correcting graph matching with a multiclass neural network. Pattern Recogn. Lett. **134**, 68–76 (2018)
12. Mysinger, M.M., Carchia, M., Irwin, J.J., Shoichet, B.K.: Directory of useful decoys, enhanced (dud-e): better ligands and decoys for better benchmarking. J. Med. Chem. **55**(14), 6582–6594 (2012)
13. Neuhaus, M., Bunke, H.: Self-organizing maps for learning the edit costs in graph matching. IEEE Trans. Syst. Man Cybernet.. Part B (Cybernetics)**35**(3), 503–514 (2005)
14. Neuhaus, M., Bunke, H.: Automatic learning of cost functions for graph edit distance. Inf. Sci. **177**(1), 239–247 (2007)
15. Rica, E., Álvarez, S., Serratosa, F.: On-line learning the graph edit distance costs. Pattern Recognit. Lett. **146**, 55–62 (2021). https://doi.org/10.1016/j.patrec.2021.02.019, https://doi.org/10.1016/j.patrec.2021.02.019
16. Rohrer, S.G., Baumann, K.: Maximum unbiased validation (MUV) data sets for virtual screening based on PubChem bioactivity data. J. Chem. Inf. Model. **49**(2), 169–184 (2009)
17. Sanders, M.P., et al.: Comparative analysis of pharmacophore screening tools. J. Chem. Inf. Model. **52**(6), 1607–1620 (2012)
18. Santacruz, P., Serratosa, F.: Error-tolerant graph matching in linear computational cost using an initial small partial matching. Pattern Recogn. Lett. **134**, 10–19 (2018)

19. Santacruz, P., Serratosa, F.: Learning the graph edit costs based on a learning model applied to sub-optimal graph matching. Neural Process. Lett. **51**, 1–24 (2019)
20. Serratosa, F.: Fast computation of bipartite graph matching. Pattern Recogn. Lett. **45**, 244–250 (2014)
21. Serratosa, F.: Redefining the graph edit distance. SN Comput. Sci. **2**(6), 1–7 (2021)
22. Skoda, P., Hoksza, D.: Benchmarking platform for ligand-based virtual screening. In: Proceedings - 2016 IEEE International Conference on Bioinformatics and Biomedicine, BIBM 2016, pp. 1220–1227 (2017). https://doi.org/10.1109/BIBM.2016.7822693
23. Xia, J., Tilahun, E.L., Reid, T.E., Zhang, L., Wang, X.S.: Benchmarking methods and data sets for ligand enrichment assessment in virtual screening. Methods **71**, 146–157 (2015)

Self-supervised Out-of-Distribution Detection with Dynamic Latent Scale GAN

Jeongik Cho⬤ and Adam Krzyzak$^{(\boxtimes)}$ ⬤

Concordia University, Montreal, QC, Canada
`krzyzak@cs.concordia.ca`

Abstract. Dynamic latent scale GAN is a learning-based GAN inversion method with maximum likelihood estimation. In this paper, we propose a self-supervised out-of-distribution detection method using the encoder of dynamic latent scale GAN. When the dynamic latent scale GAN converges, since the entropy of the scaled latent random variable is optimal to represent in-distribution data, in-distribution data are densely mapped to latent codes with high likelihood. Therefore, out-of-distribution data can only be mapped to latent codes with low likelihood. Also, since the latent random variable of GAN is i.i.d. random variable, it is easy to calculate the log-likelihood of predicted latent code. These characteristics of dynamic latent scale GAN enable the log-likelihood of the predicted latent code to be used for out-of-distribution detection. The proposed method does not require a pre-trained model, mutual information of in-distribution data, and additional hyperparameters for prediction. The proposed method showed better out-of-distribution detection performance than the previous state-of-art method.

Keywords: Out of distribution detection · Anomaly detection · Generative adversarial networks

1 Introduction

Given in-distribution (ID) data, detecting data that does not belong to the ID data is called out-of-distribution (OOD) detection (or anomaly detection). The difference between OOD detection and simple classification is that in OOD detection, only the ID data is given to the model, and the model should classify the ID data and unseen OOD data. In general, since the range of OOD data is very wide compared to the ID data in high dimensional data space, it is almost impossible to define an OOD dataset and use it for ID-OOD classification (i.e., supervised learning).

[1, 2] proposed an OOD detection method using statistics of a pre-trained model. However, since these methods require a pre-trained model, the model cannot be customized. For example, when the pre-trained model is a large model whose input is a high-resolution image, low-resolution images should be resized and input, which is inefficient. Also, large models may not be able to be used when computational performance is limited. Or, when the input resolution of the pre-trained is low, high-resolution images

© The Author(s), under exclusive license to Springer Nature Switzerland AG 2022
A. Krzyzak et al. (Eds.): S+SSPR 2022, LNCS 13813, pp. 113–121, 2022.
https://doi.org/10.1007/978-3-031-23028-8_12

should be downsampled, which degrades OOD detection performance. Furthermore, there may not be a good pre-trained model for particular data domains.

[3, 4] proposed a method that can be applied when the mutual information (e.g., the label of the image) of ID data is given. The energy score method [4] detects OOD samples using the logit of the classifier when a classifier that is trained with ID data and its corresponding labels is given. ReAct [3] improved the performance of the energy score method by clipping the feature vector (output of the layer just before the logit layer) of the classifier. However, those methods require mutual information of ID data, so they cannot be applied when mutual information is not given.

[5, 6] proposed a method when the input is an image. These methods are difficult to use for data domains other than images.

In this paper, we propose AnoDLSGAN, a self-supervised learning method for OOD detection that does not require any mutual information of ID data or a pre-trained model. AnoDLSGAN uses an encoder of dynamic latent scale GAN (DLSGAN) [7] for OOD detection. Simply, the log-likelihood of predicted latent code by the encoder of DLS-GAN is used for OOD detection. Therefore, AnoDLSGAN does not require additional hyperparameters for prediction or mutual information of ID data.

2 Out-of-Distribution Detection with DLSGAN

DLSGAN [7] is a learning-based GAN inversion method with maximum likelihood estimation of the encoder. It solves the problem that the generator loses information when training the InfoGAN [12] that predicts all latent codes through the dynamic latent scale. The encoder of DLSGAN maps input data to predicted latent codes.

We found that the likelihood of the predicted latent code of the input data predicted by DLSGAN's encoder can be used for OOD detection. There are two characteristics that allow the DLSGAN encoder to be used for OOD detection.

First is the latent entropy optimality. As DLSGAN training progresses, the entropy of scaled latent random variable decreases, and the entropy of the scaled encoder output increases. When DLSGAN is converged, the generator generates ID data with a scaled latent random variable, and the entropy of scaled latent random variable and scaled encoder output becomes optimal entropy for expressing ID data with the generator and encoder. It means that ID data generated by the generator is densely mapped to latent codes with high likelihood. Therefore, by the pigeonhole principle, OOD data can only be mapped to latent codes with low likelihood.

Secondly, elements of DLSGAN encoder output are independent of each other and follow a simple distribution (the same as latent distribution). Therefore, it is very easy to calculate the log-likelihood of predicted latent code.

The following equation shows the negative log-likelihood of the predicted latent code of input data.

$$ood\ score = -\sum\nolimits_{i=1}^{d_z} \log f\left(E_i(x)|\mu_i, v_i\right) \tag{1}$$

In Eq. (1), x and E represent the input data point and DLSGAN's encoder, respectively. $E(x)$ represents the d_z-dimensional predicted latent code of input data point x. f

represents the probability density function of the i.i.d. latent random variable Z. μ and v represent the latent mean vector and latent variance vector for the probability density function f. μ is the mean vector of latent random variable Z. v is the traced latent variance vector of DLSGAN. $E_i(x)$, μ_i, and v_i represent i-th element of $E(x)$, μ, and v, respectively.

Since each element of the encoder output $E(X)$ is independent of each other, the negative log-likelihood of the predicted latent code can be simply calculated by adding the negative log-likelihood of each element.

The *ood score* is the negative log-likelihood of the predicted latent code $E(x)$. If the *ood score* is greater than the threshold, the input data is classified as OOD data. Otherwise, it is classified as ID data.

AnoDLSGAN is a self-supervised OOD detection method, so it does not require any mutual information of ID data or a pre-trained model. Also, only one inference of encoder E is required to classify input data.

3   Experiments

3.1   Experiments Settings

We used MNIST handwritten digits dataset [8] as an ID dataset and CMNIST (Corrupted MNIST) dataset [9], FMNIST (Fashion MNIST) [10], and KMNIST (Kuzushiji MNIST) [11] dataset as OOD datasets. For the preprocessing, we added padding to the images to make the resolution 32×32 and normalized the pixel values to be between -1 and 1.

Figure 1 shows samples of ID images and OOD images. The first column of Fig. 1 shows ID images. Columns 13–23 (right part of the right white line) show far OOD images of the OOD dataset (CMNIST, FMNIST, KMNIST). Columns 2–12 (between the two white lines) show near OOD images. Near OOD images are generated by linear interpolation between far OOD images and ID images (i.e., *near OOD image = far OOD image* $\times$ *k* + *ID image* $\times$ $(1 - k)$). We used $k = 0.1$ to generate near OOD images. The near OOD images are hard to distinguish for humans without looking very closely.

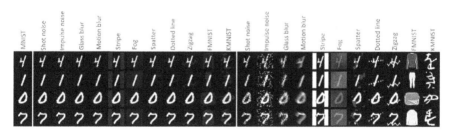

Fig. 1. Sample images from datasets. Column 1: ID images, columns 2–12: near OOD images, columns 13–23: far OOD images.

Images from the CMNIST dataset were generated by adding corruption to the images from the MNIST dataset. Therefore, each image from the CMNIST dataset has a corresponding original image from the MNIST dataset. When generating near OOD images with the CMNIST dataset, corresponding images from the MNIST dataset were used as ID images, not random images from the MNIST dataset.

We trained four types of models for OOD detection: AnoDLSGAN, InfoGAN [12], autoencoder, and classifier. Each model is trained only with the ID train dataset. AnoDLS-GAN, InfoGAN, and autoencoder were trained without labels, while the classifier was trained with labels. The reason we trained different types of models is that each OOD detection method requires a different model. For example, energy score [4] and ReAct [3] require a trained classifier for OOD detection.

Training AnoDLSGAN is the same as training DLSGAN.

Training InfoGAN is the same as training DLSGAN without a dynamic latent scale. The latent random variable of InfoGAN has a larger entropy than the optimal entropy for expressing ID data. Therefore, ID data cannot densely map to latent space. To show that AnoDLSGAN has high OOD detection performance because ID data is densely mapped to latent space, we also experimented with InfoGAN for OOD detection.

OOD detection with InfoGAN is the same as AnoDLSGAN except for the dynamic latent scale (i.e., the same OOD score function as AnoDLSGAN with traced latent variance vector is used for OOD detection).

The autoencoder is trained with reconstruction loss (mean squared error loss between an input image and reconstructed image). Reconstruction loss is also used for OOD detection with the autoencoder [13].

The classifier is trained with cross-entropy loss. Energy score with ReAct is used for OOD detection with the classifier.

Following hyperparameters were used for training models.

$$batch\ size = 32$$

$$optimizer = Adam \begin{pmatrix} learning\ rate = 0.001 \\ beta_1 = 0 \\ beta_2 = 0.99 \end{pmatrix}$$

$$epoch = 30$$

$$learning\ rate\ decay\ rate\ per\ epoch = 0.95$$

learning rate decay rate per epoch is a value multiplied by the learning rate for each epoch.

The following figures show the model architecture used in the experiments.

Input: [32, 32, 1] image	
From image layer(filter size = 128)	
Conv layer (filter size = 128)	
Conv layer (filter size = 256)	
Downsample 2 ×	
Conv layer (filter size = 256)	
Conv layer (filter size = 512)	
Downsample 2 ×	
Conv layer (filter size = 512)	
Conv layer (filter size = 1024)	
Downsample 2 ×	
Flatten layer	
Adversarial (1)	Latent vector (256)
	Classifier (10)

Fig. 2. Encoder architecture. *Adversarial*, *Latent vector*, and *Classifier* are fully connected layers, and the numbers in parentheses indicate the output dimensions.

Input: [256] latent vector
Fully connected layer (1024 × 4 × 4)
Reshape [1024,4,4]
Upsample 2 ×
Conv layer (filter size = 512)
Conv layer (filter size = 512)
Upsample 2 ×
Conv layer (filter size = 256)
Conv layer (filter size = 256)
Upsample 2 ×
Conv layer (filter size = 128)
Conv layer (filter size = 128)
To image layer(filter size = 1)

Fig. 3. Decoder architecture.

In Figs. 2 and 3, the activation functions of all layers except the output layer are leaky ReLU, and the kernel size of all convolution layers is "3 × 3". *From image layer* and *To image layer* are both convolution layers with kernel size "1 × 1". Equalized learning rate [16] is used for all layers.

DLSGAN and InfoGAN use an encoder as a discriminator and a decoder as a generator. The encoder's classifier output was not used for GAN training. NSGAN with R1 regularization [14] was used for GAN training. Also, an exponential moving average with *decay rate* = 0.999 was used for traced variance vector v. Following hyperparameters were used for GANs training.

$$\lambda_{enc} = 1$$

$$\lambda_{r1} = 0.1$$

$$Z = (Z_i)_{i=1}^{256} \overset{i.i.d.}{\sim} N\left(0, 1^2\right)$$

λ_{r1} is R1 regularization weight. The paper proposed R1 regularization used $\gamma/2$ as regularization weight, so based on that definition, γ is 0.2 when λ_{r1} is 0.1.

Autoencoder uses encoder and decoder. Adversarial output and classifier output of the encoder were not used for encoder training.

The classifier uses only an encoder. The adversarial output of the encoder was not used for classifier training.

GAN and autoencoder take more time to train than classifier because those models use both encoder and decoder for training. GAN and classifier require one encoder inference to classify one data, while autoencoder requires one encoder and decoder inference.

We evaluated the model performance using 7-fold cross-validation. Each MNIST dataset has 70k images. Therefore, the ID train dataset has 60k images, the ID test dataset has 10k images, and each OOD test dataset has 10k images in each fold. The average area under the ROC curve (AUROC) was used for OOD detection performance evaluation.

Full codes for the experiments are available at "https://github.com/jeongik-jo/Ano DLSGAN".

3.2 Experiments Results

Table 1 shows the basic model performance for each model. In Table 1, FID [15] shows the generative performance of GANs. 10k generated images and 10k ID test images for each fold were used for FID evaluation. PSNR and SSIM show the difference between test images and reconstructed images.

Table 1 shows the basic model performance. AnoDLSGAN showed better reconstruction performance (PSNR and SSIM) than InfoGAN as [7]. However, the reconstruction performance of the autoencoder was much better than GANs. The classifier showed high accuracy.

Table 1. Basic model performance

	AnoDLSGAN (ours)	InfoGAN [12]	Autoencoder	Classifier
FID [15]	5.26	6.44	–	–
PSNR	15.44	14.60	28.98	–
SSIM ($\times 100$)	56.48	50.59	96.25	–
Accuracy ($\times 100$)	–	–	–	99.49

Table 2 shows the OOD detection performance for each method. 1000 evenly spaced points between the minimum and maximum values of the *ood score* of the ID test dataset were used for the threshold value. Each value in the table is the average AUROC

Table 2. OOD detection performance for each method. Each value in the table is the average AUROC multiplied by 100

AUROC (×100)		GAN		Autoen-coder	Classifier							
		AnoDLSGAN (Ours)	InfoGAN [12]	Reconstruction [13]	Energy [3,4] t=1.0, p=0.85	Energy t=1.0, p=0.9	Energy t=1.0, p=0.95	Energy t=1.0, p=1.0	Energy t=10.0, p=0.85	Energy t=10.0, p=0.9	Energy t=10.0, p=0.95	Energy t=10.0, p=1.0
N E A R O O D	Shot noise	51.54	50.41	53.11	51.93	51.97	52.03	51.83	51.97	52.01	52.07	52.18
	Impulse noise	88.16	55.55	72.21	52.02	52.06	52.13	51.91	52.05	52.09	52.16	52.26
	Glass blur	62.28	50.65	48.37	54.30	54.42	54.59	54.18	54.42	54.52	54.68	54.95
	Motion blur	82.14	51.98	50.35	55.39	55.50	55.67	55.10	55.54	55.63	55.79	56.05
	Stripe	99.99	93.98	98.72	53.29	53.39	53.54	53.22	53.37	53.46	53.59	53.81
	Fog	97.81	59.16	75.53	57.57	57.76	58.03	57.24	57.78	57.95	58.20	58.60
	Spatter	70.50	50.92	52.31	51.04	51.06	51.10	50.99	51.06	51.08	51.12	51.17
	Dotted line	70.73	52.42	55.95	50.50	50.51	50.52	50.46	50.50	50.50	50.51	50.53
	Zigzag	89.90	56.40	63.95	51.02	51.04	51.06	50.94	51.02	51.03	51.05	51.09
	FMNIST	99.65	65.82	80.99	59.97	60.18	60.49	59.40	60.17	60.36	60.65	61.09
	KMNIST	98.84	65.24	83.20	60.50	60.71	61.00	59.84	60.69	60.88	61.16	61.59
F A R O O D	Shot noise	94.23	71.60	99.75	76.57	76.73	76.91	73.50	76.37	76.55	76.74	76.99
	Impulse noise	99.99	99.99	100.00	87.12	86.99	86.75	81.51	86.25	86.16	85.93	85.64
	Glass blur	99.76	91.01	99.81	93.42	93.74	94.08	88.07	93.66	93.95	94.25	94.48
	Motion blur	99.99	95.44	97.87	91.42	91.74	92.07	86.31	91.78	92.07	92.36	92.58
	Stripe	99.99	99.99	100.00	91.13	91.01	90.75	84.61	90.59	90.47	90.19	89.82
	Fog	99.99	99.95	99.99	99.91	99.92	99.92	92.80	99.94	99.94	99.95	99.95
	Spatter	99.49	85.79	99.41	68.03	68.16	68.29	65.87	67.70	67.87	68.02	68.19
	Dotted line	99.19	87.25	99.97	64.29	64.09	63.75	61.55	62.79	62.73	62.49	62.07
	Zigzag	99.83	95.49	99.99	70.31	70.02	69.47	66.06	67.63	67.55	67.14	66.29
	FMNIST	99.99	99.95	99.99	97.40	97.47	97.52	90.80	97.23	97.29	97.33	97.33
	KMNIST	99.93	98.20	99.99	94.63	94.47	94.19	87.67	92.96	92.87	92.60	92.21

multiplied by 100. In "Energy" of Table 2, "t" represents the temperature of the energy score [4], and "p" represents the activation percentage of ReAct [3]. When p = 1.0, it is the same as that ReAct was not applied.

In Table 2, one can see that the overall performance of AnoDLSGAN is the best. The performance of AnoDLSGAN is not significantly different from that of autoencoder reconstruction in far OOD detection, but it shows significantly better performance in near OOD detection, even if the reconstruction performance of autoencoder is much better than AnoDLSGAN's.

Additionally, OOD detection with autoencoder reconstruction shows better performance than the energy score with ReAct. Comparing p = 1.0 with the others in the energy score, one can see that there is a performance improvement when ReAct is applied as [3].

Energy score with ReAct showed good performance in easy OOD of far OOD datasets detection (FMNIST and KMNIST) but showed relatively poor performance in some

CMNIST datasets (Spatter, Dotted line, Zigzag). Also, it could hardly distinguish the near OOD images.

Also, one can see that AnoDLSGAN has clearly better near OOD detection performance than InfoGAN. This indicates that AnoDLSGAN performs better than InfoGAN because ID data is densely mapped to the latent space due to the latent entropy optimality of DLSGAN.

All methods failed to distinguish Shot noise near OOD images. Shot noise near OOD images are difficult to classify even for humans.

4   Conclusion

In this paper, we showed that the encoder of DLSGAN can be used for OOD detection. The likelihood of the predicted latent code of the input data is used for OOD detection. The latent entropy optimality and easy-to-compute latent code likelihood enable DLS-GAN's encoder to be used for OOD detection. OOD detection with AnoDLSGAN is very simple and does not require a pre-trained model, mutual information of ID data, and additional hyperparameters for prediction. AnoDLSGAN showed high OOD detection performance compared to the previous state-of-art method.

References

1. Huang, R., Geng, A., Li, Y.: On the importance of gradients for detecting distributional shifts in the wild. In: NIPS (2021). https://proceedings.neurips.cc/paper/2021/file/063e26c670d07bb7c4d30e6fc69fe056-Paper.pdf
2. Rippel, O., Mertens, P., Merhof, D.: Modeling the distribution of normal data in pre-trained deep features for anomaly detection. In: ICPR (2020). https://ieeexplore.ieee.org/abstract/document/9412109
3. Sun, Y., Guo, C., Li, Y.: ReAct: out-of-distribution detection with rectified activations. In: NIPS (2021). https://openreview.net/forum?id=IBVBtz_sRSm
4. Liu, W., Wang, X., Owens, J.D., Li, Y.: Energy-based out-of-distribution detection. In: NIPS (2020). https://proceedings.neurips.cc/paper/2020/file/f5496252609c43eb8a3d147ab9b9c006-Paper.pdf
5. Yi, J., Yoon, S.: Patch SVDD: patch-level SVDD for anomaly detection and segmentation. In: Ishikawa, H., Liu, C.-L., Pajdla, T., Shi, J. (eds.) ACCV 2020. LNCS, vol. 12627, pp. 375–390. Springer, Cham (2021). https://doi.org/10.1007/978-3-030-69544-6_23
6. Yan, X., Zhang, H., Xu, X., Hu, X., Heng, P.A.: Learning semantic context from normal samples for unsupervised anomaly detection. In: AAAI (2021). https://ojs.aaai.org/index.php/AAAI/article/view/16420/16227
7. Cho, J., Krzyzak, A.: Dynamic latent scale for GAN inversion. In: Proceedings of the 11th ICPRAM (2022). https://www.scitepress.org/Link.aspx?doi=10.5220/0010816800003122
8. LeCun, Y., Cortes, C., Burges, C.J.C.: THE MNIST DATABASE of handwritten digits. http://yann.lecun.com/exdb/mnist/
9. Mu, N., Gilmer, J.: MNIST-C: a robustness benchmark for computer vision. arXiv preprint (2019). https://arxiv.org/abs/1906.02337
10. Xiao, H., Rasul, K., Vollgraf, R.: Fashion-MNIST: a novel image dataset for benchmarking machine learning algorithms. arXiv preprint (2017). http://arxiv.org/abs/1708.07747

11. Clanuwat, T., Bober-Irizar, M., Kitamoto, A., Lamb, A., Yamamoto, K., Ha, D.: Deep learning for classical Japanese literature. arXiv preprint (2018). https://arxiv.org/abs/1812.01718

12. Chen, X., Duan, Y., Houthooft, R., Schulman, J., Sutskever, I., Abbeel, P.: InfoGAN: interpretable representation learning by information maximizing generative adversarial Nets. In: NIPS (2016). https://papers.nips.cc/paper/2016/hash/7c9d0b1f96aebd7b5eca8c3edaa19ebb-Abstract.html

13. Tensorflow autoencoder tutorial. https://www.tensorflow.org/tutorials/generative/autoen coder. Accessed 28 Apr 2022

14. Mescheder, L., Geiger, A., Nowozin, S.: Which training methods for GANs do actually converge? In: PMLR (2018). http://proceedings.mlr.press/v80/mescheder18a

15. Heusel, M., Ramsauer, H., Unterthiner, T., Nessler, B., Hochreiter, S.: GANs trained by a two time-scale update rule converge to a local Nash equilibrium. In: NIPS (2017). https://papers.nips.cc/paper/2017/hash/8a1d694707eb0fefe65871369074926d-Abstract.html

16. Karras, T., Aila, T., Laine, S., Lehtinen, J.: Progressive growing of GANs for improved quality, stability, and variation. In: ICLR (2018). https://openreview.net/forum?id=Hk99zCeAb

A Novel Graph Kernel Based on the Wasserstein Distance and Spectral Signatures

Yantao Liu[1,2], Luca Rossi[2(✉)], and Andrea Torsello[3]

[1] Beijing University of Posts and Telecommunications, Beijing, China
[2] Queen Mary University of London, London, UK
luca.rossi@qmul.ac.uk
[3] Ca' Foscari University of Venice, Venice, Italy

Abstract. Spectral signatures have been used with great success in computer vision to characterise the local and global topology of 3D meshes. In this paper, we propose to use two widely used spectral signatures, the Heat Kernel Signature and the Wave Kernel Signature, to create node embeddings able to capture local and global structural information for a given graph. For each node, we concatenate its structural embedding with the one-hot encoding vector of the node feature (if available) and we define a kernel between two input graphs in terms of the Wasserstein distance between the respective node embeddings. Experiments on standard graph classification benchmarks show that our kernel performs favourably when compared to widely used alternative kernels as well as graph neural networks.

Keywords: Graph kernel · Wasserstein distance · Spectral signature

1 Introduction

Graph-based representations have long been used as a natural way to abstract data where structure plays a key role, from biological [7] to collaboration networks [21]. The rich expressiveness of graphs comes at the cost of an increased difficulty in applying standard pattern recognition techniques, which usually require the graphs to be first embedded into a vectorial space, a procedure that is far from trivial. This is due to the lack of canonical ordering for the nodes in a graph as well as the fact that graphs do not necessarily map to vectors of fixed length, due to varying number of nodes and edges.

Graph kernels provide an elegant way to handle graph data in machine learning problems. By either explicitly or implicitly embedding the graphs into a vectorial space where a kernel measure is defined, graph kernels allow to frame the problem of learning on graphs in the context of kernel methods [6,26,27]. The common idea underpinning most of these kernels is that of decomposing each graph into a set of substructures. Input graphs are then compared in terms of the

© The Author(s), under exclusive license to Springer Nature Switzerland AG 2022
A. Krzyzak et al. (Eds.): S+SSPR 2022, LNCS 13813, pp. 122–131, 2022.
https://doi.org/10.1007/978-3-031-23028-8_13

substructures they contain. Note that the kernel trick on which these methods are based requires the kernel measure to be positive definite in order to be seen as a dot product in the embedding space. However, while most graph kernels are positive definite, some are not [13,24], and thus several learning algorithms have been proposed to deal with indefinite kernels [23]. More recently graph neural networks (GNNs) have emerged as an alternative approach where structural information is exploited by iteratively propagating node features between groups of neighbouring nodes. The features obtained after several rounds (i.e., layers) of propagation effectively provide an embedding for the graph nodes and can then be used to solve the learning task at hand [2,14,19,25]. Crucially, GNNs that follow this simple feature propagation principle have been shown to be no more powerful than the Weisfeiler-Lehman (WL) graph kernel [22].

In this paper, we propose a new graph kernel inspired by two popular shape descriptors found in computer vision [3,28], where shapes are often discretised into mesh representations and the mesh nodes are effectively embedded into a vectorial space in order to establish a set of correspondences between pairs of input shapes. Given a graph, we propose to use the Heat Kernel Signature (HKS) [28] and the Wave Kernel Signature (WKS) [3] to embed its nodes into a space that captures both local and global structural information. For a pair of graphs, this yields two distributions of points in the embedding space, allowing us to compute the kernel between the graphs in terms of the overall similarity between the two distributions. With these distributions to hand, similarly to [29], we propose to take the negative exponential of the Wasserstein distance, a widely used distance function between probability distributions. By treating each graph as a distribution of points encoding structural information at different scales, our kernel is able to capture fine grained structural information that may be missed by looking at coarser substructures as in standard graph kernels. Indeed, the experimental evaluation shows that our kernel outperforms widely used graph kernels as well as GNNs on standard graph classification benchmarks.

The reminder of the paper is organised as follows. Section 2 briefly reviews the related work, while Sect. 3 and Sect. 4 introduce the necessary background and our kernel, respectively. Section 5 discusses the experimental setup and results and finally Sect. 6 concludes the paper.

2   Related Work

Thanks to the kernel trick, any learning algorithm formulated in terms of scalar products of implicit or explicit embeddings of the inputs can be applied to data (e.g., vectors, graphs) on which a kernel is defined. More precisely, given a set X and a positive definite kernel $k : X \times X \to \mathbb{R}$, we know that there is a map $\phi : X \to H$ into a Hilbert space H, such that $k(x, y) = \phi(x)^\top \phi(y)$ for all $x, y \in X$.

In the case of graphs, the majority of kernels belongs to the family of R-convolution kernels [16], which decomposes kernels into simpler substructures and uses these to define the overall similarity between pairs of input graphs. Common examples of kernels in this family include the shortest-path kernel [6] and the graphlet kernel [27], where the difference lies in the type of substructure used to summarise the graph structure, i.e., shortest-paths and subgraphs, respectively. Information propagation kernels, on the other hand, include methods where pairs of input graphs are compared based on how information diffuses on them. Examples include random walk kernels [4,18], quantum walk kernels [5,24], and kernels based on iterative label refinements [26]. While some kernels work only on undirected and unattributed graphs, other kernels are designed to handle features as well, either discrete- or continuous-valued [26]. For a detailed review and historical perspective on graph kernels, we refer the reader to the recent survey of Kriege et al. [20].

The most closely related work to ours is that of Togninalli et al. [29], who use the WL test to propagate node feature information across the graph and then use the Wasserstein distance between the resulting embeddings. The WL embeddings can be seen as a discrete time process capturing structural information at different scales, similar in a way to the heat diffusion underpinning the HKS we propose to use in this paper, however the latter is a continuous-time process and as such allows us to naturally capture more fine grained information [28]. The WKS, on the other hand, is a different process entirely. While the WKS also encodes information at different scales by virtue of being defined as a function of the Laplacian eigenfrequencies, in contrast to the HKS, it clearly separates the influence of different frequencies (and thus scales) treating all of them equally [3].

3   Background

The proposed kernel is based on the computation of spectral signatures over the graph nodes and the Wasserstein distance between the resulting sets of points. This section introduces the relevant background.

3.1   Spectral Signatures: HKS and WKS

The HKS [28] and the WKS [3] are two types of spectral signatures [3,9,10,28] that were introduced in computer vision for the analysis of non-rigid three-dimensional shapes. The latter are often abstracted in terms of meshes connecting points sampled on the shape surfaces, however in the following text we will describe both signatures in the more general setting of graphs.

Let $G = (V, E)$ be an undirected graph with node set V and edge set E. Then $L = D - A$ is the Laplacian matrix of a graph G, where A is the adjacency matrix of G and D is the diagonal matrix with elements $D_{i,i} = \sum_j A_{i,j}$. Both the HKS and the WKS are based on the idea of using the spectrum of the graph Laplacian to characterise the structure around the nodes of a graph, however they differ in the type of process used to explore the graph structure, i.e., heat diffusion and wavefunction propagation, respectively.

With the spectral decomposition of L to hand, and assuming that at time $t = 0$ all the heat is concentrated on the node x of the graph, solving the heat equation on the graph G gives us the following equation to determine the amount of heat on x at time t [28], i.e.,

$$HKS_G(x,t) = \sum_{i=1}^{|V|} e^{-\lambda_i t} \phi_{x,i}^2 \,, \tag{1}$$

where λ_i is the i-th eigenvalue of the Laplacian and $\phi_{x,i}$ is the i-th element of the Laplacian eigenvector associated to x. The HKS of x is then constructed by evaluating Eq. 1 for different $t \in [t_{min}, t_{max}]$ [28] and effectively captures increasingly larger scale information around x for increasing values of t. In this paper we set both t_{min} and t_{max} as in [28].

The WKS, on the other hand, measures the probability of a quantum particle with a certain energy distribution to be located on a given node of the graph [3], i.e.,

$$WKS_G(x,b) = \frac{1}{C_b} \sum_{k \geq 0} e^{-\frac{(b - \log(\lambda_i))^2}{2\sigma^2}} \phi_{x,i}^2 \,, \tag{2}$$

where $C_b = \sum_{i=1}^{|V|} e^{-\frac{(b - \log(\lambda_i))^2}{2\sigma^2}}$, λ_i is the i-th eigenvalue of the Laplacian and $\phi_{x,i}$ denotes the i-th element of the eigenvector associated to x. The WKS for a point x is then constructed by evaluating Eq. 2 for different values of $b \in [b_{min}, b_{max}]$ [3]. Like the HKS, the WKS captures structural information at different scales by looking at the squared entries of the Laplacian eigenvectors, weighted according to a function of the corresponding eigenvalues. The key difference is indeed the latter function, which operates as a band-pass filter in the WKS, while it works as a low-pass filter in the HKS. This in turn enables the WKS to achieve a better separation of the frequency information. In this paper we set both b_{min} and b_{max} as in [28].

3.2 Wasserstein Distance

Originating in the context of the optimal transport theory [30], the Wasserstein distance is used to describe the distance between probability distributions on a given metric space. Given two probability distribution σ and μ on a metric space M with Euclidean distance d, the L^p-Wasserstein distance for $p \in [1, \infty]$ is defined as

$$W_p(\sigma, \mu) = \left(\inf_{\gamma(\sigma, \mu) \in \Gamma} \int_{M \times M} d(x,y)^p d\gamma(x,y) \right)^{\frac{1}{p}}, \tag{3}$$

where Γ is the set of all transportation plans $\gamma(\sigma, \mu)$ over $M \times M$ with marginals σ and μ. Similarly to [29] we only consider the case $p = 1$.

While Eq. 3 considers continuous probability distributions, in the discrete case we can rewrite it (for $p = 1$) as

$$W_1(X,Y) = \min_{P \in \Gamma(X,Y)} \langle P, D \rangle, \tag{4}$$

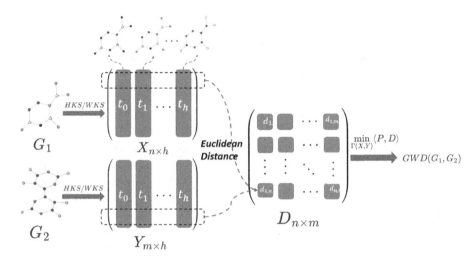

Fig. 1. Graph kernel construction pipeline. Given a pair of input graphs G_1 and G_2, for each node of the graphs we compute the corresponding HKS or WKS signature and we concatenate the resulting vector with the one-hot encoding of the node feature (if available). The kernel between the input graphs is then defined in terms of the Wasserstein distance between the set of points corresponding to G_1 and G_2.

where $X \in \mathbb{R}^{n \times m}$ and $Y \in \mathbb{R}^{q \times m}$ are two sets of vectors, $\langle ., . \rangle$ denote the Frobenius dot product, D is a matrix containing the distance between the vectors in X and Y, and $\Gamma \in (X, Y)$ is the set of transport plan matrices.

4   From Spectral Signatures to Graph Kernels

Following [29], we propose to use Eq. 4 to define a kernel between graphs. Figure 1 shows the proposed pipeline. Given two input graphs G_1 and G_2 with n and m nodes respectively, we first compute the matrices $X^{n,h}$ and $Y^{m,h}$ where the rows represent the HKS (or WKS) signatures graphs nodes and h is the number of sampled values for the time (see Eq. 1) and energy (see Eq. 2). In other words, X and Y correspond to the embeddings of the nodes in G_1 and G_2, respectively.

With these embeddings to hand, we compute the matrix D of pairwise Euclidean distances between the elements of X and Y. We then compute the kernel distance between G_1 and G_2 as

$$GWD(G_1, G_2) = \min_{P \in \Gamma(X,Y)} \langle P, D \rangle. \tag{5}$$

Finally, we compute the kernel between G_1 and G_2 as

$$GWK(G_1, G_2) = e^{-\lambda GWD(G_1,G_2)}, \tag{6}$$

where λ is a real positive number.

Table 1. Some basic statistics on the graph datasets considered.

Datasets	MUTAG	PTC-MR	PROTEINS	D&D
Avg number of vertices	17.93	14.29	39.06	284.32
Avg number of edges	19.79	14.69	72.82	715.66
Number of graphs	188	344	1113	1178
Number of classes	2	2	2	2

Taking Node Features into Account. If the input graphs are attributed, i.e., the node have associated features, for each node we generate a new embedding combining the structural and feature information as follows. Let X_i be the structural signature (HKS or WKS) computed on the i-th node of G and let F_i denote the feature on this node. If F_i is categorical, we transform it into a vector using one-hot encoding. Then, we replace the node embedding vector X_i by concatenating X_i and F_i, i.e.,

$$X_i = \text{concatenate}(w * X_i, (1 - w) * F_i),\qquad(7)$$

where $0 \leq w \leq 1$ is a parameter that allows us to adjust the relative importance of the structural and feature information.

Positive Definiteness. While we are unable to provide proof of the positive definiteness of our kernel, the results of Togninalli et al. [29] suggest that this may be the case. Indeed, we observe empirically that the kernel matrices computed in our experiments are positive definite.

Computational Complexity. It can be easily seen that the computational complexity of our kernel is dominated by eigendecomposition of the Laplacian matrix, which for a graph with n nodes is $O(n^3)$. A naive implementation of the Wasserstein distance computation would have complexity $O(n^3 \log(n))$, however we reduce it to $O(n^2)$ by applying a Sinkhorn approximation [1].

5   Experiments

In this section we describe our experiment setup and discuss the results of our experiments. Specifically, we perform a sensitivity study of the kernel parameters and then we test the performance of the kernels on a graph classification task.

5.1   Experimental Setup

We consider four widely used graph classification datasets: MUTAG [11], PTC-MR [17], PROTEINS and D&D [7]. In these datasets, chemical molecules are modelled as graphs where the vertex features (categorical) encode the atom type

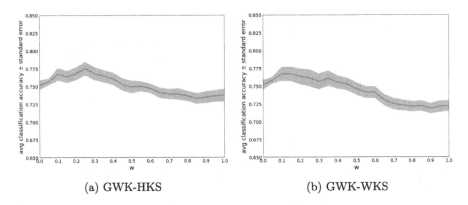

Fig. 2. Average accuracy ($\pm$ standard error) on the PROTEINS dataset for increasing values of w for the GWK-HKS (a) and the GWK-WKS (b).

and the edges encapsulate the chemical bonds between them. Further information on these dataset can be found in Table 1.

We compare the performance of our kernel against 4 widely used kernels and 4 state-of-the-art GNNs, namely 1) the Shortest Path (SP) kernel [6], 2) the Random Walk kernel (RW) [7], 3) the Weisfeiler Lehman kernel (WL) [26], 4) the Graphlet kernel (GR) [27] (with $k = 3$, $k = 4$), 5) GraphSAGE [15], 6) DiffPoll [32], 7) DGCNN [33], and 8) GIN [31]. We also compare our method against the Wasserstein Weisfeiler Lehman kernel (WWL) [29].

In the following, we indicate our kernel as GWK-HKS or GWK-WKS, to indicate the type of spectral signature used to compute the nodes embedding. For each dataset and method, we perform 10-fold cross validation and we report the average accuracy ($\pm$ standard error) over the 10 folds. The hyper-parameters as well as the SVM classifier are optimised by grid-search using a 9:1 split of the training folds. We use the GraKel implementation of the kernels[1] except for WWL, for which we report the results from [29], and we refer to [8] and [12] for the GNNs results.

5.2   Sensitivity Study

We commence by studying the effect of varying the coefficient w on the classification performance of our kernel. Figure 2 shows that the optimal accuracy is achieved by taking into account both the structural and feature information, highlighting that our kernel is indeed successfully able to capture both. We also note that for GWK-WKS the performance remains relatively stable around the optimum, suggesting a small sensitivity to the choice of w with respect to GWK-HKS, where we observe a sharper peak around $w = 0.25$.

---

[1] https://ysig.github.io/GraKeL/0.1a8/.

Table 2. Average accuracy ($\pm$ standard error) on the datasets considered. The best performing method for each dataset is highlighted in bold and the second best performing method is underlined.

Methods	MUTAG	PTC-MR	PROTEINS	D&D
SP	86.64 $\pm$ 2.57	61.94 $\pm$ 2.05	75.47 $\pm$ 1.37	**80.05 $\pm$ 1.19**
GR	66.49 $\pm$ 0.72	57.65 $\pm$ 0.46	68.16 $\pm$ 0.91	N.A
RW	83.48 $\pm$ 3.10	57.94 $\pm$ 0.73	N.A	N.A
WL	81.35 $\pm$ 1.90	61.47 $\pm$ 1.34	75.66 $\pm$ 1.17	79.03 $\pm$ 1.23
WWL	87.25 $\pm$ 1.50	**66.31 $\pm$ 1.21**	74.28 $\pm$ 0.56	<u>79.69 $\pm$ 0.50</u>
DiffPool	81.35 $\pm$ 1.86	55.87 $\pm$ 2.73	73.13 $\pm$ 1.49	75.00 $\pm$ 1.11
GIN	78.13 $\pm$ 2.88	56.72 $\pm$ 2.66	73.05 $\pm$ 0.90	75.30 $\pm$ 0.92
DGCNN	85.06 $\pm$ 2.50	53.50 $\pm$ 2.71	74.31 $\pm$ 1.03	76.60 $\pm$ 1.36
GraphSAGE	77.57 $\pm$ 4.22	59.87 $\pm$ 1.91	73.11 $\pm$ 1.27	72.90 $\pm$ 0.63
GWK-HKS	88.30 $\pm$ 1.79	56.40 $\pm$ 2.46	<u>76.35 $\pm$ 0.87</u>	77.97 $\pm$ 1.35
GWK-WKS	**89.88 $\pm$ 1.39**	<u>62.79 $\pm$ 1.74</u>	**76.89 $\pm$ 1.02**	78.95 $\pm$ 1.10

5.3 Graph Classification Results

Table 2 shows the results of the graph classification experiment, where we highlight the highest average accuracy for each dataset in bold and the second highest one is underlined. GKW-WKS outperforms all competing methods on two out of four datasets and remains highly competitive in the remaining two, while GKW-HKS is the second best performing method on MUTAG and PROTEIN. The WWL kernel is the best performing approach on the PTC-MR dataset, while the SP kernel outperforms both our methods and WWL (although the difference is not statistically significant) on D&D. It should be noted however that the SP kernel is significantly more computationally intensive than our kernel, with a computational complexity of $O(n^4)$. Also, note that the results of the WWL kernel in [29] were computed without making use of the Sinkhorn approximation. While avoiding this approximation can result a slightly higher accuracy, it also leads to a computational complexity of $O(n^3 \log n)$. Finally, note that when the execution of a kernel on a dataset fails, we indicate the corresponding result in Table 2 as not available.

6 Conclusion

In this paper we proposed a novel graph kernel based on two widely used spectral signatures (HKS and WKS) and the Wasserstein distance. The kernel is able to capture both structural and node feature information of the graph and outperforms (or performs comparably to) alternative methods on standard graph classification datasets. Future work will explore the use of alternative structural signatures, more effective ways to encode the node features, as well as new ways to combine structural and node feature information.

References

1. Altschuler, J., Niles-Weed, J., Rigollet, P.: Near-linear time approximation algorithms for optimal transport via sinkhorn iteration. In: Advances in Neural Information Processing Systems, vol. 30 (2017)
2. Atwood, J., Towsley, D.: Diffusion-convolutional neural networks. In: Advances in Neural Information Processing Systems, pp. 1993–2001 (2016)
3. Aubry, M., Schlickewei, U., Cremers, D.: The wave kernel signature: a quantum mechanical approach to shape analysis. In: 2011 IEEE International Conference on Computer Vision Workshops (ICCV Workshops), pp. 1626–1633. IEEE (2011)
4. Bai, L., Hancock, E.R.: Graph kernels from the Jensen-Shannon divergence. J. Math. Imaging Vis. **47**(1), 60–69 (2013)
5. Bai, L., Rossi, L., Torsello, A., Hancock, E.R.: A quantum Jensen-Shannon graph kernel for unattributed graphs. Pattern Recogn. **48**(2), 344–355 (2015)
6. Borgwardt, K.M., Kriegel, H.P.: Shortest-path kernels on graphs. In: Fifth IEEE International Conference on Data Mining (ICDM 2005), pp. 8-pp. IEEE (2005)
7. Borgwardt, K.M., Ong, C.S., Schönauer, S., Vishwanathan, S., Smola, A.J., Kriegel, H.P.: Protein function prediction via graph kernels. Bioinformatics **21**(suppl_1), i47–i56 (2005)
8. Cosmo, L., Minello, G., Bronstein, M., Rodolà, E., Rossi, L., Torsello, A.: Graph kernel neural networks. arXiv preprint arXiv:2112.07436 (2021)
9. Cosmo, L., Minello, G., Bronstein, M., Rodolà, E., Rossi, L., Torsello, A.: 3D shape analysis through a quantum lens: the average mixing kernel signature. Int. J. Comput. Vis. 1–20 (2022)
10. Cosmo, L., Minello, G., Bronstein, M., Rossi, L., Torsello, A.: The average mixing kernel signature. In: Vedaldi, A., Bischof, H., Brox, T., Frahm, J.-M. (eds.) ECCV 2020. LNCS, vol. 12365, pp. 1–17. Springer, Cham (2020). https://doi.org/10.1007/978-3-030-58565-5_1
11. Debnath, A.K., Lopez de Compadre, R.L., Debnath, G., Shusterman, A.J., Hansch, C.: Structure-activity relationship of mutagenic aromatic and heteroaromatic nitro compounds. correlation with molecular orbital energies and hydrophobicity. Journal of medicinal chemistry 34(2), 786–797 (1991)
12. Errica, F., Podda, M., Bacciu, D., Micheli, A.: A fair comparison of graph neural networks for graph classification. arXiv preprint arXiv:1912.09893 (2019)
13. Fröhlich, H., Wegner, J.K., Sieker, F., Zell, A.: Optimal assignment kernels for attributed molecular graphs. In: Proceedings of the 22nd International Conference on Machine Learning, pp. 225–232 (2005)
14. Gilmer, J., Schoenholz, S.S., Riley, P.F., Vinyals, O., Dahl, G.E.: Neural message passing for quantum chemistry. In: International Conference on Machine Learning, pp. 1263–1272. PMLR (2017)
15. Hamilton, W., Ying, Z., Leskovec, J.: Inductive representation learning on large graphs. In: Advances in Neural Information Processing Systems, vol. 30 (2017)
16. Haussler, D.: Convolution kernels on discrete structures. Technical report, Department of Computer Science, University of California (1999)
17. Helma, C., King, R.D., Kramer, S., Srinivasan, A.: The predictive toxicology challenge 2000–2001. Bioinformatics **17**(1), 107–108 (2001)
18. Kashima, H., Tsuda, K., Inokuchi, A.: Marginalized kernels between labeled graphs. In: Proceedings of the 20th International Conference on Machine Learning (ICML 2003), pp. 321–328 (2003)

19. Kipf, T.N., Welling, M.: Semi-supervised classification with graph convolutional networks. In: Proceedings of the 5th International Conference on Learning Representations, ICLR 2017 (2017)
20. Kriege, N.M., Johansson, F.D., Morris, C.: A survey on graph kernels. Appl. Netw. Sci. **5**(1), 1–42 (2020)
21. Lima, A., Rossi, L., Musolesi, M.: Coding together at scale: github as a collaborative social network. In: Eighth International AAAI Conference on Weblogs and Social Media (2014)
22. Morris, C., et al.: Weisfeiler and leman go neural: higher-order graph neural networks. In: Proceedings of the AAAI Conference on Artificial Intelligence, pp. 4602–4609 (2019)
23. Ong, C.S., Mary, X., Canu, S., Smola, A.J.: Learning with non-positive kernels. In: Proceedings of the Twenty-First International Conference on Machine Learning, p. 81 (2004)
24. Rossi, L., Torsello, A., Hancock, E.R.: Measuring graph similarity through continuous-time quantum walks and the quantum Jensen-Shannon divergence. Phys. Rev. E **91**(2), 022815 (2015)
25. Scarselli, F., Gori, M., Tsoi, A.C., Hagenbuchner, M., Monfardini, G.: The graph neural network model. IEEE Trans. Neural Networks **20**(1), 61–80 (2008)
26. Shervashidze, N., Schweitzer, P., Van Leeuwen, E.J., Mehlhorn, K., Borgwardt, K.M.: Weisfeiler-Lehman graph kernels. J. Mach. Learn. Res. **12**(9) (2011)
27. Shervashidze, N., Vishwanathan, S., Petri, T., Mehlhorn, K., Borgwardt, K.: Efficient graphlet kernels for large graph comparison. In: Artificial Intelligence and Statistics, pp. 488–495. PMLR (2009)
28. Sun, J., Ovsjanikov, M., Guibas, L.: A concise and provably informative multiscale signature based on heat diffusion. In: Computer Graphics Forum, vol. 28, pp. 1383–1392. Wiley Online Library (2009)
29. Togninalli, M., Ghisu, E., Llinares-López, F., Rieck, B., Borgwardt, K.: Wasserstein Weisfeiler-Lehman graph kernels. In: Advances in Neural Information Processing Systems, vol. 32 (2019)
30. Villani, C.: Optimal Transport: Old and New, vol. 338. Springer, Heidelberg (2009). https://doi.org/10.1007/978-3-540-71050-9
31. Xu, K., Hu, W., Leskovec, J., Jegelka, S.: How powerful are graph neural networks? arXiv preprint arXiv:1810.00826 (2018)
32. Ying, Z., You, J., Morris, C., Ren, X., Hamilton, W., Leskovec, J.: Hierarchical graph representation learning with differentiable pooling. In: Advances in Neural Information Processing Systems, vol. 31 (2018)
33. Zhang, M., Cui, Z., Neumann, M., Chen, Y.: An end-to-end deep learning architecture for graph classification. In: Thirty-Second AAAI Conference on Artificial Intelligence (2018)

Discovering Respects for Visual Similarity

Olivier Risser-Maroix[(✉)], Camille Kurtz, and Nicolas Loménie

Université Paris Cité, LIPADE, Paris, France
orissermaroix@gmail.com

Abstract. Similarity is a fundamental concept in artificial intelligence and cognitive sciences. Despite all the efforts made to study similarity, defining and measuring it between concepts or images remains challenging. Fortunately, measuring similarity is comparable to answering *"why"*/*"in which respects"* two stimuli are similar. While most related works done in computer sciences try to measure the similarity, we propose to analyze it from a different angle and retrospectively find such respects. In this paper, we provide a pipeline allowing us to find, for a given dataset of image pairs, *what* are the different concepts generally compared and *why*. As an indirect evaluation, we propose generating an automatic explanation of clusters of image pairs found using a model pretrained on texts/images. An experimental study highlights encouraging results toward a better comprehension of visual similarity.

Keywords: Visual similarity · Respects · Image representation · Clustering

1 Introduction

The notion of similarity is a key component of our cognitive system. It is used as a heuristic in different tasks such as learning concepts or reasoning in the presence of new information. In parallel, similarity is considered as a cornerstone in artificial intelligence (AI) [1], used in many areas such as machine learning, and information retrieval, but also in related fields such as data analysis, psychometrics or cognitive sciences. Even if the notion of similarity between concepts or images has been studied for a while in human perception, yet it is still difficult to define [2–5]. Due to the task complexity, similarity suffered a lot of criticisms, such as being too subjective. Authors of [6] reproach that the similarity of A to B is an ill-defined and useless concept until "in what respects" A is similar to B can be stated. They argued that one must specify in what respects two things are similar. Authors of [7] found that *when researchers ask people "How similar are X and Y?" It is as if people are answering the question "How are X and Y similar?"* making thus the procedure of fixing such *respects* a critical aspect of similarity. In this paper, we are interested in finding such *respects*.

In [3], the authors propose to split the similarities shared by two concepts over two axes: the first is relational and not always visible (e.g., *job* and *jail* are related because it is difficult to escape from both of them); the second axis corresponds

A. Krzyzak et al. (Eds.): S+SSPR 2022, LNCS 13813, pp. 132–141, 2022.
https://doi.org/10.1007/978-3-031-23028-8_14

to the attributes shared between two concepts (e.g. *zebra* and *jail* because of the *stripes*) and may be more visual. If two concepts share both relations and attributes, we deal with a *literal similarity match* (e.g. *prison* and *jail*). Here we are interested in the *non-literal* similarity where respects are visual.

During the last decade, cognitive and computer scientists started to examine the similarities and discrepancies between human and machine perception and how neural architectures may help in the comprehension of human judgments of similarity. Such a similarity can be either guided by general concepts [8–10] or performed by visual correspondences [11–15]. As stated by [16], strategies based on visual stimuli yield good results with simple images [12] or when focusing on a narrow class [8,13,17,18]. Only a few works [10,19,20] proposed to link the similarity assessment task to the task of finding the respects. [19,20] worked on narrow image domains and on *literal* similarity by using the *Caltech-UCSD Birds 200* and *Stanford Cars* datasets. Here the respect is the semantic similarity. In parallel, authors from [10] used the *THINGS* dataset composed of 1854 concepts. They did not work directly from images but from a more abstract representation obtained from human judgments derived from a *select the one-odd-out* task. By being broader and more focused on *non-literal* similarity, the *Totally Looks Like* [16] dataset is more challenging and illustrates well the complexity of the task (see e.g. in Fig. 2). In this dataset, the respects used to make comparisons are multiples, such as color, texture, shape, etc. Recently, [21,22] proposed comparable approaches to deal with such complex examples of cases and datasets, without taking into account the respects.

In this paper, we aim to go further in the study of visual similarity. We propose to work from semantic and visual embeddings to discover: *what* can be compared and *why* it is compared. The *what* is often denoted as a tuple of *source* and *target*. Whereas the *why* corresponds to the respects. For example in Fig. 2 (left), the first pair corresponds to a man and a cat (*what*) compared together because of their eyebrows (*why*). To achieve this, we start from the pipeline proposed by [21] to compute visual embeddings, and [23], a neural network trained on images and texts, to obtain semantics embeddings from images. Our contributions are then (1) to adapt a clustering algorithm to find the *what* and (2) a method based on NMF (non-negative matrix factorization) to discover the *why*. Finally, we exploit the zero-shot ability of [23] to get an automatic interpretation of the clusters discovered, and we observe correlations with human thoughts, which can be considered as an indirect evaluation of the respects for visual similarity highlighted for a dataset.

2 Method

2.1 Dataset Selection

We used the *Totally Looks Like* (*TLL*) image dataset [16], a stimulating example for this research area. *TLL* consists of 6k+ image pairs coming from human propositions, and image similarity is determined by visual cues from the image contents (rather than semantic categories). The images are related to various domains such as photography, cartoons, etc., making the task more difficult but

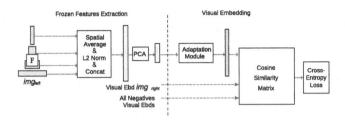

Fig. 1. Visual features adaptation pipeline [21]. Each image pair (left and right) is passed through a frozen feature extractor (here ResNet50x4 from [23]). After the visual adaptation, a similarity matrix between each image from the left and each image from the right is built. The objective function is to bring closer images from a given pair than any other image from the left/right set.

more reflective of the human ability to link semantically unrelated concepts. The process of human visual similarity combines numerous layers of analysis, such as color, texture, shape, etc. Thus, one could argue that similarity is too subjective and an ill-posed problem. Nevertheless, it has been found in [5] that when the task is specified, humans remain consistent in their judgments [4]. This statement was also verified on *TLL* with human experiments [16]. In this case, the respects depend on the task of finding visual connections from unrelated images. However, *TLL* suffers from class imbalance. As suggested in [21], such a dataset can be decomposed into two subsets: *TLL_face* contains 1817 pairs where faces are compared together; *TLL_obj* contains 4199 pairs, with all other kinds of comparisons (faces are still present but not compared only because of the facial similarity). [22] considered all the *TLL* and took this bias to their advantage by using a *FaceNet* network pre-trained on the *VGG-Face2* dataset. In [13], facial similarity is treated separately from general image similarity since humans have dedicated neural processing for faces, different than for other scenes. Thus, as proposed in [21], we only consider the *TLL_obj*.

Another interesting dataset is *THINGS* coupled with human judgments from [10]. We will refer to this combination as *SPOSE*. [10] propose to embed 1854 concepts, each represented by a single image, into a 49 (sparse) dimensions vector. By asking the participants to find commonality through a set of images representing a concept, they found these dimensions to be highly interpretable. In Fig. 3, humans have found these semantically different objects activating a given dimension which may highlight the concept of *disk*. Each image is associated with a behavioral embedding, while each dimension is associated with a variable number of words humans give when asked to find commonalities between images that best represent each dimension. We will consider *SPOSE* to validate the correlation between the zero-shot interpretation and human judgments.

2.2   Image Representation

The features extraction step is highly influenced by the chosen extractor. For a given architecture, the dataset scales and content domains, as well as the training objective, will have direct impacts on the re-usability of the features. From the model provided by [21], we replaced the default feature extractor by ResNet50x4, a network pre-trained using CLIP [23]. CLIP is a framework that learns contrastively visual concepts from language supervision [23]. Since the CLIP objective function is biased toward the semantic of images, we denote those embeddings: $x_{sem} \in \mathbb{R}^{640}$. Due to their simplicity and features highly correlated with human judgments on diverse tasks, convolutional neural networks have become popular in cognitive sciences [24,25]. Pre-trained features have been found to be highly correlated with human judgments on different tasks such as typicality rating or similarity assessment on natural stimuli [26]. However, those results may still be highly improved by adapting the features into a more psychological space [8,14,15,18,22] to better correlate with human judgments for each given task requirement (semantic, perceptual, or visual similarity). The whole pipeline generally consists of a frozen features extractor and an adaptation module to adapt representations.

In our case, we adapted the features from the images of *TLL_obj* to a visual space with the method from [21], whose the adaptation module is composed of a linear layer followed by a ReLU activation (Fig. 1). Visual features $x_{vis} = adaptation(x_{sem})$ are denoted: $x_{vis} \in \mathbb{R}_{+}^{1024}$. We used the CLIP ResNet50x4 CNN as a feature extractor instead of the default ResNet50 pre-trained on ImageNet and used the following hyper-parameters for training: temperature $\sigma = 15$, default Adam, mirror augmentation. This feature extractor and those hyper-parameters increased their asymmetric recall score at 1 to an average of 0.39 on 20 runs with a 75–25 train test split. Once hyper-parameters were found, for better generalization for our final task, we trained on 90% of the dataset instead of the default 75%.

Based on these image embeddings, the next step is to go further in the analysis and the comprehension of visual similarity to discover: *what* can be compared and *why* it is compared (the respects).

2.3   What

With similarity arise the questions of *what can be compared* and *why things are compared*. In the case of *literal* similarity, such questions are trivial: two images are compared since they share both relations and attributes. The *literal* similarity can be considered as a tautology and is poorly informative [6]. On another hand, *non-literal* similarity (defined as a form of metaphor) needs to be informative. For example comparing a *dahu* to a *dahu* does not help us to define a *dahu*. But by comparing a *dahu* to a chamois arises more information on the morphology of what a *dahu* may look like. The large diversity of comparisons made in the *TLL_obj* dataset can help psychologists to understand the choices made by humans to build pairs. The similarity is asymmetrical, but in the *TLL_obj*, pairs are not provided with a referent and a referee.

In order to understand what can be compared in a dataset, we propose a modification of k-Means to cluster image pairs (Algorithm 1). In our case, we face pairs of images and assume that each image of a given pair should belong

Fig. 2. (left) Example of *what* can be compared. Here cats are compared to human because of their poses, facial expression, etc. (right) Example of *why* things can be compared. Here all persons, animals and pareidolia have similar facial expressions. This may be one *respect* for their similarity.

to different clusters. Thus we could discover the different objects which could be compared together by clustering the displacement from the cluster of one image to the cluster of the second image. However, the pairs are not made in a specific order (objects of a given semantic class could be left or right placed).

To bypass this issue, we propose to cluster the semantic displacement (SD) between the two image embeddings (abbreviated as *ebds* in Algorithm 1). SD corresponds to the normalized difference between each image of a given pair. By clustering those vectors, we aim to get the different displacements used to create pairs. This idea is similar to [27] where the authors proposed to re-balance dataset classes by adding the displacements found with a simple k-means to real points. However they did not check whether these displacements are consistent/interpretable. While they performed clustering on the displacement vectors for a given class, we propose to extend the procedure to all images. The displacement from A to B is the strict opposite of the B to A one, they should belong to the same cluster. As in the original k-means they would be affected to different clusters, we propose to solve this issue by modifying the affectation step and the centroids updating step by considering the absolute value of the cosine similarity of each semantic displacement (SD) vector to the centroids (C). Here,

---

Algorithm 1: "what"-clustering

Input:
    $ebds_{left} \leftarrow \ell_2(ebds_{left})$, $ebds_{right} \leftarrow \ell_2(ebds_{right})$, K // Number of clusters
  $SD \leftarrow \ell_2(ebds_{left} - ebds_{right})$;
  $C \leftarrow \text{init\_centroids}(SD, K)$;
while *not converged* **do**
    // Cluster affectation
    $sim\_to\_C \leftarrow \cos\_sim(C, SD)$;
    $abs\_sim \leftarrow \text{abs}(sim\_to\_C)$;
    $sign\_sim \leftarrow \text{sign}(sim\_to\_C)$;
    $\hat{c} \leftarrow \text{argmax}(abs\_sim, 0)$;
    // Centroids update
    for *i in range(K)* **do**
      $C_i \leftarrow \ell_2(\text{mean}(SD_{\hat{c}=i} \times sign\_sim_{i,\hat{c}=i}))$
    end
end

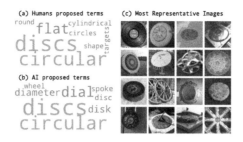

Fig. 3. Example of most typical images representing a given dimension from [10]. Common words found by humans and AI are highlighted in green, see Sect. 3.1.

0 becomes the smallest similarity possible: the semantic displacement vector is orthogonal to the centroids and not related at all. During the centroids update, the centroid C_i should not be overwritten by the simple mean of each cluster instance. For example, if there is as much positive signed similar instances as negatively signed one, the centroid could collapse to the null vector (e.g. cat to dog + dog to cat = do not move). Furthermore, using the naive mean in our case does not make sense. We propose thus to multiply each instance of a centroid C_i by the sign of the similarity between the semantic displacement and C_i.

By doing so, we can reorder pairs of a given cluster in consistent ones. As illustrated in Fig. 2 (left block), after automatic re-ordering all humans are on the left while cats are on the right.

2.4 Why

According to [7], answering the question of *"how similar are X and Y?"* it is as if people are answering the subtly different question *"how are X and Y similar?"*. Finding the diverse reasons of *"why two images are compared"* (finding the respects) is thus a critical aspect of similarity comparisons. We propose here to check whether the learned visual embeddings allow to discover those respects. To this end, we propose a method to find such reasons in a topics modeling fashion. Non-negative matrix factorization (NMF) [28] has been used successfully to analyze data from various domains such as audio source separation, face recognition, etc. In our case, we propose to adapt the NMF to pairs in order to find clusters of respects. Let us consider an image pair. After adaptation to the visual space, the embeddings of the images are defined on $\mathbb{R}_+^{1024}$ and ℓ_2 normalized. Thus, we consider each dimension similar to the quantity of how a given word is used to describe the concept. So, the reason of the similarity between images would be the selection of the words in the first image AND the second. We propose to use the *fuzzy T-norm* $*$ (equivalent to Hadamard product $\odot$ in vectorized form) to model this logical AND. We ℓ_2 normalize the output to get comparable vectors. Pairs p are computed as: $p = \ell_2(x_{vis,left} \odot x_{vis,right})$. We feed the NMF algorithm with this representation for topic discovery.

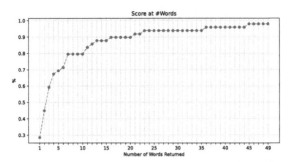

Fig. 4. Percentage of clusters having at least one word in common between the human (fix) and AI proposed (variable, x-axis), see Sect. 3.1.

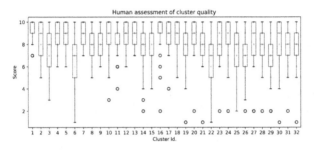

Fig. 5. Human assessment of *why* cluster quality experiment (Sect. 3.2). We observe that 28 out of 32 clusters have a quality score greater or equal to 8.

For a set of pairs representation p_i stacked in a matrix P, the NMF is optimized with the following equation: $\text{NMF}(P, k) = \text{argmin}_{\hat{P}_k} \parallel P - \hat{P}_k \parallel_F^2$, subject to $\hat{P}_k = HM, \forall ij, J_{ij}, W_{ij} \geq 0$ with $H \in \mathbb{R}^{n \times k}$ and $W \in \mathbb{R}^{k \times m}$ reducing dimensionality to rank k.

Figure 2 (right) illustrates a result obtained with this approach. Here all persons, animals have open mouths and similar facial expressions.

3   Evaluations

3.1   Automatic Interpretability and Human Validation

Thanks to the proposed approaches, we automatically built *what* and *why* clusters from the *TLL_obj* dataset. In this preliminary study, and according to the dataset's content, we fixed the number of clusters to 32. We aim in the following to automatically validate the cluster interpretability. To do so, we propose to employ the zero-shot ability of CLIP [23]. For each cluster, we search the n most typical pairs. The typicality of pairs for k-means is assessed by using the distance to the centroids: closer is the pair, more typical is the ranked pair. For the *why* clusters, since the NMF does not provide centroids, we measure typicality

Fig. 6. Human assessment of cluster quality experiment (Sect. 3.2). Example of *why* cluster #20 and word cloud associated to participants' answers. On 33 answers, 29 explicitly mentioned the word *color* while some others used terms such as *rainbow*. As a comparison, the term *shape* was used only 7 times.

in a similar way as [26]. We assume that typicality ratings are related to the strength of a model's classification response to the category of interest. Here, for each factor of the NMF, we took the pairs most activating the corresponding dimension. From those pairs, we extracted semantics embeddings from each image pair (right and left) with CLIP. For the *what* clusters, after automatically re-ordering pairs as describer earlier, we averaged *left* and *right* semantic embeddings separately. This allows us to have embeddings representing concepts X and Y such that X looks like Y or Y looks like X. In parallel, we average all semantic embeddings to find the respect for each factor of the NMF. From now, the steps are the same for both *what* and *why*. CLIP was trained on 400 million text-image pairs and is approaching, in zero-shot, the performances of the original supervised ResNet-50 on ImageNet. Label candidates are proposed by using *prompt engineering*. A given label name W is surrounded by text in natural language such as: *"this is a picture of a W"* then passed to the text-encoder of CLIP. The list of candidates is ranked by computing the similarity to the encoded image. The one with the highest score is picked. We followed this procedure by using the 11.200 most used English words. Top words were picked and displayed in a word cloud representation (Fig. 2).

To check whether this procedure is able to find common concepts of images which can be unrelated, we use results from human experiments done by [10] with the THINGS dataset combined in the dataset presented earlier as SPOSE. To validate how our word proposition pipeline could agree with humans for abstract concepts (Fig. 3), we extracted semantic embeddings from the most typical images of each conceptual dimension proposed by [10] and averaged them. We checked if at least one common word was found and the top-k words proposed by AI. Results are shown in Fig. 4. We observe that with only 7 words suggested by CLIP in a zero-shot setting, this allows 80% of the clusters to have at least one word in common with the ones proposed by humans from more than 10k choices. From Fig. 3, we observe that only two of them overlap with the same number of words suggested by humans and by machine. However, one can note that this metric does not well handle synonymy, and words automatically suggested are closely related to the common concept of these images: "circularity".

3.2 Human Assessment of Cluster Quality

While we found visual clusters of *respects* meaningful, we run an evaluation campaign to validate the proposed method for the *why* clusters. Since our work is exploratory, we do not have ground truth labels for an automatic evaluation of our clusters. Thus we propose to assess the quality of the *why* clusters found from human judgments. We generated 32 *why* clusters as in Fig. 2 (right). For each cluster, we asked 33 participants to find which may be the common *respect* in the cluster and then to evaluate the cluster according to the inferred rule on a Likert scale ranging from 1 (the cluster does not follow the supposed rule) to 10 (all visible pairs follow the cluster supposed rule). While the inferred rule is hard to process (Fig. 6), the score given for each cluster is used to assess the quality of the given cluster. Scores are reported in Fig. 5. At the end of the experiment, we asked the participants to assess their work. Except for two outliers, participants are pretty confident in the quality of their answers.

4 Conclusion

This paper discusses on few aspects of the non-literal similarity applied to the case of images. As original and ongoing work, we proposed three complementary methods. Their common objective is to make the analysis easier and get insights on similarity from data. The first method aims to discover the *sources and targets* (*what*), involved in *non-literal* similarity by adapting a clustering algorithm. The second one relies on a pipeline to discover different *respects* (*why*) involved in similarity. Some were simple to anticipate, such as color, while others were non-trivial to find, such as shape, facial expression, or even hair-cut. The last method benefits from the zero-shot ability of a model jointly trained on text and images to help to give sense to the different *what* and *why* clusters discovered. An experimental study with these methods highlights encouraging results to start larger-scale human experiments with the ultimate goal of better understanding the complex notion of visual similarity.

References

1. Rissland, E.: AI and similarity. IEEE Intell. Syst. **21**, 39–49 (2006)
2. Tversky, A., Gati, I.: Studies of similarity. In: Cognition and Categorization (1978)
3. Gentner, D., Markman, A.B.: Structure mapping in analogy and similarity. Am. Psychol. **52**, 45–56 (1997)
4. Tirilly, P., Mu, X., Huang, C., Xie, I., Jeong, W., Zhang, J.: On the consistency and features of image similarity. In: IIiX, pp. 164–173 (2012)
5. Liu, T., Cooper, L.A.: The influence of task requirements on priming in object decision and matching. Mem. Cognit. **29**, 874–882 (2001)
6. Goodman, N.: Seven strictures on similarity. In: Problems and Projects (1972)
7. Medin, D.L., Goldstone, R.L., Gentner, D.: Respects for similarity. Psychol. Rev. **100**, 254–274 (1993)

8. Peterson, J.C., Abbott, J.T., Griffiths, T.L.: Evaluating (and improving) the correspondence between deep neural networks and human representations. Cogn. Sci. **42**, 2648–2669 (2018)

9. Zheng, C.Y., Pereira, F., Baker, C.I., Hebart, M.N.: Revealing interpretable object representations from human behavior. In: ICLR (2019)

10. Hebart, M.N., Zheng, C.Y., Pereira, F., Baker, C.I.: Revealing the multidimensional mental representations of natural objects underlying human similarity judgements. Nat. Hum. Behav. **4**, 1173–1185 (2020)

11. Kubilius, J., Bracci, S., Op de Beeck, H.P.: Deep neural networks as a model for human shape sensitivity. PLoS Comput. Biol. **12**, e1004896 (2016)

12. Pramod, R.T., Arun, S.P.: Do computational models differ systematically from human object perception? In: CVPR, pp. 1601–1609 (2016)

13. Sadovnik, A., Gharbi, W., Vu, T., Gallagher, A.C.: Finding your lookalike: measuring face similarity rather than face identity. In: CVPR Workshops, pp. 2345–2353 (2018)

14. Zhang, R., Isola, P., Efros, A.A., Shechtman, E., Wang, O.: The effectiveness of deep features as a perceptual metric. In: CVPR, pp. 586–595 (2018)

15. German, J.S., Jacobs, R.A.: Can machine learning account for human visual object shape similarity judgments? Vis. Res. **167**, 87–99 (2020)

16. Rosenfeld, A., Solbach, M.D., Tsotsos, J.K.: Totally looks like - how humans compare, compared to machines. In: ACCV, pp. 282–297 (2018)

17. Hamilton, M., Fu, S., Freeman, W.T., Lu, M.: Conditional image retrieval. CoRR, abs/2007.07177 (2020)

18. Attarian, M., Roads, B.D., Mozer, M.C.: Transforming neural network visual representations to predict human judgments of similarity. In: SVRHM Workshop (2020)

19. Chen, C., Li, O., Barnett, A., Su, J., Rudin, C.: This looks like that: deep learning for interpretable image recognition. CoRR, abs/1806.10574 (2018)

20. Nauta, M., Jutte, A., Provoost, J.C., Seifert, C.: This looks like that, because ... explaining prototypes for interpretable image recognition. CoRR, abs/2011.02863 (2020)

21. Risser-Maroix, O., Kurtz, C., Loménie, N.: Learning an adaptation function to assess image visual similarities. In: ICIP, pp. 2498–2502 (2021)

22. Takmaz, A., Probst, T., Pani Paudel, D., Van Gool, L.: Learning feature representations for look-alike images. In: CVPR Workshops (2019)

23. Radford, A., Kim, J.W., Hallacy, C., Ramesh, A., et al.: Learning transferable visual models from natural language supervision. In: ICML, vol. 139, pp. 8748–8763 (2021)

24. Cichy, R.M., Kaiser, D.: Deep neural networks as scientific models. Trends Cogn. Sci. **23**, 305–317 (2019)

25. Zhuang, C., et al.: Unsupervised neural network models of the ventral visual stream. Proc. Natl. Acad. Sci. **118**, 237–250 (2021)

26. Lake, B., Zaremba, W., Fergus, R., Gureckis, T.: Neural networks predict category typicality ratings for images. In: CogSci, pp. 1–15 (2015)

27. Hariharan, B., Girshick, R.: Low-shot visual recognition by shrinking and hallucinating features. In: ICCV, pp. 3018–3027 (2017)

28. Lee, D.D., Seung, H.S.: Learning the parts of objects by non-negative matrix factorization. Nature **401**, 788–791 (1999)

Graph Regression Based on Graph Autoencoders

Sarah Fadlallah[ID], Carme Julià[ID], and Francesc Serratosa[✉][ID]

Universitat Rovira i Virgili, Tarragona, Catalonia, Spain
{sarah.fadlallah,carme.julia,francesc.serratosa}@urv.cat

Abstract. We offer in this paper a trial of encoding graph data as means of efficient prediction in a parallel setup. The first step converts graph data into feature vectors through a Graph Autoencoder (G-AE). Then, derived vectors are used to perform a prediction task using both a Neural Network (NN) and a regressor separately. Results for graph property prediction of both models compared to one another and baselined against a classical graph regression technique i.e. Nearest Neighbours, showed that using embeddings for model fitting has a significantly lower computational cost while giving valid predictions. Moreover, the Neural Network fitting technique outperforms both the regression and Nearest Neighbours methods in terms of accuracy. Hence, it can be concluded that using a non-linear fitting architecture may be suitable for tasks similar to representing molecular compounds and predicting their energies, as results signify the G-AE's ability to properly embed each graph's features in the latent vector. This could be of particular interest when it comes to representing graph features for model training while reducing the computational cost.

Keywords: Graph embedding · Autoencoders · Graph regression · Graph convolutional networks · Neural networks · Nearest neighbor · Molecular descriptors · Atomisation energy

1 Introduction

A natural way to represent information in a structured form is as a graph. A graph is a data structure describing a collection of entities, represented as nodes, and their pairwise relationships, represented as edges. With modern technology and the rise of the internet, graphs are everywhere: social networks, the world wide web, street maps, knowledge bases used in search engines, and even chemical molecules are commonly represented as a set of entities and relations between them, as pointed out in [8].

There are two main applications for working with graphs: graph classification and graph regression. The former consists in obtaining a set of features that allow to discriminate between graphs of different classes, while the latter aims at predicting a global graph property. This paper addresses the second case.

© The Author(s), under exclusive license to Springer Nature Switzerland AG 2022
A. Krzyzak et al. (Eds.): S+SSPR 2022, LNCS 13813, pp. 142–151, 2022.
https://doi.org/10.1007/978-3-031-23028-8_15

The classical graph regression applications are based on the Nearest Neighbour algorithm for regression, in which the distance between elements is the graph edit distance [11]. While this can be a valid approach for graph regression, it imposes significant runtime restrictions, since the cost of calculating the distance between old and new points increases with the number of observations. For this reason, this paper evaluates graph regression approaches based on the graph autoencoder model presented in [8].

Specifically, the current paper proposes a two-step graph regression approach based on graph autoencoders. Initially, G-AEs are used to embed the graph in a vector of features. Then, the obtained embeddings were used to predict a global graph property through a fitting technique. To fit said embeddings, linear regression and a shallow Neural Network were employed. Figure 1 summarises the scheme of the two-step proposed approach.

It should be highlighted that the decoder is only used in the G-AE learning process, since the goal of the G-AE is not to generate new graphs, but to obtain a good embedding of each of them in the encoder module.

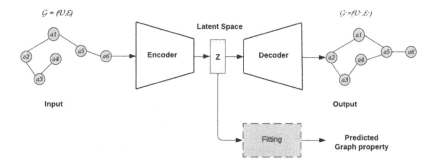

Fig. 1. General scheme of the proposed two-step graph regression approach

With the growing tendency to represent chemical compounds as a graph using a vector representation [4], we decided that our method can be applied to predict the atomisation energy needed for drug-target interaction calculations. In the next section, we briefly explain the main ideas and intuition behind graph embedding and graph regression, introducing the methods and algorithms that compose our graph regression. Then, in Sect. 3 we explain the approach in more detail, and in Sect. 4 we analyse the practical application. Finally, we conclude the paper in Sect. 5.

2 Related Work

2.1 Graph Embedding and Graph Regression

Graphs can be an effective method for visualising and characterising instances of data with complex structures and rich attributes [13]. To capture the interrelationships between a system and its components, e.g. social networks, and biological structures, graph embedding can be utilised for representing the local features of a graph as a vector. This of course, requires the selection of a well-suited set of representatives as feature vectors. Once achieved, classical statistical learning machinery can be used on graph-based input for pattern recognition problems [5].

For the reconstruction of graphs from a vector of features, weighted K-Nearest Neighbours (K-NN) and -neighborhood graphs are two of the most common techniques used [12]. One downside of regression with K-NN and similar methods however, is the calculation of distances, which increases the computational cost in correspondence to the number of data points.

2.2 Autoencoders and Graph Autoencoders

Graph-based machine learning techniques have witnessed increased attention in the recent years [14]. The utilization of Convolutional Neural Networks (CNN) to exploit the richness of graph representations has produced efficient methods for classification and feature extraction tasks, e.g., the architecture presented in [3]. In this approach, classical CNNs were generalized to graphs by using Graph Signal Processing tools with a strictly localised filter for efficient learning and evaluation, while keeping the computational complexity low. This work influenced the development of a convolution network architecture that can scale linearly in the number of graph edges to encode both the structure of a given graph and the features of its nodes [8]. This encoding can be achieved by learning the hidden layer representations in which they are embedded.

Later on, Graph Variational Autoencoders (G-VAE) were presented [7]. Their proposed scheme aimed at predicting missing edges in graph-structured data. Their experiment was initially compared to two common baseline methods, Deep Walk and Spectral Embedding. The two proposed flavours of graph gutoencoder (variational and non-variational) achieved good performance results. Moreover, when including node features for testing the variational graph autoencoder on known datasets, i.e., Citseer, PubMed, and Cora, their model outperformed the aforementioned techniques. The formulation used in this paper is detailed in the Sect. 3 section.

2.3 Prediction of Chemical Compound Properties

The search through the Chemical Compound Space is increasing in complexity due to the recent progress in high-throughput screening, combinatorial chemistry and molecular biology. With this progress, the challenge arises when it comes to testing synthetic molecules in a short period of time in the pharmaceutical industry [2]. With the parallel advancement of statistical tools through machine learning, a plethora of research exploring various techniques is available to aid in the problem of chemical compound space search. Researchers direct their efforts into high accuracy prediction of different molecular properties from existing molecular structures, e.g., calculating the ground-state energy of a given molecule [6]. Algorithms such as the Kernel Ridge Regression and Multilayer

Neural Networks were optimised for potential energy prediction. While accuracy of predictions has improved, the question of having efficient representations of invariant properties is still open [9].

3   The Method

A graph $\mathbf{G(X,A)}$, with n number of nodes, is represented by a node attribute matrix $\mathbf{X}$, and an adjacency matrix $\mathbf{A}$. Attributes on the nodes are a vector of real numbers and edges do not have attributes. $\mathbf{A}_{i,j} = 1$ means that there is an edge between the ith node and the jth node. $\mathbf{A}_{i,j} = 0$ means otherwise. Graph edges are undirected, thus $\mathbf{A}_{i,j} = \mathbf{A}_{j,i}$.

GAEs are composed of two main modules: an encoder and a decoder. The encoder embeds input graphs through a graph convolution network as defined in [8] and returns a latent vector $\mathbf{Z}$. Equation 1 show the encoder function:

$$\mathbf{Z} = \mathbf{GCN}(\mathbf{X}, \mathbf{A}) = \sigma\left(\tilde{A}ReLU\left(\tilde{A}XW_0\right)W_1\right) \tag{1}$$

where $\tilde{A}$ is a symmetrically normalised adjacency matrix computed from $\mathbf{A}$, while $\mathbf{W_0}$ and $\mathbf{W_1}$ are the weight matrices for each layer, which are learned through a learning algorithm. Note that $\sigma\left(\cdot\right)$ is the sigmoid function and $ReLU$ is the classical non-negative linear equation.

The decoder is defined as Eq. 2:

$$\mathbf{A}^* = \sigma\left(\mathbf{Z}\mathbf{Z}^T\right) \tag{2}$$

where $\mathbf{T}$ means the transpose matrix. Note $\mathbf{A}^*$ represents the reconstructed adjacency matrix.

Since the aim of the graph autoencoder is to reconstruct the adjacency matrix such that being similar to the original adjacency matrix, the learning algorithm minimises the mean square distance between these matrices by Eq. 3,

$$\mathbf{L}(\mathbf{A}, \mathbf{A}^*) = min \sum_{i,j=1}^{n} (A_{ij} - \hat{A}_{ij})^2 \tag{3}$$

Finally, the general minimisation criteria is the sum of Eq. 3 given all graphs in the dataset.

The aim of our method is to predict general graph properties accurately, while taking into consideration computational cost. This is achieved through two steps. In the first one, we explore the potential of G-AEs introduced in [7] as means for extracting a representative vector $\mathbf{Z}$. In the second one, we apply classical machine learning techniques, such as linear regression (option 1) or shallow neural networks (option 2), to predict the general graph properties given the embedding of the graph.

Figure 2 shows our detailed scheme (see Fig. 1 for the general scheme). In the upper part, we see the graph autoencoder defined in [7] and summarised in Sect. 3 represented as a graph convolutional network (GCN). In the lower part, we see two options for the fitting module, one based on a classical linear regression while the other was based on a neural network. The encoder of the graph autoencoder is defined as a GCN (Eq. 1) and the decoder as a sigmoid function (Eq. 2).

The learning algorithm is composed of two steps. In the first step, we learn the weights of the graph autoencoder W_0 and W_1 as explained in Sect. 2.2. In the second step, we do not modify these weights and we learn the weights of the linear regression (option 1) or the neural network (option 2). To do so, we first compute the latent vector Z of each graph (applying the encoder and the learned weights W_0 and W_1). Then, given these vectors and the global properties of the graphs that we have in the training set, we deduce the weights of the fitting modules (linear regression or neural network).

Note the decoder (Eq. 2) is only used to learn the G-AE weights. It is not used neither in the process of learning the fitting modules nor in the process of predicting the global properties. Thus, it is needed to reconstruct the adjacency matrix A^* while learning W_0 and W_1 but it is not needed to reconstruct G^*.

4   Motive and Practical Application

We have applied our method to predict the atomisation energy, which is needed for drug-target interaction and binding calculations. This binding energy is used to assess how potent a suggested compound of a pharmaceutical drug will be. While there are some methods that predict said energy, many have the drawback of being computationally costly given not so large compounds. Therefore, speeding up the process of feature engineering and model training can be of immense aid when it comes to scanning the chemical search space and other problems that are time sensitive.

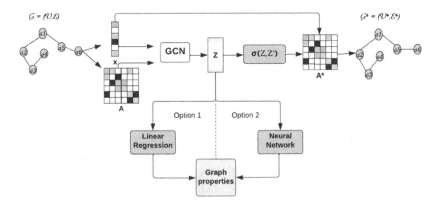

Fig. 2. Embedding and learning scheme of the two-step proposed approach with a G-AE and two prediction models

4.1   Database

The proposed architecture was tested on the quantum mechanics database. This database, named **QM7**, is a subset of a larger database containing molecules of up to 23 atoms per compound, with the total of 7165 molecular instances for the small subset. While the original dataset contains information about different features, e.g., Coulomb Matrices, atomisation energies, and atomic charge, only a selected set of the molecular features was used for training the model. The dataset was preprocessed to generate graphs where edges resembled the connected atoms in a molecule, while the atomic energy and the Cartesian coordinates of the atoms were registered as features of the node, Each compound can be represented as an attributed graph. No additional attributes were set on the edges.[1], the first 5000 compounds were used for learning purposes and the rest of them were used for testing.

4.2   Architecture Configuration

The GAE has a latent space with dimensions: 24×3. Note that 23 is the maximum number of atoms per compound and we have added the bias. Moreover, the dimensions of the weights of the autoencoder are W0: 4×100 and W1: 100×3. Additionally, for the fitting module, we have used two different architectures: a classical linear regression module and a neural network. Figure 3 shows the evolution of the graph autoencoder and neural network loss functions given the training set through the 20 epochs of the learning algorithm.

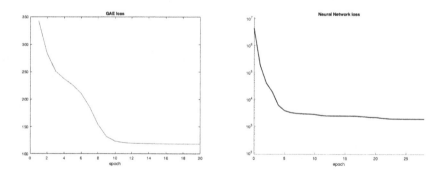

Fig. 3. Loss of the graph autoencoder (left) and the neural network (right) given the training set for the 20 epochs

The neural network has one hidden layer. The input of this network is the vectorised latent vector of the graph autoencoder. Thus, it has a length of 24×3 $+1 = 73$. (1 is for the bias). The number of neurons of the hidden layer is 35. The neural network output is a real number, which is the predicted Energy. Finally,

---

[1] http://quantum-machine.org/datasets/.

the activation function for the hidden layer is a sigmoid function, whereas it is a linear function for the output layer. We used 20 epochs and the back-propagation algorithm to learn the weights.

The linear regression module receives the vectorised latent vector, similarly to the neural network and generates only one output value. The number of weights is 73.

4.3   Energy Prediction

Figure 4 (Left side figures) show predicted energy (vertical axis) versus the real energy (horizontal axis) achieved through linear regression, a neural network, and the Nearest Neighbour models respectively. We realise that in the case of regression, the predictions seem to be quite accurate although the neural network obtains a better prediction as a lower mean square error (MSE) was achieved.

This result indicates that, on the one hand, the GAE is able to properly embed the features of each graph in the latent vector. On the other hand, the latent domain is not linear with respect to the energy of the compound since it is better to use a non-linear fitting architecture as a neural network.

For further analysis of the method performance, the right side figures in Fig. 4 show the histogram of the absolute difference between the predicted energy and the real one. As supposed, the bars indicating the error score of the neural network are more concentrated in the left side of the histogram, as opposed to the case of linear regression. Furthermore, the tie in the neural network is also shorter than it is in the linear regression. The Nearest Neighbour algorithm had the highest MSE value amongst the tested approaches.

4.4   Runtime Analysis

Fitting the regression model on a training set of 5000 graphs took approximately 5 h of computing. The learning of the Neural Network needed about 2 min, while predictions obtained through the Nearest Neighbour algorithm took over 12 h of computing. Neural Networks show evident superiority in terms of computating time.

Experiments were carried out using Matlab 2021 on an Intel(R) Xeon(R) W - 2123 CPU @ 3.60 GHz processor, with 16 GB of RAM.

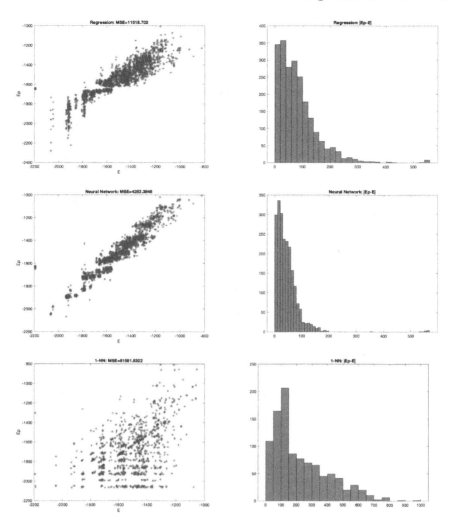

Fig. 4. Predicted energy (left column) (vertical axis) versus real energy (horizontal axis), and absolute error score (right column) computed via the fitting techniques: linear regression, neural network, and nearest neighbour (top to bottom).

5   Conclusions and Future Work

We have tested a method to perform graph regression based on a graph autoencoder and a fitting module. The encoder of the graph autoencoder is based on a graph convolutional network, while the decoder is based on a sigmoid function. In addition, we have proposed two options for the fitting module. One was composed of a classical linear regression and the other one was a neural network. This method has been applied to a well-known database obtaining promising results.

We tested a classical graph regression based on a K-Nearest Neighbour method for regression in which the distance between elements was the graph edit distance. This distance was computed by the bipartite algorithm [10] and edit costs were learned by algorithm in [1]. Results show that classical methods perform poorly on a large set of graphs, in addition to them suffering from significant runtime issues.

As a future work, we want to define a learning algorithm such that the graph autoencoder and neural network weights are learned at once, which may result in better predictions. Moreover, we will test our method on different databases for further validation.

Acknowledgements. This research is supported by the Universitat Rovira i Virgili through the Martí Franquès grant.

References

1. Algabli, S., Serratosa, F.: Embedding the node-to-node mappings to learn the graph edit distance parameters. Pattern Recogn. Lett. **112**, 353–360 (2018)
2. Böhm, H., Schneider, G., Kubinyi, H., Mannhold, R., Timmerman, H.: Virtual Screening for Bioactive Molecules, Volume 10. Methods and Principles in Medicinal Chemistry, Wiley (2000). https://books.google.co.cr/books?id=mcrwAAAAMAAJ
3. Defferrard, M., Bresson, X., Vandergheynst, P.: Convolutional Neural Networks on Graphs with Fast Localized Spectral Filtering. arXiv e-prints arXiv:1606.09375 (2016)
4. Garcia-Hernandez, C., Fernández, A., Serratosa, F.: Ligand-based virtual screening using graph edit distance as molecular similarity measure. J. Chem. Inf. Model. **59**(4), 1410–1421 (2019)
5. Gibert, J., Valveny, E., Bunke, H.: Graph embedding in vector spaces by node attribute statistics. Pattern Recogn. **45**(9), 3072–3083 (2012)
6. Hansen, K., et al.: Assessment and validation of machine learning methods for predicting molecular atomization energies. J. Chem. Theory Comput. **9**(8), 3404–3419 (2013). https://doi.org/10.1021/ct400195d. pMID: 26584096 26584096 26584096
7. Kipf, T.N.: Deep learning with graph-structured representations. Ph.D. thesis, University of Amsterdam (2020)
8. Kipf, T.N., Welling, M.: Semi-supervised classification with graph convolutional networks. CoRR abs/1609.02907 (2016). http://arxiv.org/abs/1609.02907
9. Montavon, G., et al.: Learning Invariant Representations of Molecules for Atomization Energy Prediction, vol. 1, pp. 449–457. Massachussetts Institute of Technology Press (2012)
10. Serratosa, F.: Fast computation of bipartite graph matching. Pattern Recogn. Lett. **45**, 244–250 (2014)
11. Serratosa, F., Cortés, X.: Graph edit distance: moving from global to local structure to solve the graph-matching problem. Pattern Recogn. Lett. **65**, 204–210 (2015)
12. Shekkizhar, S., Ortega, A.: Graph construction from data by non-negative kernel regression. In: ICASSP 2020 - 2020 IEEE International Conference on Acoustics, Speech and Signal Processing (ICASSP), pp. 3892–3896 (2020). https://doi.org/10.1109/ICASSP40776.2020.9054425

13. Wu, Z., Pan, S., Chen, F., Long, G., Zhang, C., Yu, P.S.: A comprehensive survey on graph neural networks. IEEE Trans. Neural Netw. Learn. Syst. **32**(1), 4–24 (2021). https://doi.org/10.1109/TNNLS.2020.2978386

14. Zhou, J., et al.: Graph neural networks: a review of methods and applications (2018). https://doi.org/10.48550/arXiv.1812.08434. https://arxiv.org/abs/1812.08434

Distributed Decision Trees

Ozan İrsoy[1]([✉]) [iD] and Ethem Alpaydın[2] [iD]

[1] Department of Computer Science, Cornell University, Ithaca, NY 14853, USA
oirsoy@cs.cornell.edu
[2] Department of Computer Science, Özyeğin University,
Çekmeköy 34794 Istanbul, Turkey
ethem.alpaydin@ozyegin.edu.tr

Abstract. In a budding tree, every node is part internal node and part leaf. This allows representing the tree in a continuous parameter space and training it with backpropagation, like a neural network. Unlike a traditional tree whose construction is composed of two distinct stages of growing and pruning, "bud" nodes grow into subtrees or are pruned back dynamically during learning. In this work, we extend the budding tree and propose the distributed tree where the children use different and independent splits; hence, multiple paths in a tree can be traversed at the same time. In a traditional tree, the learned representations are local, that is, activation makes a soft selection among all the root-to-leaf paths in a tree, but the ability to combine multiple paths of the distributed tree gives it the power of a distributed representation, as in a traditional perceptron layer. Our experimental results show that distributed trees perform comparably or better than budding and traditional hard trees.

Keywords: Decision trees · Hierarchical mixture of experts · Local vs distributed representations

1 Introduction

The decision tree is one of the most widely used models for supervised learning. Consisting of internal decision nodes and terminal leaf nodes, it implements a hierarchical decision function [3,13,14]. The decision nodes act as input-dependent gates and the response is read from the leaves.

Different decision tree architectures differ in the way the decision nodes are defined and/or how the tree is constructed. The basic decision tree is *hard* in the sense that a decision node chooses one of the branches. It is also *univariate*, that is, only one input attribute is used in making a decision.

In this work[1], we are going to discuss a number of extensions of the basic hard univariate tree; in Sect. 2, we will discuss the multivariate tree, the soft

---

[1] This submission is a revised version of our original report arXiv:1412.6388 from 2014, which has not been published in any conference/journal since then. Given the current need for interpretable/explainable alternatives to neural networks, we have decided to pick up this line of research again. O. İrsoy is now with Bloomberg L.P.

© The Author(s), under exclusive license to Springer Nature Switzerland AG 2022
A. Krzyzak et al. (Eds.): S+SSPR 2022, LNCS 13813, pp. 152–162, 2022.
https://doi.org/10.1007/978-3-031-23028-8_16

tree, and the budding tree. In Sect. 3, we will propose a new type of tree, namely the distributed budding tree. We give experimental results in Sects. 4 and 5, and conclude in Sect. 6.

2 Different Tree Architectures

2.1 Hard Decision Trees

Given input $x = [1, x_1, \ldots, x_d]$, the response at node m is defined recursively:

$$y_m(x) = \begin{cases} \rho_m & \text{if } m \text{ is a leaf} \\ y_{ml}(x) & \text{else if } g_m(x) > 0 \\ y_{mr}(x) & \text{else if } g_m(x) \leq 0 \end{cases} \quad (1)$$

If m is an internal node, the decision is forwarded to the left or right subtree depending on the outcome of the gating $g_m(x)$ (In this work, we assume that all input attributes are numeric and we build binary trees.) If m is a leaf node, for regression $\rho_m \in \mathbb{R}$ returns the scalar response value; for binary classification $\rho_m \in [0, 1]$ returns the probability of belonging to the positive class, and for tasks requiring multidimensional outputs (e.g. multiclass classification, vector regression), ρ_m is a vector.

In a univariate tree, the gating uses a single input dimension:

$$g_m(x) = x_{j(m)} - c_m \quad (2)$$

where $j(m)$ denotes the attribute index $(1 \ldots d)$ used in node m and c_m is the corresponding threshold value against which the value of that attribute is compared. We choose one of the two branches depending on the sign of the difference. Decision tree induction algorithms, e.g., C4.5 [13], allow us find the best attribute and threshold at each node using the subset of the training data reaching that node (i.e., satisfying all the conditions on the nodes above m).

The *multivariate tree* [12,16] is a generalization where all input attributes are used in gating. In a linear multivariate tree, we have

$$g_m(x) = w^T x \quad (3)$$

A univariate decision defines a split boundary that is orthogonal to one of the axes but the linear multivariate node allows arbitrary *oblique* splits. If we relax the linearity assumption on $g_m(\cdot)$, we get the *multivariate nonlinear tree*; for example, in [4], $g_m(\cdot)$ is defined as a multilayer perceptron. In an *omnivariate tree*, a node can be univariate, linear, or nonlinear multivariate [15].

Regardless of the gating and the leaf response functions, inducing the optimal decision tree is a difficult problem [14]. Finding the smallest decision tree that perfectly classifies a given set of input is NP-hard [5]. Thus typically, decision trees are constructed greedily.

Essentially, decision tree induction consists of two steps:

1. Growing the tree: Starting from the root node, at each node m, we search for the best decision function $g_m(\boldsymbol{x})$ that splits the data that reaches the node m. If the split provides an improvement in terms of a measure (e.g. entropy), we keep the split, and recursively repeat the process for the two children ml and mr. If the split does not provide any improvement, then m is kept as a leaf and ρ_m is assigned accordingly.
2. Pruning the tree: Once the tree is grown, we can check if reducing the tree complexity by replacing subtrees with leaf nodes leads to an improvement on a separate development set. This is done to avoid overfitting and improve generalization of the tree.

2.2   Soft Decision Trees

The hierarchical mixture of experts [10] define a soft decision tree where the gating nodes are linear multivariate and the splitting is soft; that is, instead of choosing one of the two children, both are chosen with different probabilities. This allows the overall output of the tree to be continuous and we can use backpropagation to update all of the tree parameters, in the gating nodes and the leaves.

More formally, a soft decision tree models the response as the following recursive definition:

$$y_m(\boldsymbol{x}) = \begin{cases} \rho_m & \text{if } m \text{ is a leaf} \\ g_m(\boldsymbol{x})y_{ml}(\boldsymbol{x}) + (1 - g_m(\boldsymbol{x}))y_{mr}(\boldsymbol{x}) & \text{otherwise} \end{cases} \qquad (4)$$

where $g_m(\boldsymbol{x}) = \sigma(\boldsymbol{w}_m^T \boldsymbol{x})$ where $\sigma(\cdot)$ is the sigmoid (logistic) function.

Soft decision trees are grown incrementally in a similar fashion to hard trees [8]. Every node is recursively split until a stopping criterion is reached, and we use gradient-descent to learn the splitting hyperplane of the parent and the response values of its two children.

In a previous study, we have shown how convolutional preprocessing can be added to soft decision trees thereby achieving an architecture that can compete with deep multilayer network [1].

2.3   Budding Trees

Budding trees [9] generalize soft trees further by softening the notion of *being a leaf* as well. Every node (called a *bud* node) is part internal node and part leaf. For a node m, the degree of how much of m is a leaf is defined by the *leafness* parameter $\gamma_m \in [0, 1]$. This allows the response function to be continuous with respect to the parameters, including the structure of a tree.

The response of bud node m is recursively defined:

$$y_m(\boldsymbol{x}) = (1 - \gamma_m)\big[g_m(\boldsymbol{x})y_{ml}(\boldsymbol{x}) + (1 - g_m(\boldsymbol{x}))y_{mr}(\boldsymbol{x})\big] + \gamma_m\rho_m \qquad (5)$$

The recursion ends when a node with $\gamma_m = 1$ is encountered. (In a traditional decision tree, γ_m is binary; it is 0 for an internal node and 1 for a leaf.)

The augmented error that we minimize is

$$E' = E + \lambda \sum_m (1 - \gamma_m) \text{ subject to } \gamma_m \in [0, 1], \forall m \tag{6}$$

where E is the original regression/classification error calculated using the output of the tree on a labelled training set, and the second term is a regularization/penalty term forcing nodes to be leaves.

Note that while backpropagating, not only we update all the decision node weights and the leaf values, but we also update γ_m, that is, we are also updating the structure of the tree. Initially, a node starts with its $\gamma_m = 1$ and if it is updated to be less, we physically split the node and grow a subtree with its gating function and children (with their own γ starting from 1). Similarly, a previously grown subtree that turned out be useless can be pruned back because of the regularization term.

Another difference is that any training instance updates all the parameters of all the tree and so the growth of any subtree affects the rest of the tree.

3   Distributed Budding Trees

Even though the soft tree (and also the budding tree) provide a means for hierarchical representation learning, the resulting representations are local. The overall response is a convex combination of all the leaves. This provides a soft selection among the leaves that is akin to a hierarchical softmax, and limits the representational power—If we use hard thresholds instead of soft sigmoidal thresholds, it essentially selects one of the leaves (paths). Thus, it partitions the input space into as many regions as the number of leaves in the tree, which results in a representation power that is linear in the number of nodes.

To overcome this limit, we extend the budding tree to construct a distributed tree where the two children of a node are selected by different and independent gating functions. Observe that the source of locality in a budding tree comes from the convexity among leaves and selecting one child more means selecting the other child less. We relax this constraint in the distributed budding tree by untying the gating function that controls the paths for the two subtrees:

$$y_m(\boldsymbol{x}) = (1 - \gamma_m)\big[g_{ml}(\boldsymbol{x})y_{ml}(\boldsymbol{x}) + g_{mr}(\boldsymbol{x})y_{mr}(\boldsymbol{x})\big] + \gamma_m \rho_m \tag{7}$$

where $g_{ml}(\boldsymbol{x}) = \sigma(\boldsymbol{w}_{ml}^T \boldsymbol{x})$ and $g_{mr}(\boldsymbol{x}) = \sigma(\boldsymbol{w}_{mr}^T \boldsymbol{x})$ are the conditions for the left and right children and $\boldsymbol{w}_{ml}, \boldsymbol{w}_{mr}$ respectively are the untied linear split parameters of the node m for the left and right splits. Hence the conditions for left and right subtrees are independent—we get the budding (and soft) tree if $g_{mr}(\boldsymbol{x}) \equiv 1 - g_{ml}(\boldsymbol{x})$.

With this definition, a tree no longer generates local representations, but distributed ones. The selection of one child is independent of the selection of the other, and for an input, multiple paths can be traversed (see Fig. 1). Intuitively, a distributed tree becomes similar to a *hierarchical sigmoid* layer as opposed to

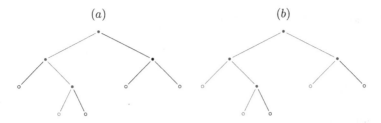

(a)   (b)

Fig. 1. (a) Operation of a budding tree which (softly) selects a single path across the tree. (b) Operation of a distributed tree where it can (softly) select multiple paths.

a hierarchical softmax. In the case of hard thresholds, more than one leaf node (path) can be selected for an instance, implying any one of 2^L possibilities where L is the number of leaves (as opposed to exactly one of L for the traditional tree). This results in a representation power exponential in the number of nodes.

The distributed budding tree still retains the hierarchy that exists in traditional hard/soft decision trees and budding trees. Essentially, each node m can still veto its entire downward subtree by not being activated. The activation of a node m means that there is at least one leaf in the subtree that is relevant to this particular input.

Previously we have also applied the same idea of continuous construction by softly adding/removing complexity to multilayer perceptrons and we have proposed two methods; in tunnel networks, it is done at the level of hidden units and in budding perceptrons, it works at the level of hidden layers [7].

4   Experiments

In this section, we report quantitative experimental results for regression, binary, and multiclass classification tasks.

We use ten regression (*ABAlone, ADD10, BOSton, CALifornia, COMp, CONcrete, puma8FH, puma8FM, puma8NH, puma8NM*), ten binary (*BREast, GERman, MAGic, MUSk2, PIMa, POLyadenylation, RINgnorm, SATellite47, SPAmbase, TWOnorm*) and ten multiclass (*BALance, CMC, DERmatology, ECOli, GLAss, OPTdigits, PAGeblock, PENdigits, SEGment, YEAst*) data sets from the UCI repository [2], as in [9]. We compare distributed trees with budding trees and the C4.5 for regression and classification tasks. Linear discriminant tree (LDT) which is a hard, multivariate tree [16] is used as an additional baseline for the classification tasks.

We adopt the following experimental methodology: We first separate one third of the data set as the test set over which we evaluate the final performance. With the remaining two thirds, we apply 5×2-fold cross validation. Hard trees (including the linear discriminant tree) use the validation set as a pruning set. Distributed and budding trees use the validation set to tune the learning rate and λ. Statistical significance is tested with the paired t-test for the performance measures, and the Wilcoxon Rank Sum test for the tree sizes,

Table 1. Regression results.

	MSE			Size		
	C4.5	Bud	Dist	C4.5	Bud	Dist
ABA	54.13	41.61	41.79	44	35	**24**
ADD	24.42	4.68	**4.56**	327	35	**27**
BOS	34.21	21.85	**18.28**	19	19	33
CAL	31.18	24.01	**23.26**	300	94	**47**
COM	3.61	1.98	1.97	110	**19**	29
CON	26.89	15.62	**15.04**	101	38	43
8FH	41.69	37.83	37.89	47	**13**	21
8FM	6.89	5.07	**5.02**	164	17	15
8NH	39.46	**34.25**	34.53	77	**27**	43
8NM	6.69	3.67	3.61	272	37	37

both with significance level $\alpha = 0.05$. The values reported are results on the test set not used for training or validation (model selection). Significantly best results are shown in boldface in the figures.

Table 1 shows the mean squared errors and the number of nodes of the C4.5, budding tree, and distributed tree on the regression data sets. We see that the distributed trees perform significantly better on five data sets (*add10, boston, california, concrete, puma8fm*), whereas the budding tree performs better on one (*puma8nh*), the remainder four being ties. In terms of tree sizes, both the distributed tree and the budding tree have three wins (*abalone, add10, california* and *comp, puma8fh, puma8nh*, respectively), the remaining four are ties. Note that at the end of the training, because of the stochasticity of training, both distributed trees and budding trees have nodes that are almost leaf (having $\gamma \approx 1$). These nodes can be pruned to get smaller trees with negligible change in the overall response function.

Table 2 shows the percentage accuracy of C4.5, LDT, budding and distributed trees on binary classification data sets. In terms of accuracy, LDT has four wins (*german, pima, polyadenylation, twonorm*), distributed tree has five wins (*magick, musk2, ringnorm, satellite, spambase*) and the remaining one (*breast*) is a tie. On its four win, LDT produces very small trees (and on three, it produces the smallest trees). This suggests that with proper regularization, it is possible to improve the performance of budding and distributed trees. In terms of tree sizes, LDT has six wins (*breast, polyadenylation, ringnorm, satellite, spambase, twonorm*) and C4.5 has one (*german*) with the remaining three ties.

Table 3 shows the percentage accuracy of C4.5, LDT, budding and distributed trees on multiclass classification data sets. The distributed tree is significantly better on five data sets (*balance, dermatology, optdigits, pendigits, segment*), and the remaining five are ties. In terms of tree sizes, again LDT produces smaller trees with four wins (*balance, cmc, glass, optdigits*), and budding tree has one win (*pendigits*).

Table 2. Binary classification results

	Accuracy				Size			
	C4.5	LDT	Bud	Dist	C4.5	LDT	Bud	Dist
BRE	93.29	95.09	95.00	95.51	7	**4**	12	23
GER	70.06	**74.16**	68.02	70.33	1	3	42	39
MAG	82.52	83.08	86.39	**86.64**	53	38	122	40
MUS	94.54	93.59	97.02	**98.24**	62	11	15	35
PIM	72.14	**76.89**	67.20	72.26	8	5	68	35
POL	69.47	**77.45**	72.57	75.33	34	3	61	97
RIN	87.78	77.25	88.51	**95.06**	93	3	61	117
SAT	84.58	83.30	86.87	**87.91**	25	9	38	41
SPA	90.09	89.86	91.47	**93.29**	36	**13**	49	23
TWO	82.96	**98.00**	96.74	97.64	163	3	29	25

Table 3. Multiclass classification results

	Accuracy				Size			
	C4.5	LDT	Bud	Dist	C4.5	LDT	Bud	Dist
BAL	61.91	88.47	92.44	**96.36**	6	3	29	28
CMC	50.00	46.65	53.23	52.87	25	3	28	51
DER	94.00	93.92	93.60	**95.84**	16	11	11	20
ECO	77.48	81.39	83.57	83.83	10	11	24	58
GLA	56.62	53.38	53.78	55.41	21	9	21	55
OPT	84.85	93.73	94.58	**97.13**	121	**31**	40	92
PAG	96.72	94.66	96.52	96.63	24	29	37	27
PEN	92.96	96.60	98.14	**98.98**	170	66	**54**	124
SEG	94.48	91.96	95.64	**96.97**	42	33	33	76
YEA	54.62	56.67	59.32	59.20	25	22	41	42

5   Visualization

In this section, we visualize trees learned on synthetic and real data.

Synthetic data is designed as a one-dimensional regression dataset where the response is a sinusoidal of the input with a small additive noise, and the input is limited to the interval $[-3, 3]$. 300 samples are drawn for both training and validation sets. Trees are displayed in Fig. 2.

We observe that both type of trees use smoothly sloped sigmoid gating functions to interpolate between two endpoints to incrementally build up the sinusoidal shape. Since distributed trees have two (untied) gates, its decision nodes are able to (softly) split the space into not two but three regions. Therefore

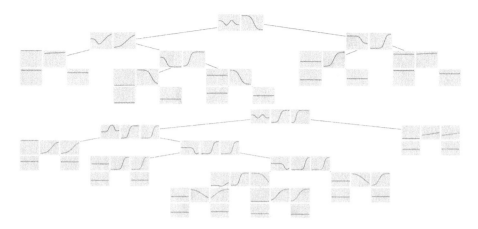

Fig. 2. Budding (top) and distributed (bottom) trees trained on toy data. Red curves display response by the subtree at that node. Blue curves display the gating function at that node. For distributed trees, blue and purple curves display (untied) left and right gating functions. (Color figure online)

for some nodes we observe curves transitioning between three pieces, creating shapes with more than one break point, while using flat children.

In addition to synthetic data we use the MNIST handwritten digit database [11] to visualize trees as shown in Fig. 3. Each node is visualized by the average value of all instances that fall into that node. In a hard tree, this is trivial to define, however in soft, budding and distributed trees every instance falls into every possible node with some nonzero value, however small. Therefore we compute a notion of a *soft membership* which is the multiplication of all gating values through the particular path, and use that to compute a weighted average instead. Additionally, distributed decision trees have a notion of *not taking either of the path to children*, finishing the computation at the given node as if it was a leaf. For each internal node, we measure this by multiplying the soft membership so far with $(1 - g_{ml})(1 - g_{mr})$, and display the resulting average right below each plot. Nodes also carry a transparency measured by the *non-leafness* of all the ancestors in the path $(1 - \gamma_m)$ multiplied, since the more each ancestor tends to becoming a leaf the less impact the node has in the overall output.

For both trees we see interesting hierarchies: For instance, budding tree splits apart digit 4 from similar looking input, and then into two, digits of 7 and 9. Distributed tree has several variations of digit 3 spreading apart into branches from the same ancestor. Additionally, we see that distributed tree occasionally early stops at an internal node by shutting off both children, for instance digit 4 is recognized this way at the first level, below the root.

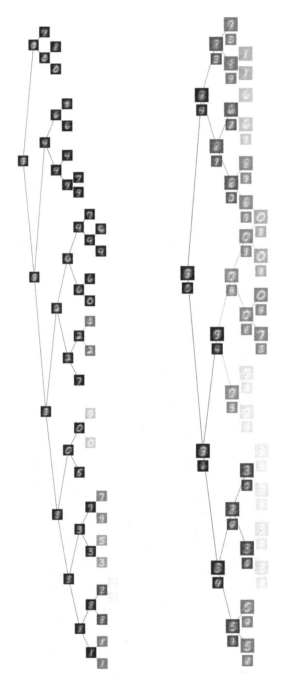

Fig. 3. Budding (left) and distributed (right) trees trained on MNIST, displayed vertically (root is at the left, growing towards the right). Each node displays the average
input falling into that node (path), weighted by the soft memberships assigned by successive compositions of gating. Distributed trees show an additional image that belongs
to neither left or right subtree, measured by how *closed* both gates are.

6   Conclusions

After reviewing soft and budding trees, we propose a distributed tree model that overcomes the locality of traditional trees and can learn distributed representations of the data. It does this by allowing multiple root-to-leaf path (soft) selections over a tree structure, as opposed to selection of a single path as done by budding (soft) and traditional (hard) trees. This increases their representational power from linear to exponential in the number of leaves. Quantitative evaluation on several data sets shows that this increase is indeed helpful in terms of predictive performance.

Previously we have proposed using decision-tree autoencoders [6] and we believe that distributed trees too can be considered as alternative to layers of perceptrons for deep learning in that they can learn hierarchical distributed representations of the input in its different levels.

Even though the selection of left and right subtrees is independent in a distributed tree, it still preserves the hierarchy in its tree structure as in traditional decision trees and budding trees. This is because an activation close to zero for a node has the ability to veto its entire subtree, and an *active* node means that it believes that there is some relevant node down in that particular subtree. Visualizing the decisions and paths learned by soft and budding trees help towards the interpretation of the trained models.

References

1. Ahmetoğlu, A., İrsoy, O., Alpaydın, E.: Convolutional soft decision trees. In: Kůrková, V., Manolopoulos, Y., Hammer, B., Iliadis, L., Maglogiannis, I. (eds.) ICANN 2018. LNCS, vol. 11139, pp. 134–141. Springer, Cham (2018). https://doi.org/10.1007/978-3-030-01418-6_14

2. Bache, K., Lichman, M.: UCI machine learning repository (2013). http://archive.ics.uci.edu/ml

3. Breiman, L., Friedman, J.H., Olshen, R.A., Stone, C.J.: Classification and Regression Trees. Wadsworth & Brooks, Pacific Grove (1984)

4. Guo, H., Gelfand, S.B.: Classification trees with neural network feature extraction. IEEE Trans. Neural Netwo. **3**, 923–933 (1992)

5. Hancock, T., Jiang, T., Li, M., Tromp, J.: Lower bounds on learning decision lists and trees. Inf. Comput. **126**(2), 114–122 (1996)

6. İrsoy, O., Alpaydın, E.: Autoencoder trees. In: Asian Conference on Machine Learning (2015)

7. İrsoy, O., Alpaydın, E.: Continuously constructive deep neural networks. IEEE Trans. Neural Netwo. Learn. Syst. **31**(4), 1124–1133 (2020)

8. İrsoy, O., Yıldız, O.T., Alpaydın, E.: Soft decision trees. In: International Conference on Pattern Recognition (2012)

9. İrsoy, O., Yıldız, O.T., Alpaydın, E.: Budding trees. In: International Conference on Pattern Recognition (2014)

10. Jordan, M.I., Jacobs, R.A.: Hierarchical mixtures of experts and the EM algorithm. Neural Comput. **6**(2), 181–214 (1994)

11. LeCun, Y., Cortes, C.: The MNIST database of handwritten digits (1998)

12. Murthy, S.K., Kasif, S., Salzberg, S.: A system for induction of oblique decision trees. J. Artiff. Intell. Res. **2**, 1–32 (1994)
13. Quinlan, J.R.: C4.5: Programs for Machine Learning. Morgan Kaufmann Publishers Inc., San Francisco (1993)
14. Rokach, L., Maimon, O.: Top-down induction of decision trees classifiers-a survey. IEEE Trans. Syst. Man Cybern. Part C Appl. Rev. **35**(4), 476–487 (2005)
15. Yıldız, O.T., Alpaydın, E.: Omnivariate decision trees. IEEE Trans. Neural Netw. **12**(6), 1539–1546 (2001)
16. Yıldız, O.T., Alpaydın, E.: Linear discriminant trees. Int. J. Pattern Recogn. Artif Intell. **19**(03), 323–353 (2005)

A Capsule Network for Hierarchical Multi-label Image Classification

Khondaker Tasrif Noor[1]([envelope])[ID], Antonio Robles-Kelly[1][ID], and Brano Kusy[2][ID]

[1] School of IT, Deakin University, Waurn Ponds, Geelong, VIC, Australia
`knoor@deakin.edu.au`
[2] CSIRO Data 61, Queensland Centre for Advanced Technologies, Pullenvale, QLD, Australia

Abstract. Image classification is one of the most important areas in computer vision. Hierarchical multi-label classification applies when a multi-class image classification problem is arranged into smaller ones based upon a hierarchy or taxonomy. Thus, hierarchical classification modes generally provide multiple class predictions on each instance, whereby these are expected to reflect the structure of image classes as related to one another. In this paper, we propose a multi-label capsule network (ML-CapsNet) for hierarchical classification. Our ML-CapsNet predicts multiple image classes based on a hierarchical class-label tree structure. To this end, we present a loss function that takes into account the multi-label predictions of the network. As a result, the training approach for our ML-CapsNet uses a coarse to fine paradigm while maintaining consistency with the structure in the classification levels in the label-hierarchy. We also perform experiments using widely available datasets and compare the model with alternatives elsewhere in the literature. In our experiments, our ML-CapsNet yields a margin of improvement with respect to these alternative methods.

Keywords: Hierarchical image classification · Capsule networks · Deep learning

1 Introduction

Image classification is a classical problem in computer vision and machine learning where the aim is to recognise the image features so as to identify a target image class. Image classification has found application in areas such as face recognition [25], medical diagnosis [32], intelligent vehicles [26] and online advertising [12] amongst others. Note that these classification tasks are often aimed at a "flat" class-label set where all the classes are treated equally, devoid of a taxonomical or hierarchical structure between them. This contrasts with hierarchical classification tasks, which require multiple label predictions per instance. Moreover, these should be consistent with the hierarchical structure of the class-label set under consideration.

A. Krzyzak et al. (Eds.): S+SSPR 2022, LNCS 13813, pp. 163–172, 2022.
https://doi.org/10.1007/978-3-031-23028-8_17

In this paper, we propose a multi-label image classification model (ML-CapsNet) for hierarchical image classification based on capsule networks [20]. We note that capsule networks (CapsNets) can learn both, the image features and their transformations. This allows for a natural means to recognition by parts. Recall that capsule networks (CapsNets) use a set of neurons to obtain an "activity vector". These neurons are grouped in capsules, whereby deeper layers account for several capsules at the preceding layer by agreement. Capsules were originally proposed in [10] to learn a feature of instantiation parameters which are robust to variations in position, orientation, scale and lighting. The assertion that CapsNets can overcome viewpoint invariance problems was further explored in [20], where the authors propose a dynamic routing procedure for routing by agreement. Hinton et al. propose in [11] a probabilistic routing approach based upon the EM-algorithm [4] so as to learn part-whole relationships. Building upon the EM routing approach in [11], Bahadori [1] proposes a spectral method to compute the capsule activation and pose.

It is not surprising that, due to their viewpoint invariance, their capacity to address the "Picasso effect" in classifiers and their robustness to input perturbations when compared to other CNNs of similar size [9], CapsNets have been the focus of great interest in the computer vision and machine learning communities. CapsNets have found applications in several computer vision tasks such as text classification [2], 3D data processing [33], target recognition [18] and image classification [28]. They have also been used in architectures such as Siamese networks [16], generative adversarial networks [23] and residual networks (ResNets) [13].

Despite the interest of the community in CapsNets, to our knowledge, they have not been employed for hierarchical multi-label classification (HMC). Since hierarchical multi-label classification can be viewed as a generalisation of multi-class problems with subordinate, not exclusive classes, it is a challenging problem in machine learning and pattern recognition that has attracted considerable attention in the research community. It has been tackled in a number of ways, spanning from kernel methods [19] to decision trees [24] and, more recently, artificial neural networks [27]. Nonetheless this attention and the fact that image HMC has been applied to the annotation of medical images [8], these methods often do not focus on images, but rather on problems such as protein structure prediction [24], data-dependent grouping [22] or text classification [15].

Further, image hierarchical multi-label classification methods are relatively few elsewhere in the literature. This is even more surprising since it is expected that incorporating hierarchies in the model would improve generalisation, particularly when the training data is limited. Image HMC methods often employ the hierarchical semantic relationships of the target classes so as to improve visual classification results, making use of the hierarchical information as a guide to the classifier. Along these lines, word hierarchies have been applied to provide consistency across multiple datasets [17], deliver a posterior confidence estimate [3] and to optimise the accuracy-specificity trade-off in visual recognition [5]. In [7], order-preserving embeddings are used to model label-to-label and image-to-image hierarchical interactions. In [34], the authors propose a Branch-CNN (B-CNN), a CNN architecture for hierarchical data organised as a tree. The B-CNN

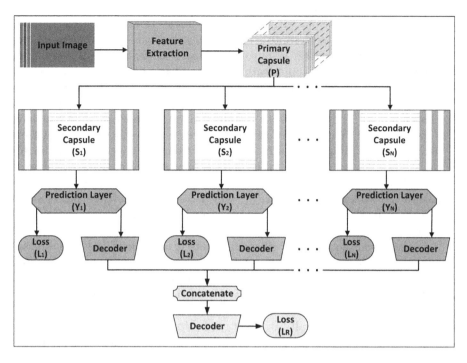

Fig. 1. Architecture of our multi-label capsule network. Our Network has multiple secondary capsules, each of these corresponding to a class-hierarchy. The output of these is used for class prediction and the reconstruction of the input image using the decoder outputs as shown in the figure.

architecture is such that the model outputs multiple predictions ordered from coarse to fine along concatenated convolutional layers. In [30], the authors propose a hierarchical deep CNN (HD-CNN) that employs component-wise pre-training with global fine-tuning making use of a multinomial logistic loss.

Here, we present a CapsNet for image HMC. Our choice of capsules is motivated by their capacity to learn relational information, seeking to profit from the capacity of CapsNets to model the semantic relationships of image features for hierarchical classification. Since CapsNets were not originally proposed for HMC tasks, we propose a modified loss function that considers the consistency of the predicted label with that endowed by the hierarchy under consideration. Further, our reconstruction loss is common to all the secondary capsule decoders, making use of the combined predictions for the hierarchies and further imposing consistency on the HMC classification task.

2 Hierarchical Multi-label Capsules

Our ML-CapsNet uses a capsule architecture to predict multiple classes based on a hierarchical/taxonomical label tree. The overall architecture of ML-CapsNet is presented in Fig. 1. Note our ML-CapsNet has a feature extraction block aimed at converting pixel intensities into local features that can be used as inputs for the primary capsule network. In our architecture, the feature map is reshaped accordingly to allow for the input features to be passed on to the primary capsule network $\mathcal{P}$, which is shared amongst all the secondary ones. These secondary capsule networks, denoted $\mathcal{S}_n$ in the figure, correspond to the nth hierarchy in the class-label set. Thus, each of the hierarchical levels, from coarse-to-fine, have their own secondary capsule networks and, therefore, there are as many of these as hierarchies. For simplicity, each of these secondary capsules sets contain K_n capsules, where K_n corresponds to the number of classes in the hierarchy indexed n. Note that, as shown in the figure, each of the secondary capsule networks $\mathcal{S}_n$ has their own prediction layer $\mathcal{Y}_n$, which computes the class probabilities using a logit function. Thus, the $\mathcal{Y}_n$ layers provide predictions for the K_n classes corresponding to the nth hierarchy class level.

It is also worth noting that, since CapsNets train making use of the reconstruction loss, here we use an auxiliary decoder network to reconstruct the input image. To this end, we make use of the activity vector delivered by the secondary capsule networks $\mathcal{S}_n$ for each hierarchy. We do this by combining the decoder outputs for each class-hierarchy into a final one via concatenation as shown in Fig. 1. As a result, the overall loss function L_T for our ML-CapsNet is a linear combination of all prediction losses and the reconstruction loss defined as follows

$$L_T = \tau L_R + (1 - \tau) \sum_{n=1}^{N} \lambda_n L_n \qquad (1)$$

where n is the index for hierarchical level in the level tree, N is the total number of levels, λ_n a weight that moderates the contribution of each class hierarchy to the overall loss and τ is a scalar that controls the balance between the classification loss L_n and the reconstruction one L_R. In Eq. 1 the reconstruction loss is given by the L-2 norm between the input instance x and the reconstructed one $\hat{x}$. Thus, the loss becomes

$$L_R = ||x - \hat{x}||_2^2 \qquad (2)$$

For L_n we employ the hinge loss given by

$$L_n = T_n \max(0, m^+ - ||\mathbf{v}_k||)^2 + \gamma(1 - T_n)\max(0, ||\mathbf{v}_k|| - m^-)^2 \qquad (3)$$

where $T_n = 1$ if and only if the class indexed k is present, m^+, m^- and γ are hyper-parameters and $\mathbf{v}_k$ is the output vector for the capsule for the kth class under consideration.

3   Experiments

We have performed experiments making use of the MNIST [6], Fashion-MNIST [29], CIFAR-10 and CIFAR-100 [14] datasets. We have also compared our results with those yielded by the B-CNN approach proposed in [34] and used the CapsNet in [20] as a baseline. Note that, nonetheless the CapsNet in [20] is not a hierarchical classification model, it does share with our approach the routing procedure and the structure of the primary and secondary capsules.

3.1   Implementation Details and Datasets

We have implemented our ML-CapsNet on TensorFlow and, for all our experiments, have used the Adam optimiser with TensorFlow's default settings. For all the experiments we use exponential decay to adjust the learning rate value after every epoch, where the initial learning rate is set to 0.001 with a decay rate of 0.995.

As mentioned earlier, our network has a common feature extraction block and primary capsule $\mathcal{P}$ shared amongst the secondary capsules $\mathcal{S}_n$. In our implementation, the feature extraction block is comprised by two convolutional layers for the MNIST dataset and 5 convolutional layers for Fashion-MNIST, CIFAR-10 and CIFAR-100 datasets. We do this based on the complexity of the dataset so that the feature extraction block can adapt accordingly. For all the convolutional layers we have used 3×3 filters with zero-padding, $ReLu$ activations and gradually increased the number of filters from 32 filters for the first to 512 in the following layers, i.e. 64, 128, 256 and 512. For all our results, we have used an 8-dimensional primary capsule and all the secondary capsules $\mathcal{S}_n$ are 16-dimensional with dynamic routing [20]. In all our experiments we have set m^+, m^- and γ to 0.9, 0.1 and 0.5, respectively. The value of τ in Eq. 1 has been set to 0.0005.

We have assigned coarse and medium classes to the datasets to construct a hierarchical label tree. In this setting, coarse labels are a superclass of several corresponding medium-level labels for the dataset and these labels in turn are a superclass for the fine labels prescribed for the dataset under consideration. As a result, each instance will have multiple labels in the hierarchy, i.e. one per level. Recall that the MNIST and Fashion-MNIST dataset contains 28×28 grey-scale images. Both datasets have $60,000$ training and $10,000$ testing images. For the MNIST, which contains images of handwritten digits, we have followed [34] and added five coarse classes for the dataset. The Fashion-MNIST dataset is similar to the MNIST dataset and contains images of fashion products for ten fine classes. We use the coarse to medium hierarchy presented in [21], which

Table 1. Accuracy yielded by our ML-CapsNet, the B-CNN [34] and the baseline [20] on the MNIST and Fashion-MNIST datasets. The absolute best are in bold.

Model	Accuracy (%)				
	MNIST		Fashion-MNIST		
	Coarse	Fine	Coarse	Medium	Fine
CapsNet	–	99.3	–	–	91.2
B-CNN	99.2	99.28	99.57	95.52	92.04
ML-CapsNet	**99.65**	**99.5**	**99.73**	**95.93**	**92.65**

Table 2. Accuracy yielded by our ML-CapsNet, the B-CNN [34] and the baseline [20] on the CIFAR-10 and CIFAR-100 datasets. The absolute best are in bold.

Model	Accuracy (%)					
	CIFAR-10			CIFAR-100		
	Coarse	Medium	Fine	Coarse	Medium	Fine
CapsNet	–	–	70.42	–	–	34.93
B-CNN	95.63	86.95	84.95	71.04	61.94	55.52
ML-CapsNet	**97.52**	**89.27**	**85.72**	**78.73**	**70.15**	**60.18**

manually adds a coarse and medium level for the dataset. In [34], the coarse level has two classes, and the medium level has six classes. The CIFAR-10 and CIFAR-100 dataset consist 32×32 colour images and have $50,000$ training images and $10,000$ testing images in 10 and 100 fine classes, respectively. Here, we use the class-hierarchy used in [34], which adds an additional coarse and medium levels for the dataset whereby the coarse level contains two classes and the medium contains seven classes. The CIFAR-100 dataset is similar to CIFAR-10, except it has 100 fine and 20 medium classes. We have used the hierarchy in [34], which groups the medium and fine classes into eight coarse ones.

3.2   Experimental Setup

As mentioned earlier, in order to compare our ML-CapsNet with alternatives elsewhere in the literature, we have used the CapsNet as originally proposed in [20] as a baseline and compared our results with those yielded by the B-CNN [34]. In all our experiments, we normalise the training and testing data by subtracting the mean and dividing by the standard deviation. For all the datasets, the B-CNN follows the exact model architecture and training parameters used by the authors in [34]. For training the B-CNN model on the Fashion-MNIST dataset, we followed the architecture for the MNIST dataset and added an additional branch for the fine level.

When training our ML-CapsNet on the CIFAR-10 and CIFAR-100 datasets, we apply MixUp data augmentation [31] with $\alpha = 0.2$. Also, recall that the λ_n values in Eq. 1 govern the contribution of each hierarchy level to the overall loss function. Thus, for our ML-CapsNet, we adjust the λ_n values as the training progresses so as to shift the importance of the class-level hierarchy from course-to-fine. As a result, on the MNIST dataset we set the initial λ_n to $0.90, 0.10$ for the coarse and fine levels, respectively. The value of λ_n is then set to $0.10, 0.90$ after 5 epochs and $0.02, 0.98$ after 10. For the Fashion-MNIST dataset the initial loss weight values for λ_n are $0.98, 0.01, 0.01$ for the coarse, medium and fine hierarchies. The value of λ_n after 5 epochs becomes $0.10, 0.70, 0.20$, after 10 epochs shifts to $0.07, 0.10, 0.83$, at 15 epochs is set to $0.05, 0.05, 0.90$ and, finally, at 25 epochs assumes the value of $0.01, 0.01, 0.98$. For the CIFAR-10 dataset, the initial λ_n values are $0.90, 0.05$ and 0.05 for the coarse, medium

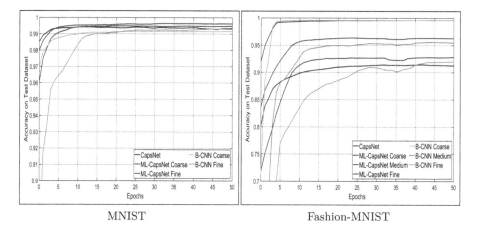

MNIST   Fashion-MNIST

Fig. 2. Accuracy as a function of training epoch for all the models under consideration. The left-hand panel shows the plots for the MNIST dataset whereas the right-hand panel corresponds to the Fashion-MNIST dataset.

and fine class-hierarchy levels. The λ_n then changes to 0.10, 0.70 and 0.20 after Epoch 5 and, at Epoch 11 is set to 0.07, 0.20 and 0.73. At Epoch 17 is set to 0.05, 0.15 and 0.80, taking its final value of 0.05, 0.10 and 0.85 at Epoch 24. For the CIFAR-100 dataset the initial values are 0.90, 0.08 and 0.02, which then change to 0.20, 0.70 and 0.10 after 7 epochs. After Epoch 15 we set these to 0.15, 0.30 and 0.55. At Epoch 22 we modify them to be 0.10, 0.15 and 0.75 and, at Epoch 33, they take their final value of 0.05, 0.15 and 0.80.

3.3 Results

We now turn our attention to the results yielded by our network, the CapsNet as originally proposed in [20] and the B-CNN in [34] when applied to the four datasets under consideration. In Table 1 we show the performance yielded by these when applied to the MNIST and Fashion-MNIST dataset. Similarly, in Table 2, we show the accuracy for our network and the alternatives when applied to the CIFAR-10 and CIFAR-100 datasets. In the tables, we show the accuracy for all the levels of the applicable class-hierarchies for both, our network and the B-CNN. We also show the performance on the CapsNet in [20] for the fine class-hierarchy. We do this since the baseline is not a hierarchical classification one, rather the capsule network as originally proposed in [20] and, therefore, the medium and coarse label hierarchies do not apply.

Note that, for all our experiments, our network outperforms the alternatives. It is also worth noting that, as compared with the baseline, it fairs much better for the more complex datasets of CIFAR-10 and CIFAR-100. This is also consistent with the plots in Figs. 2 and 3, which show the model accuracy on the dataset under consideration for every one of their corresponding class-hierarchies. Note that in several cases, our network not only outperforms B-CNN, but also

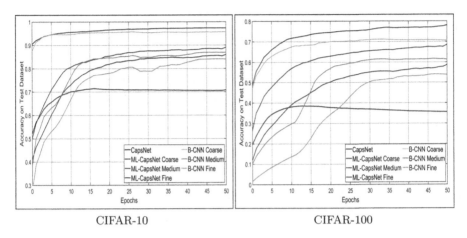

CIFAR-10   CIFAR-100

Fig. 3. Accuracy as a function of training epoch for all the models under consideration. The left-hand panel shows the plots for the CIFAR-10 dataset whereas the right-hand panel corresponds to the CIFAR-100 dataset.

converges faster. It is also important to note that, as compared to the baseline, our ML-CapsNet employs the hierarchical structure of the class-set to impose consistency through the loss. This appears to introduce a structural constraint on the prediction that helps improve the results.

This is consistent with the reconstruction results shown in Fig. 4. In the figure, we show the reconstruction results for the classes in the Fashion-MNIST for both, our ML-CapsNet and the capsule network as originally proposed in [20]. In the figure, from top-to-bottom, the rows show the input image, the image reconstructed by our network and that yielded by that in [20]. Note the images reconstructed using our network are much sharper, showing better detail and being much less blurred. This is somewhat expected, since the reconstruction

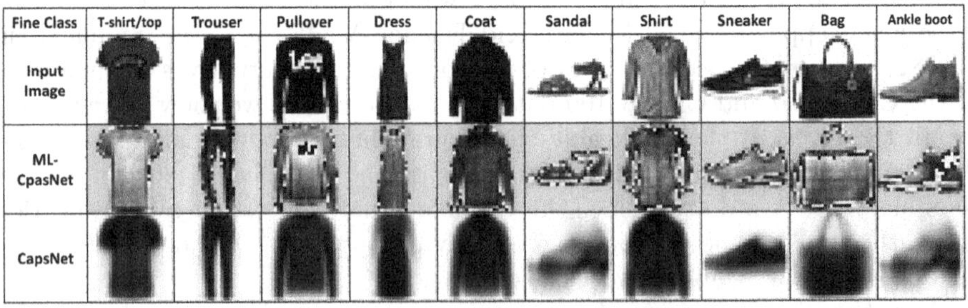

Fine Class	T-shirt/top	Trouser	Pullover	Dress	Coat	Sandal	Shirt	Sneaker	Bag	Ankle boot
Input Image										
ML-CpasNet										
CapsNet										

Fig. 4. Sample reconstructed images for the Fashion-MNIST dataset. From top to bottom we show the input image, the image reconstructed by our ML-Capsnet and that reconstructed using the capsule network in [20].

error is also used by the loss and, hence, better reconstruction should yield higher accuracy and vice versa.

4 Conclusions

In this paper, we have presented a capsule network for image classification, which uses capsules to predict multiple hierarchical classes. The network presented here, which we name ML-CapsNet, employs a shared primary capsule, making use of a secondary one for each class-label set. To enforce the multi-label structure into the classification task, we employ a loss which balances the contribution of each of the class-sets. The loss proposed here not only enforces consistency with the label structure, but incorporates the reconstruction loss making use of a common encoder. We have shown results on four separate widely available datasets. In our experiments, our ML-CapsNet outperforms the B-CNN [34] and the classical capsule network in [20]. As expected, it also delivers better reconstructed images than those yielded by the network in [20].

References

1. Bahadori, M.T.: Spectral capsule networks. In: International Conference on Learning Representations Workshops (2018)
2. Chen, B., Huang, X., Xiao, L., Jing, L.: Hyperbolic capsule networks for multi-label classification. In: Proceedings of the 58th Annual Meeting of the Association for Computational Linguistics, pp. 3115–3124 (2020)
3. Davis, J., Liang, T., Enouen, J., Ilin, R.: Hierarchical classification with confidence using generalized logits. In: 2020 25th International Conference on Pattern Recognition (ICPR), pp. 1874–1881 (2021)
4. Dempster, A., Laird, N., Rubin, D.: Maximum-likelihood from incomplete data via the EM algorithm. J. R. Stat. Soc. Ser. B Methodol. **39**, 1–38 (1977)
5. Deng, J., Krause, J., Berg, A.C., Fei-Fei, L.: Hedging your bets: optimizing accuracy-specificity trade-offs in large scale visual recognition. In: 2012 IEEE Conference on Computer Vision and Pattern Recognition, pp. 3450–3457 (2012)
6. Deng, L.: The MNIST database of handwritten digit images for machine learning research [best of the web]. IEEE Signal Process. Mag. **29**(6), 141–142 (2012)
7. Dhall, A., Makarova, A., Ganea, O., Pavllo, D., Greeff, M., Krause, A.: Hierarchical image classification using entailment cone embeddings. In: Computer Vision and Pattern Recognition Workshops, pp. 836–837 (2020)
8. Dimitrovski, I., Kocev, D., Loskovska, S., Džeroski, S.: Hierarchical annotation of medical images. Pattern Recogn. **44**(10), 2436–2449 (2011)
9. Hahn, T., Pyeon, M., Kim, G.: Self-routing capsule networks. In: Advances in Neural Information Processing Systems, vol. 32 (2019)
10. Hinton, G.E., Krizhevsky, A., Wang, S.D.: Transforming auto-encoders. In: International Conference on Artificial Neural Networks, pp. 44–51 (2011)
11. Hinton, G.E., Sabour, S., Frosst, N.: Matrix capsules with EM routing. In: International Conference on Learning Representations (2018)
12. Hussain, Z., et al.: Automatic understanding of image and video advertisements. In: Computer Vision and Pattern Recognition, pp. 1705–1715 (2017)

13. Jampour, M., Abbaasi, S., Javidi, M.: Capsnet regularization and its conjugation with resnet for signature identification. Pattern Recogn. **120**, 107851 (2021)
14. Krizhevsky, A., Hinton, G., et al.: Learning multiple layers of features from tiny images (2009)
15. Meng, Y., Shen, J., Zhang, C., Han, J.: Weakly-supervised hierarchical text classification. In: Proceedings of the AAAI Conference on Artificial Intelligence, vol. 33, pp. 6826–6833 (2019)
16. Neill, J.O.: Siamese capsule networks. arXiv E-preprints (2018)
17. Redmon, J., Farhadi, A.: Yolo9000: better, faster, stronger. In: Computer Vision and Pattern Recognition, pp. 7263–7271 (2017)
18. Ren, H., Yu, X., Zou, L., Zhou, Y., Wang, X., Bruzzone, L.: Extended convolutional capsule network with application on SAR automatic target recognition. Signal Process. **183**, 108021 (2021)
19. Rousu, J., Saunders, C., Szedmak, S., Shawe-Taylor, J.: Kernel-based learning of hierarchical multilabel classification models. J. Mach. Learn. Res. **7**, 1601–1626 (2006)
20. Sabour, S., Frosst, N., Hinton, G.E.: Dynamic routing between capsules. In: Advances in Neural Information Processing Systems, vol. 30 (2017)
21. Seo, Y., Shin, K.S.: Hierarchical convolutional neural networks for fashion image classification. Expert Syst. Appl. **116**, 328–339 (2019)
22. Ubaru, S., Dash, S., Mazumdar, A., Günlük, O.: Multilabel classification by hierarchical partitioning and data-dependent grouping. In: Advances in Neural Information Processing Systems (2020)
23. Upadhyay, Y., Schrater, P.: Generative adversarial network architectures for image synthesis using capsule networks. arXiv E-preprint (2018)
24. Vens, C., Struyf, J., Schietgat, L., Džeroski, S., Blockeel, H.: Decision trees for hierarchical multi-label classification. Mach. Learn. **73**(2), 185–214 (2008)
25. Wang, M., Deng, W.: Deep face recognition: a survey. Neurocomputing **429**, 215–244 (2021)
26. Wang, Z., Zhan, J., Duan, C., Guan, X., Lu, P., Yang, K.: A review of vehicle detection techniques for intelligent vehicles. IEEE Trans. Neural Netw. Learn. Syst. (2022)
27. Wehrmann, J., Cerri, R., Barros, R.: Hierarchical multi-label classification networks. In: International Conference on Machine Learning, pp. 5075–5084 (2018)
28. Xiang, C., Zhang, L., Tang, Y., Zou, W., Xu, C.: MS-CapsNet: a novel multi-scale capsule network. IEEE Signal Process. Lett. **25**(12), 1850–1854 (2018)
29. Xiao, H., Rasul, K., Vollgraf, R.: Fashion-MNIST: a novel image dataset for benchmarking machine learning algorithms. arXiv preprint arXiv:1708.07747 (2017)
30. Yan, Z., et al.: HD-CNN: hierarchical deep convolutional neural networks for large scale visual recognition. In: International Conference on Computer Vision, pp. 2740–2748 (2015)
31. Zhang, H., Cisse, M., Dauphin, Y.N., Lopez-Paz, D.: mixup: beyond empirical risk minimization. In: International Conference on Learning Representations (2018)
32. Zhang, Z., Xie, Y., Xing, F., McGough, M., Yang, L.: MDNet: a semantically and visually interpretable medical image diagnosis network. In: Computer Vision and Pattern Recognition, pp. 6428–6436 (2017)
33. Zhao, Y., Birdal, T., Deng, H., Tombari, F.: 3D point capsule networks. In: Computer Vision and Pattern Recognition (2019)
34. Zhu, X., Bain, M.: B-CNN: branch convolutional neural network for hierarchical classification. arXiv E-prints pp. arXiv-1709 (2017)

Monte Carlo Dropout for Uncertainty Analysis and ECG Trace Image Classification

Md. Farhadul Islam(✉) ⓘ, Sarah Zabeen ⓘ, Md. Humaion Kabir Mehedi ⓘ,
Shadab Iqbal ⓘ, and Annajiat Alim Rasel ⓘ

BRAC University, Dhaka, Bangladesh
{md.farhadul.islam,sarah.zabeen,humaion.kabir.mehedi,
shadab.iqbal,annajiat.alim.rasel}@g.bracu.ac.bd

Abstract. Cardiovascular diseases (CVDs), such as arrhythmias (abnormal heartbeats) are the prime cause of mortality across the world. ECG graphs are utilized by cardiologists to indicate any unexpected cardiac activity. Deep Neural Networks (DNN) serve as a highly successful method for classifying ECG images for the purpose of computer-aided diagnosis. However, DNNs can not quantify uncertainty in predictions, as they are incapable of discriminating between anomalous data and training data. Hence, a lack of reliability in automated diagnosis and the potential to cause severe decision-making issues is created, particularly in medical practises. In this paper, we propose an uncertainty-aware ECG classification model where Convolutional Neural Networks (CNN), combined with Monte Carlo Dropout (MCD) is employed to evaluate the uncertainty of the model, providing a more trustworthy process for real-world scenarios. We use ECG images dataset of cardiac and covid-19 patients containing five categories of data, which includes COVID-19 ECG records as well as data from other cardiovascular disorders. Our proposed model achieves 93.90% accuracy using this dataset.

Keywords: ECG · COVID-19 · Cardiovascular diseases · Uncertainty analysis · Convolutional Neural Network · Monte Carlo Dropout

1 Introduction

Deep Learning (DL) models have been immensely popularized over the last few years due to their contribution to various fields of machine learning based systems. However, DNNs can not measure the uncertainty which ends up as a drawback in real world applications of DNNs, especially those concerning medical image processing. ECG Classification is a classical example of medical image and signal processing where DNNs are utilized. An ECG is obtained by the electrodes attached to different parts of the body, which record potential changes in the periodic beat of the heart using an electrocardiograph or a vector electrocardiograph. ECGs can greatly assist in helping us to effectively diagnose

© The Author(s), under exclusive license to Springer Nature Switzerland AG 2022
A. Krzyzak et al. (Eds.): S+SSPR 2022, LNCS 13813, pp. 173–182, 2022.
https://doi.org/10.1007/978-3-031-23028-8_18

and classify any existing heart disease. Diseases relating to the heart or blood vessels are collectively called CVDs. In 2019, the American Heart Association announced that CVDs are the major cause of death. It accumulated over 17.6 million deaths in 2016 and the number is estimated to reach 23.6 million by 2030 [1]. A patient suffering from myocardial infarction has a greater chance of recovery, the sooner the abnormal cardiac activity is detected by an ECG [2]. An automated ECG diagnostic system holds the capability to examine the general populace and provide a valuable second opinion for health care practitioners.

It is reported that COVID-19 infection can possibly lead to severe myocarditis in previously healthy patients [3]. Despite the acute myocardial damage, the majority of the COVID-19 patients recover without any obvious cardiac issues. Currently, the world is emerging from the depths of the pandemic so the focus of health care shifts to finding out whether cardiac monitoring in COVID-19 survivors is necessary or not.

Uncertainty analysis is imperative in guaranteeing better, definitive results, and accordingly, making them more applicable for real-life scenarios [4]. For estimating uncertainty, Bayesian neural networks (BNN) [5,6] are a probabilistic variant of neural networks that are fundamentally suitable. The variational inference [6] is generally applied to calculate the posterior model by using variational (such as Gaussian distribution) distribution. However, this approach is not very convenient and does not improve the accuracy. Hence A method especially suitable for BNNs which may not always outperform other state-of-the-art models, is the MCD method. Gal and Ghahramani [7] introduced this idea of using dropouts to determine the model's uncertainty. They noticed that, while dropouts are generally applied to prevent overfitting, it may very well be implemented to depict a model's rough estimate of the weight's posterior. Besides, the convenient implementation of MCD does not affect the model performance.

Considering the risks of classifying ECG images, we propose an Uncertainty-aware CNN model using MCD to safely continue the autonomous diagnosis process. Our main contributions are:

- Using CNN to classify five distinct categories of ECG images including COVID-19.
- Using MCD to analyze the uncertainty of the model
- Finding uncertain samples, for safer diagnosis

2   Literature Review

Many studies on ECG classification have been conducted over the years, given the importance of the problem. Hong et al. [8] provide a good review of ECG. We only discuss about the relevant studies here.

2.1   ECG Classification

In most of the ECG classification studies, the input is either a 1-dimensional numerical ECG signal values [9] or a 2-dimensional image of the ECG signal

[10]. In our study, we train a 2-dimensional CNN on 2-dimensional images of ECG signals for disease classification. Irmak et al. [11] present a CNN model to detect COVID-19 with the help of ECG trace images. They achieve 83.05% five-class classification accuracy using their approach. Sobahi et al. [12] propose an attention-based 3-dimensional CNN model with residual connections (RC). Their approach reaches 92.0% accuracy on four-class classification. Attallah et al. [13] develops five distinct DL models. The predictions of three machine learning classifiers is merged using a classification system, which is developed following the implementation of a feature selection approach. The model achieves 91.7% accuracy in three-class classification.

2.2   Uncertainty Estimation in Medical Image Analysis

Studies regarding uncertainty in DNN prototypes are conducted in order to increase the reliability of the prediction methods. Evaluation of uncertainty in medical applications has been frequently applied for diagnosis of diseases, such as covid-19 [14], tuberculosis [15], and cancer [16] etc. Ghoshal et al. [17] used drop-weights based BNN to quantify uncertainty in DL techniques. They show that the predictive accuracy of the model was notably associated with its uncertainty. Milanés-Hermosilla et al. [18] implement Shallow CNN and with an ensemble model to classify motor imagery. Besides, they use MCD to find the uncertainty of their predictions, which make their model more reliable.

3   Proposed Methodology

We describe our proposed methodology in this section. First, we discuss about the dataset collection and analysis, data processing, CNN model development and finally MCD technique for uncertainty analysis.

3.1   Dataset

We use ECG Images dataset of Cardiac and COVID-19 Patients [19], which has 1937 distinct patient records. Data is gathered using ECG Device 'EDAN SERIES-3' placed in different health institutes across Pakistan. The dataset is a 12-lead based standard ECG images for different patients with five labels such as, **COVID-19, Abnormal Heartbeat (HB), Myocardial Infarction (MI), Previous History of MI (PMI)**, and **Normal Person**. The majority of cardiac abnormalities cause only slight variations in the ECG data. Such as, a peak-to-peak interval or a particular wave. These differences are frequently used to categorize abnormalities. Figure 1 shows some samples from the dataset.

3.2   Data Preprocessing

We have preprocessed the ECG images from the dataset by converting them from RGB (three channel) to grayscale (one channel) images and resizing them into

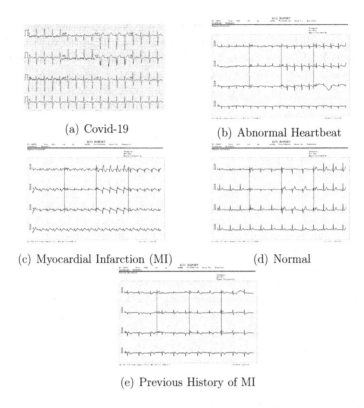

(a) Covid-19   (b) Abnormal Heartbeat

(c) Myocardial Infarction (MI)   (d) Normal

(e) Previous History of MI

Fig. 1. Sample of each classes from the dataset

70×70 resolution. We use the cropped version of this dataset, where the images are cropped according to the ECG leads [20]. This requires less computation cost and better classification performance. Figure 2 exhibits the cropping process.

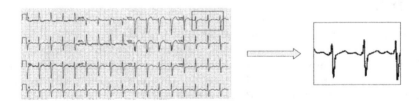

Fig. 2. Processed data

3.3   CNN Architecture

Our CNN model starts with an input image with the resolution of 70×70 pixels. Then we use six 2-dimensional convolutional layers and with 3×3 kernel size. Also, to avoid overfitting in the model, we added a batch normalization layer after each convolutional layer. Batch normalization is used, so that we can use a

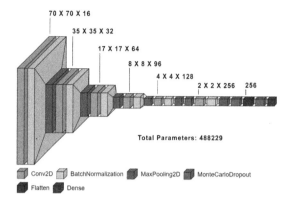

70 X 70 X 16

35 X 35 X 32

17 X 17 X 64

8 X 8 X 96

4 X 4 X 128

2 X 2 X 256   256

Total Parameters: 488229

Conv2D   BatchNormalization   MaxPooling2D   MonteCarloDropout

Flatten   Dense

Fig. 3. Visual representation of proposed deep CNN architecture

higher learning rate as it allows every layer of the network to do learning more independently. Besides that, we use a max pooling layer after each batch normalization layer with a pool size of 2 × 2 to reduce the computational cost. For all convolutional layers, we use default strides which is (1,1). As the activation function, we use Relu as the non-saturation of its gradient accelerates the convergence of stochastic gradient descent (SGD) compared to the Sigmoid/Tanh function. After that, we flatten the values to single dimension and use 20% dropout from that layer. The purpose of using dropout in our proposed model is to make it robust and also to remove any simple dependencies between the neurons. Moreover, the dropout layer is also used as the MCD layer. Although batch normalization and dropout are regularization techniques, each of them have different tasks as well other than regularization. During training, dropout modifies the standard deviation of the distribution, whereas it has no effect on the distribution during validation. Unlike dropout, Batch normalization does not do the same. Softmax activation function is added at the end of the output since here all the nodes are classified. The input to the function is transformed into a value between 0 and 1. We use MCD in the last two convolutional block for the same purpose as mentioned before. The model consists of 488229 parameters. Figure 3 is the visual representation of our proposed architecture.

3.4 Monte Carlo Dropout

The general purpose of dropout is to decrease the model complexity and prevent overfitting [21]. During the training phase, each neuron's output, in the dropout layer, is multiplied by a binary mask which is obtained from a Bernoulli distribution. Some of the neurons are set to zero this way. Afterwards, the neural network (NN) is used at test phase. The dropout technique can be used as an approximation of probabilistic Bayesian models in deep Gaussian processes, which was presented by Gal and Ghahramani [7]. MCD is a technique of performing numerous stochastic forward passes in a neural network using activated dropout throughout the testing phase to generate an ensemble of predictions that may reflect uncertainty estimations. If we are given a trained neural

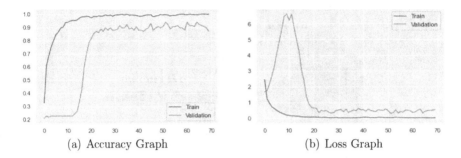

(a) Accuracy Graph   (b) Loss Graph

Fig. 4. Training and validation graph for CNN-MCD method

network model with dropout f_{nn}. To derive the uncertainty for one sample x we collect the predictions of T inferences with different dropout masks. Here $f_{nn}^{d_i}$ represents the model with dropout mask d_i. So we obtain a sample of the possible model outputs for sample x as

$$f_{nn}^{d_0}(x), ..., f_{nn}^{d_T}(x) \tag{1}$$

We obtain an ensemble prediction by computing the mean and the variance of this sample. The prediction is the mean of the model's posterior distribution for this sample and the estimated uncertainty of the model regarding x.

$$Predictive\,Posterior\,Mean,\, p\; =\; \frac{1}{T}\sum_{i=0}^{T} f_{nn}^{d_i}(x) \tag{2}$$

$$Uncertainty,\, c\; =\; \frac{1}{T}\sum_{i=0}^{T} [f_{nn}^{d_i}(x) - p]^2 \tag{3}$$

We do not alter the dropout NN model itself but simply collect the results of stochastic forward passes from the model. This is done in order to assess the prognostic mean and uncertainty of the model. Consequently, this data can be implemented with the existing NN models trained with dropout.

Table 1. Results obtained from CNN-MCD method

Accuracy	Precision	Sensitivity	Loss	F1 score	AUC score
93.9%	94.0%	93.0%	0.031	93.5%	97.6%

4   Experimental Results

4.1   Experimentation

The model has been trained and tested grayscale images. Converting colored images from the dataset into grayscale images reduced the size or number of

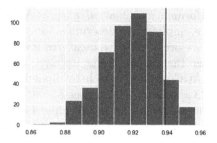

Fig. 5. Distributions of the monte carlo prediction accuracies and prediction accuracy of the ensemble (Red). (Color figure online)

unnecessary features of the images. The images are randomly divided into 8:1:1 training, validation, and test datasets, which also maintains the ratio of disease and non-disease classes in every sub-datasets. After an in-depth experiment, we use 32 training samples per iteration, and set the epochs value to 70. We set the learning rate to 1e−4. Because of using batch normalization layers, this learning rate gave us optimal results. We utilize all test samples and make 1500 predictions for each sample (Monte Carlo Sampling). This is necessary for measuring the uncertainty resulting from the predicted class-wise softmax score distribution of 1500 predictions from the test samples.

Table 2. Uncertain samples and their predictions

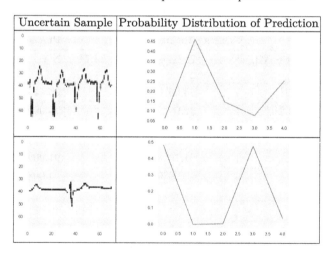

4.2   Result Analysis and Discussion

Following any image classification task, evaluation of the quality of the model depends on performance evaluation metrics. For quantitative evaluation, renowned metrics of performance evaluation, such as Accuracy, F1-Score, Sensitivity, and Precision, are employed to determine the performance of the CNN-MCD approach. There are other effective measures for performance evaluation, such as The Area Under the Curve of the Receiver Operating Characteristic (ROC), also known as AUC of ROC which is applied too. Table 1 depicts the results of the CNN-MCD approach. We obtained an accuracy of 0.939; precision is 0.940. Sensitivity is measured to be 0.930 while Loss is found to be 0.031. The overall F1 Score is 0.935 and the AUC Score is 0.976. Because of using a real-world dataset, CNN-MCD approach performs poorly with less number of epochs, after 20 epochs the results get more stable as shown in Fig. 4. We stop the training process after 70 epochs.

Figure 5 shows the distributions of the MCD predictions and prediction of the ensemble, which is acquired by computing the average and the variance of possible model outputs. We find the uncertain samples using mean and variance of the predictions. We show two uncertain samples with the probability distributions of their predictions in Table 2. This helps us understand the dataset and indicates any existing issues within the model. In the graph, classes 0 to 4 are titled covid-19, HB, MI, Normal, PMI respectively. The first sample successfully predicts the label as 'HB', but from the distribution we can see that, the model is quite uncertain about this prediction as it also confuses the input as 'PMI'. The second sample fails to predict the label. The model is very uncertain about labelling its predictions as 'covid-19' or 'PMI'. Hence, MCD helps the model to point out uncertainty in a way that prevents uncertain diagnosis.

Table 3. Comparison among different models.

Method	Classes	Accuracy	Precision	Sensitivity
ResNet-50, InceptionV3, VGG-16 [11]	5	78.08%, 79.35%, 83.74%	NA	NA
CNN Model of [11]	5	83.05%	96.12%	95.39%
Attention-based CNN with Residual Connections [12]	4	92.00%	95.99%	82.00%
ECG-BiCoNet [13]	3	91.70%	91.90%	95.90%
Proposed CNN-MCD	5	93.90%	94.00%	93.00%

In terms of model performance, our proposed model outperforms all the existing models in multi-class classification shown in Table 3. Popular models such as VGG-16, ResNet-50, InceptionV3 achieves accuracy from 79%-83%. Other methods mentioned in Table 3 conducts their classification on three or four classes while our five-class classification accuracy is higher than those. Even though our model could not outperform in precision and recall in some cases, it is very close to those results.

5   Conclusion and Future Work

We present the uncertainty of a CNN model in ECG Classification using MCD, which shows the uncertainty and improves the predictive accuracy of the CNN for safer and better analysis of quantitative metrics. Besides, it keeps the model structure the same as before and adds minimal cost to inference time. We use MCD method to do such which is very significant to decrease the risk factors in cardiovascular disease diagnosis. Our proposed CNN-MCD method not only surpass state-of-the-art models in a huge margin but also allows users to find uncertain samples to reduce risk factors. Our proposed method can be used simply by taking an image of the ECG. This will help to diagnose patients quickly if the specialist is not currently available. The MCD method used in our proposed model helps to identify out-of-scope samples which can be prevented as these are very risky to deal with. Besides trace images, the growing popularity of smart watches allows another bright idea to be explored, which is using a MCD model in smart watches which are embedded with ECG trackers. This will allow smart watches to provide more reliable and real time predictions. Smart watches can inform experts for dealing with uncertain predictions since many smart watches have calling or messaging feature.

References

1. Benjamin, E.J., et al.: Heart disease and stroke statistics-2019 update: a report from the american heart association. Circulation **139**, e56–e528 (2019). https://www.ahajournals.org/doi/abs/10.1161/CIR.0000000000000659
2. Lown, B., Klein, M.D., Hershberg, P.I.: Coronary and precoronary care. Am. J. Med. **46**, 705–724 (1969). https://www.sciencedirect.com/science/article/pii/0002934369900229. ISSN 0002-9343
3. Tajbakhsh, N., et al.: Convolutional neural networks for medical image analysis: fine tuning or full training? IEEE Trans. Med. Imaging **35**, 1 (2016)
4. Begoli, E., Bhattacharya, T., Kusnezov, D.F.: The need for uncertainty quantification in machine-assisted medical decision making. Nat. Mach. Intell. **1**, 20–23 (2019)
5. Neal, R.: Bayesian Learning via Stochastic Dynamics. In: Hanson, S., Cowan, J., Giles, C. (eds.) Advances in Neural Information Processing Systems, vol. 5. Morgan-Kaufmann (1992). https://proceedings.neurips.cc/paper/1992/file/f29c21d4897f78948b91f03172341b7b-Paper.pdf
6. Gal, Y.: Uncertainty in deep learning. Ph.D. thesis, University of Cambridge (2016)
7. Gal, Y., Ghahramani, Z.: Dropout as a Bayesian approximation: representing model uncertainty in deep learning. In: Balcan, M.F., Weinberger, K.Q. (eds.) Proceedings of the 33rd International Conference on Machine Learning, New York, USA, 20–22 June 2016, pp. 1050–1059, vol. 48. PMLR (2016). https://proceedings.mlr.press/v48/gal16.html
8. Hong, S., Zhou, Y., Shang, J., Xiao, C., Sun, J.: Opportunities and challenges of deep learning methods for electrocardiogram data: a systematic review. Comput. Biol. Med. **122**, 103801 (2020). https://www.sciencedirect.com/science/article/pii/S0010482520301694. ISSN 0010-4825

9. Attia, Z., et al.: An artificial intelligence-enabled ECG algorithm for the identification of patients with atrial fibrillation during sinus rhythm: a retrospective analysis of outcome prediction. Lancet **394**, 861–867 (2019)

10. Jun, T.J., et al.: ECG arrhythmia classification using a 2-D convolutional neural network (2018). https://arxiv.org/abs/1804.06812

11. ırmak, E.: COVID-19 disease diagnosis from paper-based ECG trace image data using a novel convolutional neural network model. Phys. Eng. Sci. Med. **45**, 167–179 (2022)

12. Sobahi, N., Sengur, A., Tan, R.-S., Acharya, U.R.: Attention-based 3D CNN with residual connections for efficient ECG-based COVID-19 detection. Comput. Biol. Med. **143**, 105335 (2022). https://www.sciencedirect.com/science/article/pii/S0010482522001275. ISSN 0010-4825

13. Attallah, O.: ECG-BiCoNet: an ECG-based pipeline for COVID-19 diagnosis using Bi-Layers of deep features integration. Comput. Biol. Med. **142**, 105210 (2022). https://www.sciencedirect.com/science/article/pii/S0010482522000026. ISSN 0010-4825

14. Ghoshal, B., Tucker, A.: On cost-sensitive calibrated uncertainty in deep learning: an application on COVID-19 detection. In: 2021 IEEE 34th International Symposium on Computer-Based Medical Systems (CBMS), pp. 503–509 (2021)

15. Ul Abideen, Z., et al.: Uncertainty assisted robust tuberculosis identification with Bayesian convolutional neural networks. IEEE Access **8**, 22812–22825 (2020)

16. Jungo, A., et al.: On the Effect of Inter-observer Variability for a Reliable Estimation of Uncertainty of Medical Image Segmentation (2018). https://arxiv.org/abs/1806.02562

17. Ghoshal, B., Tucker, A.: Estimating Uncertainty and Interpretability in Deep Learning for Coronavirus (COVID-19) Detection (2020). https://arxiv.org/abs/2003.10769

18. Milanés-Hermosilla, D., et al.: Monte Carlo dropout for uncertainty estimation and motor imagery classification. Sensors **21**, 7241 (2021). https://www.mdpi.com/1424-8220/21/21/7241. ISSN 1424-8220

19. Khan, A.H., Hussain, M., Malik, M.K.: ECG images dataset of cardiac and COVID-19 patients. Data Brief **34**, 106762 (2021). https://www.sciencedirect.com/science/article/pii/S2352340921000469. ISSN 2352-3409

20. Nkengue, M.J.: ECG_Image_Cropped Version 4 (2022). https://www.kaggle.com/datasets/marcjuniornkengue/ecg-image-cropped

21. Srivastava, N., Hinton, G., Krizhevsky, A., Sutskever, I., Salakhutdinov, R.: Dropout: a simple way to prevent neural networks from overfitting. J. Mach. Learn. Res. **15**, 1929–1958 (2014). http://jmlr.org/papers/v15/srivastava14a.html

One-Against-All Halfplane Dichotomies

George Nagy$^{(\boxtimes)}$ ⓘ and Mukkai Krishnamoorthy ⓘ

Rensselaer Polytechnic Institute, Troy, N.Y. 12180, USA
nagy@ecse.rpi.edu, moorthy@cs.rpi.edu

Abstract. Given M vectors in N-dimensional attribute space, it is much easier to find M hyperplanes that separate each of the vectors from all the others than to solve M arbitrary linear dichotomies with approximately equal class memberships. An explanation of the rapid growth with M and N of the number of separable one-against-all linear halfplane dichotomies is proposed in terms of convex polyhedra in a hyperspherical shell. The counterintuitive surge is illustrated by averaged results on pseudo-random integer arrays obtained by Linear Programming and Neural Networks. Although the initial motivation arose from seemingly arbitrary rankings of scientists and universities, this project is not directed at any application.

Keywords: Linear separability · Halfplane dichotomy · Ranking

1 Introduction

We consider datasets of M homogenous patterns, i.e., sets of observations on objects that do not naturally fall into two or more categories. Each pattern is, as usual, represented by a feature vector of N elements. Examples of such datasets include census records, fact sheets for countries, cities, schools, universities or hospitals, and collections of publication data for scientists. In the small collections we have in mind, M may range from 10 to 10^6, and N from 1 to 100.

Our objectives are to explore, theoretically and experimentally, (1) why so many of the M patterns can be linearly separated from all the others when $M >> N > 3$; (2) equivalently, why most patterns can be ranked first among all the patterns by a linear weighting function; (3) whether feature difference vectors facilitate finding the weight vectors for such one-against-all dichotomies; and (4) how the required weight vectors can be obtained with either linear programming or a perceptron-type neural network. We introduced some of these issues in an ICPR 2022 submission [1].

There are, of course, other ways to analyze pattern matrices beside linear separation or ranking, including descriptive statistics, higher moments of the empirical distributions, and classification with various degrees and types of supervision. Data may be analyzed to reveal groups or clusters that are more like each other in some respect than like other groups. The expectation in such exercises is that each class is grouped somewhat compactly in feature space, with every class bounded by non-intersecting linear or nonlinear surfaces. We are, however, interested only in how many patterns in a given

A. Krzyzak et al. (Eds.): S+SSPR 2022, LNCS 13813, pp. 183–192, 2022.
https://doi.org/10.1007/978-3-031-23028-8_19

set can be linearly separated from all the others. We believe that the answer is counterintuitive because of the difference of volume-surface relations in hyperspace from our experience in the 3-D world.

In the remainder of this paper, we present a brief literature review (Sect. 2); examine the prevalence of one-against-all halfplane dichotomies in pseudo-random arrays (Sect. 3); relate the number of such dichotomies to surface-to-volume ratios in hyperspace (Sect. 4), comment on the merits of difference vectors (Sect. 5), compare Linear Programming and Neural Network solutions (Sect. 6); return briefly to MeFirst ranking (Sect. 7); and summarize our putative contributions (Sect. 8).

2 Prior Work

Most of this work is based on theory established at least 50 years ago. Our terminology and notation mirrors those of venerable textbooks on pattern recognition like DHS, Fukunaga, K&K, and Bishop [2–5]. These include material about weight vectors and separating planes in hyperspace, feature normalization and column augmentation, and some aspects of dynamic and linear programming and of single-layer neural networks for which we feel compelled to cite primary sources.

One-class classification (OCC), typically encountered in anomaly detection or separation of signal from noise, is really a two-class discriminations where one class (normal, signal, inlier) is well defined by one or more clusters of samples, and the other class (abnormal, anomalous, noise, outlier), usually with far fewer samples, does not exhibit any compactness characteristic in feature space [3]. Any pattern far enough from the normal class (according to some metric) falls into this abnormal class. One-shot learning is different: here it is desired to improve the classifier after seeing each new labeled sample [6].

Multi-category classification can be reduced to either a large binary discrimination (via Kessler's Construction [7]), or to a set of dichotomies with the same number of features, as is customary for Support Vector Machines. We mention these paradigms because they are easy to confuse with our main subject of one-against-all dichotomies.

The ease of one-against-all linear separation may be contrasted with Hughes' demonstration that the fraction of linearly separable pairs of sample sets (i.e., *halfplane dichotomies*) of M points in general positions in N-dimensional space diminishes rapidly as M exceeds $2N + 1$ (the *capacity* of a hyperplane) [8].

Hilbert's and Coxeter's illustrations of 4-dimensional polyhedra offer a gentle introduction to hyperspace [9, 10]. Formulas for the volume and surface area of an n-sphere are listed in [11]. Bishop provides a good explanation of why with rising dimensionality samples are increasingly concentrated in a thin shell [5]. Kernel methods map nonlinear boundaries into hyperplanes in higher dimensions [12].

Although Linear Programming (LP) is designed to optimize a linear function subject to a set of linear constraints [13, 14], we used it only to obtain weights for linear separation. Both the default Interior Point algorithm [15] built into MATLAB and into open-source GNU Octave, and the Dual Simplex algorithm [16], find a solution, if one exists, in polynomial time, and halt if the constraints cannot be satisfied.

Neural networks (NNs) have also been used to determine linear separability. The original "fixed-increment perceptron error-correction" procedure provably converges to a solution if one exists [17, 18]. Replacing the Heaviside step activation by a differentiable function led to gradient descent methods that accelerate convergence. However, the bound on the number of iterations can still be computed only from a known solution. It is impossible to tell whether the weight vector is still approaching the solution cone or the input data is linearly inseparable. This holds also for other algorithms for training a single layer. Therefore the maximum allowed number of epochs or training cycles must be specified. Unlike LP, a single-layer neural network can never confirm linear inseparability. We programmed an elementary α-perceptron according to the 1966 perceptron software manual [19], and ran a single-layer feed-forward network configured for gradient descent from the MATLAB Deep Learning toolbox [20].

The wide-ranging and often harmful effects of ranking human endeavor are critically examined in [21] and [22]. OpenAI forbids such use of its software [23].

3   One-Against-All Halfplane Dichotomies

We confine our attention to classifying each pattern in the dataset against every other pattern. For M patterns, we "train" M classifiers to perform M binary classifications. Each classifier is trained on all M patterns, but with a different pattern singled out for the positive class. Since each of the M classifiers knows the identity of the distinguished pattern, it could achieve 100% correct performance via table look-up. However, we seek only to isolate each pattern by a linear weighting of its features.

Since we don't have any test set (which dispenses us from the vexatious concern for generalization), how do we measure performance? Our performance metric for the M classifiers is the fraction of linearly separable dichotomies. Suppose, for example, that M = 200. If an algorithms finds 180 weight vectors such that each separates one pattern from the remaining 199, then the metric is 0.9.

Let us look at a two-dimensional example where we can plot both the patterns and the weight vectors. If one had to find seven one-against all linearly separable vectors in two-space, they would have to lie on the vertices of a convex polygon, like the four different sets of 7 points in Fig. 1. (Only one of the polygons – for the + set – is traced explicitly in the Figure.) Any points inside the polygon would not be separable from all the other points. Each point set gives rise to 7 linear dichotomies. Only the line separating one of the +'s from the other six is shown. Note that most of the points fall in an annulus (or *shell* in higher N). The next section suggests that the prevalence of one-against all linear separability is due to the increasing concentration of points in the shell.

For a fixed N, the number Fsep(M, N) of one-against-all halfplane dichotomies is a rising function of the number of patterns M that flattens out when M is large enough. We denote its asymptotic vale as M^. The function depends critically on the number of attributes N (i.e., the dimensionality of the feature space). The values of the function change by orders of magnitude, but its general behavior remains the same.

We believe that our experimental exploration of the parameter space for N up to 12 and M up to 10,000 reveals most of the interesting behavior of this function. We conducted our experiments on M $\times$ N arrays of pseudo-random integers (randi(R, M, N)) with R = 1000. (To keep coincidences low enough to avoid affecting the overall behavior, the range R must satisfy $R^N >> M$.) Depending on the variability

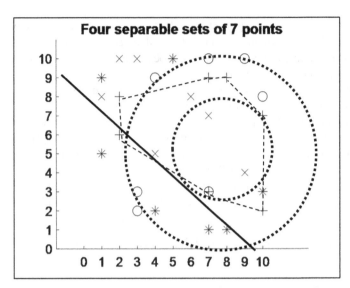

Fig. 1. Four sets of one-against-all linearly separable points at the vertices of convex polygons. One of the polygons is shown with dashed lines, The solid black line separates one of the +s from the other six +s These four (○, +, *, ×) were the only separable sets of 7 points generated by 200 pseudo-random trials with a 1–10 range of integers. The dotted circles bound an annulus that contains most of the points.

of the observations, we ran each setting with 10 or 100 trials and recorded the average Fsep and its standard deviation. For exploring the behavior of Fsep for larger values of M, we sampled only every 5th, 10th, 100th or 1000th M, as shown in the figures. Even so, several of the experiments ran for over ten hours on our vintage 2.4 GHz Dell Optiplex. The run-time is roughly proportional to $T \times M^3$, where T is the number of trials. In Sect. 6, we tabulate some results, including timing, on small and large attribute arrays.

Initially, Fsep(M, N) = M for any N, because in N-space, any N + 1 points in general positions are linearly separable. In 3-D, these points form the vertices of a tetrahedron, but in higher dimensions they are difficult to visualize. The linear growth of Fsep with M extends rapidly with N. With N = 2, the increase moderates as soon as M = 5 and the curve is almost flat by M = 100 (Fig. 2). With N = 5, the slope is still about 0.75 at M = 100 (Fig. 3). But with N = 12, the fraction of linearly separable dichotomies is 90% even at M = 6400 (Fig. 4). We lack the computer resources to explore it further. For a fixed M as a function of N, Fsep necessarily plateaus at Fsep(M, N) = M after a linear rise (Fig. 5).

4   A Geometric Perspective

Although an infinite number of distinct points located on the surface of an N-dimensional sphere are one-against-all separable, their number reaches a limiting value in any realistic scenario where the points are subject to perturbation. Both this asymptotic magnitude

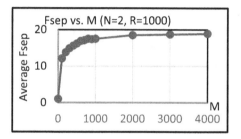

Fig. 2. Fsep, the number of separable patterns, vs. M, with N = 2

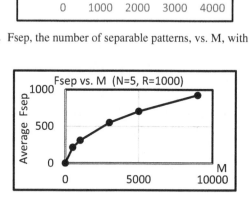

Fig. 3. Fsep, the number of separable patterns, vs. M, with N = 5

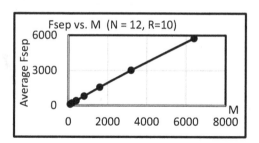

Fig. 4. Fsep, the number of separable patterns, vs. M, with N = 12

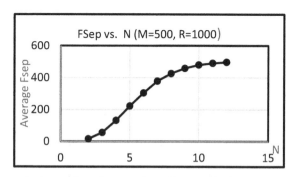

Fig. 5. Increase in the number of one-against-all separable dichotomies (FSep) with N

M^ and the value of M where it prevails increase rapidly with N. A possible cause is the following. The ratio of the volume of a shell of thickness ε to that of a unit hypersphere is $1-(1-ε)^N$. If, for example, N = 5 and ε = 0.2 (a shell of thickness equal to 20% of the radius), then the volume of the shell is 67% of the volume of the sphere. Therefore most of the randomly generated points would fall in the shell, as already suggested by Fig. 1 (where N is only 2). They are all separable only if they constitute the vertices of a convex polyhedron.

For an additional point to be separable from the rest without altering the convexity of any existing vertex, it would have to fall near the center of one of the faces of the polyhedron, as suggested by the 2-D example in Fig. 6. Since every new separable point decreases the area of the facets, the space available for new separable pints approaches zero at a rate decreasing with M. For any M, with large-enough N it is possible to generate sets of M points in general positions that are one-against-all linearly separable. We believe that these notions generalize to N dimensions and look forward to suggestions from the Workshop participants on turning arguments into a proofs.

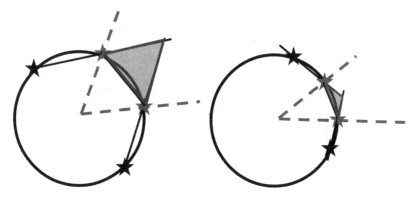

Fig. 6. How near is near enough? Suppose that there are already 4 point located on a circle (hypersphere). Where in the red sector can we add a new point? The space available for a new point shrinks as the points in the shell get closer to each other. Space available for new separable points in the red sector of the convex hull is shown in blue. (Color figure online)

5   Pairwise Attribute Difference Vectors

We must at this point introduce some notation. Our data vector is the M × N array A. Let a_k, k = 1 to M, be the row vectors of A. Let w_k be the M column vectors of the N × M weight matrix W. The task is to find M weight vectors w_k such that $a_k * w_k > a_k * w_k$ for all for all k ≠ k*. In other words, we must solve M one-against all dichotomies.

Stated in terms of *pairwise difference vectors* $d_{k*,k} = a_{k*} - a_k$, $d_{k*,k} > 0$ for all k ≠ k*. The difference vectors define homogenous halfspace dichotomies as illustrated in Fig. 7. The constraints for a linear programming solution are naturally framed in terms of difference vectors. Our experiments indicate that difference vectors also lead to faster convergence of single-layer neural networks. A possible reason is that in the direct solution, the distinguished vector a_{k*} can change the weight vector only once per epoch. In the proposed alternative, every vector plays an equal role in training.

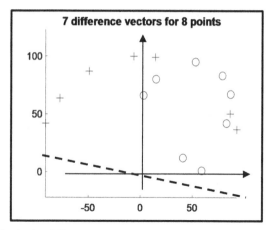

Fig. 7. Normalized pairwise difference vectors (+) for 8 one-against-all separable attribute vectors (o) form a homogenous halfplane dichotomy, bounded by the dashed line. For better visibility, the unit-length difference vectors were scaled (multiplied) by 100.

6 Linear Programming and Neural Networks

To find the weights of the separating vector, we used the difference vectors as the required constraints and minimized the sum of the weights as the objective function. We found equivalent, but not identical, solutions with the MATLAB default Interior Point algorithm and the Dual Simplex algorithm. The LP program was run for each of the M one-against-all dichotomies of the pseudo-random $M \times N$ integer arrays folded into a loop for the specified number of trials (10 or 100).

Because the networks in the MATLAB Deep Learning toolbox have so many invisible built-in functions (including their *Legacy Perceptron*), we programmed a classical α-perceptron in m-code. As shown below, it required far fewer iterations with the input preprocessed into pairwise difference vectors than directly on the rows of the attribute array. We report only the fastest configuration of the Alpha Perceptron, with dimensionality augmentation after normalization. As proved in [1], the weights obtained with such preprocessing also separate the original array.

We also experimented with a single layer Levenberg–Marquardt feedforward network. Since the feedforward net has a smooth (*hyperbolic tangent sigmoid*) activation function, we rounded its output to 0 or 1. We set the performance criterion to *least mean squares*. After missing a few dichotomies with MaxEpochs = 10, this net reached the correct value of Fsep = 496 on the 500 × 12 arrays.

The general trend of the performance comparisons are shown in Table 1. In every experiment, we checked the output weights directly on the original array. The runtimes reported were obtained with *tic-toc*, which is not immune to Windows background activity. Our objectives for the experiments were only to show that one-against-all dichotomies are easier for both LP and NN than arbitrary divisions of M patterns into approximately equal classes, and that such dichotomies benefit from pairwise difference input.

Table 1. Performance of LP and NN on 8×2 and 500×12 pseudo-random attribute arrays.

Method	M	N	MaxEpochs	Trials	Fsep	Time (s)
LP	8	2	N/A	100	5.5	7
Alpha Perceptron (diffs)	8	2	100	100	5.5	0.3
Alpha Perceptron (rows)	8	2	100,000	100	.4.9	239
Levenberg–Marquardt	8	2	10	100	5.4	7
LP	500	12	N/A	10	496	63
Alpha Perceptron (diffs)	500	12	100	10	496	9
Levenberg–Marquardt	500	12	10	10	496	564

7 Ranking

What originally led us to one-against-all dichotomies was MeFirst Ranking. A ranking of 100,000 scientists with their citation attributes was published in October 2020 in PLoS BIOLOGY [24, 25]. We wanted to demonstrate the existence of linear weighting functions that would allow most of these scientists to claim first rank in a large subgroup (defined, for example, by institution or nationality). In [1], we explored that dataset, and another of the 1000 "top" universities [26]. Rather than repeating those results here, we observe only that while the relationship of the number of separable entries to M and N, Fsep(M, N), was much the same as what we found on pseudo-random integer arrays, separating the real data took more work. Although for $M = 500$ and $N = 12$ both sets yield Fsep = 496, it takes the Alpha Perceptron 1,000,000 instead of 100 epochs to complete the task. The LP, however, shows no increase in runtime.

Table 2 is a small example with small integer coefficients that shows clearly the equivalence of linear separability and ranking. It also illustrates the method we used in all our experiments to verify the weights of the separating vector. We note that if ties for first place area not considered admissible, an additional check is required. With the range of integer arrays used in our simulations (1000), coincidences among the 1000^2 ($N = 2$) or 1000^{12} (for $N = 12$) possible entries are rare.

Because attribute vector normalization (usually to unity) and column augmentation (via the addition of a constant attribute and a threshold weight) are often used in statistical and neural network classification, in [1] we gave sufficient and necessary conditions for MeFirst ranking and one-against-all separability in terms of normalized or augmented attribute vectors. We close with an informal statement of sufficient and necessary conditions in geometric terms.

Proposition: *Each of a set of M points in N-space are one-against-all separable from the remaining M-1 points if and only if each point constitutes a distinct convex vertex of the convex hull of all M points.*

Note that points on the edges or faces, which are by convention considered part of the convex hull, are excluded, as are coincident points.

Table 2. The highest value of each weighted attribute vector is on the diagonal of the product matrix. For example, the product of $(1, 8, 8) \times (0, 1, 1)$ is the highest value (16) in the fifth column.

Attributes A			Weights W^T			Weighted sum of Attributes A $\times$ W						
9	1	1	1	0	0	9	1	1	10	2	10	11
1	9	1	0	1	0	1	9	1	10	10	2	11
1	1	9	0	0	1	1	1	9	2	10	10	11
8	8	1	1	1	0	8	8	1	16	9	9	17
1	8	8	0	1	1	1	8	8	9	**16**	9	17
8	1	8	1	0	1	8	1	8	9	9	16	17
6	6	6	1	1	1	6	6	6	12	12	12	18

8 Summary

This paper presents a problem that we believe has received little attention. We investigated the behavior of one-against-all halfplane dichotomies of pseudo-random integer arrays and found that it is very different from that of similar dichotomies with approximately balanced populations. We showed that for a limited range of M patterns in N-dimensional feature space, linear separability can be determined with either Linear Programming or single-layer Neural Networks. In contrast to balanced populations, the preferred input for both methods is an array of pairwise attribute difference vectors.

We found that one-against-all linear separability in any dimensionality N increases rapidly with M, with the slope changing gradually from unity to zero. With fixed M, the number of linearly separable halfplane dichotomies increases monotonically with N from N + 1 to its maximum of M.

We suggested that this surprising behavior is due to the concentration of high-dimensional patterns in a hyperspherical shell where they form the vertices of a convex polyhedron. Convex vertices are linearly separable from all other points regardless of whether these are on the surface or inside the entire point set.

References

1. Nagy, G., Krishnamoorthy, M.: MeFirst ranking and multiple dichotomies via linear programming and neural networks. In: ICPR 2022 (2022). Accepted
2. Duda, R.O., Hart, P.E., Stork, D.G.: Pattern Classification. Wiley, New York (2001)
3. Fukunaga, K.: Statistical Pattern Recognition, 2nd edn. Academic Press, Cambridge (1990)
4. Theodoridis, S., Koutroumbas, K.: Pattern Recognition. Academic Press, Cambridge (2009)
5. Bishop, C.M.: Pattern Recognition and Machine Learning. Springer, New York (2006). ISBN 978-0387-31073-2
6. Li, F.-F., Fergus, R., Perona, P.: One-shot learning of object categories. IEEE Trans. Pattern Anal. Mach. Intell. **28**(4), 594–611 (2006)
7. Nilsson, N.J.: Learning Machines: The Foundations of Trainable Pattern-Classifying Systems. McGraw-Hill, New York (1965)

8. Hughes, G.F.: On the mean accuracy of statistical pattern recognizers. IEEE Trans. Inf. Theory IT **4**(1), 55–63 (1968)
9. Hilbert, D., Cohn-Vossen, S.: Geometry and the Imagination, Chap. III. Chelsea Publishing, Hartford (1952)
10. Coxeter, H.S.M.: Introduction to Geometry, Chap. 22. Wiley, Hoboken (1989)
11. Woldfram MathWorld, Hypersphere. https://mathworld.wolfram.com/Hypersphere.html
12. Shawe-Taylor, J., Cristianini, N.: Kernel Methods for Pattern Analysis. Cambridge University Press, Cambridge (2004)
13. Dantzig, G.B.: Maximization of a linear function of variables subject to linear inequalities. In: Koopmans, T.C. (ed.) Activity Analysis of Production and Allocation. Wiley & Chapman-Hall (1951)
14. Gass, S.L.: Linear Programming, 5th edn. Dover, Downers Grove (2003)
15. Karmarkar, N.: A new polynomial-time algorithm for linear programming. In: Proceedings of the Sixteenth Annual ACM Symposium on Theory of Computing, STOC 1984, p. 302 (1984)
16. Nabli, H.: An overview on the simplex algorithm. Appl. Math. Comput. **210**(2), 479–489 (2009)
17. Rosenblatt, F.: The perceptron: a probabilistic model for information storage and organization in the brain. Psychol. Rev. **65**, 386 (1958)
18. Rosenblatt, F.: Principles of Neurodynamics. Spartan Books, New York (1962)
19. Barker, T.: A Computer Program for Simulation of Perceptrons and Similar Neural Networks: User's Manual. CSRP Report #8. Cornell University, Ithaca, NY (1966)
20. MATLAB: Deep Learning Toolbox. https://www.mathworks.com/products/deep-learning.html. Accessed 12 Jan 2021
21. Muller, J.Z.: The Tyranny of Metrics. Princeton University Press, Princeton (2018)
22. Biangioli, M., Lippman, A.: Gaming the Metrics. MIT Press, Cambridge (2020)
23. Johnson, S.: The Writing on the Wall, The New York Times Magazine, 17 April 2022 (2022)
24. Ioannidis, J.P.A., Boyack, K.W., Baas, J.: Updated science-wide author databases of standardized citation indicators. PLoS Biol. **18**(10), e3000918 (2020). https://doi.org/10.1371/journal.pbio.3000918
25. Mendelay: Data for "Updated science-wide author databases of standardized citation indicators". https://data.mendeley.com/datasets/btchxktzyw/2. Accessed 10 Apr 2021
26. QS World University Ratings. https://group.intesasanpaolo.com/content/dam/portalgroup/nuove-immagini/sociale/2022_QS_World_University_Rankings_1.pdf. Accessed 30 Nov 2021

Fast Distance Transforms in Graphs and in Gmaps

Majid Banaeyan$^{(\boxtimes)}$ (ID), Carmine Carratù (ID), Walter G. Kropatsch (ID),
and Jiří Hladůvka (ID)

Pattern Recognition and Image Processing Group, TU Wien, Vienna, Austria
{majid,krw,jiri}@prip.tuwien.ac.at

Abstract. Distance Transform (DT) as a fundamental operation in pattern recognition computes how far inside a shape a point is located. In this paper, at first a novel method is proposed to compute the DT in a graph. By using the edge classification and a total order [1], the spanning forest of the foreground is created where distances are propagated through it. Second, in contrast to common linear DT methods, by exploiting the hierarchical structure of the irregular pyramid, the geodesic DT (GDT) is calculated with parallel logarithmic complexity. Third, we introduce the DT in the nD generalized map (n-Gmap) leading to a more precise and smoother DT. Forth, in the n-Gmap we define n different distances and the relation between these distances. Finally, we sketch how the newly introduced concepts can be used to simulate gas propagation in 2D sections of plant leaves.

Keywords: nD distance transform · Generalized maps · Irregular pyramids · Parallel processing · Logarithmic complexity · Geodesic distance transform (GDT)

1 Introduction

The distance transform [5] computes for every pixel/voxel of an image/object how far it is from the closest obstacle, boundary, or background. While any valid metric may be involved in the computation of distance transforms, in topological data structures like graph, combinatorial maps [12], or generalized maps (n-Gmap) [7] often the shortest path between the obstacle/boundary and a given point is used. In this study, we first investigate the distance transform (DT) in graphs and then extend it to generalized map. We define different distances for every dimension (1D, 2D,..., nD) in the n-Gmap. This would be useful in many applications. In particular, in study of gas exchange through airspace of a leaf, computing the distances from stomata is very crucial to understand the different diffusion processes needed for photosynthesis.

Supported by the Vienna Science and Technology Fund (WWTF), project LS19-013, https://waters-gateway.boku.ac.at/.

Computing the DT and propagating the distances is an iterated local operation [8,15]. While local processes (e.g., convolution and mathematical morphology) are important in early vision, they are not suitable for higher level vision, such as symbolic manipulation and feature extraction where both local and global information is needed [9]. Therefore we exploit the advantage of the hierarchical structure of the pyramid [10] that encodes both local and global information similar to the human visual system [14].

In the pyramid there are two directions of processes: bottom-up and top-down. In the bottom-up (fine to coarse) process the information of the input data (e.g. intensity, color, texture) is transformed into global information. In the top-down (coarse to fine) process the global information such as the shape and the size of objects are refined into the base level of the pyramid. Therefore, the main idea of using hierarchical structure in computing the DT is to investigate the connectivity of a connected component in the local and general view within the pyramid. We will show that the connectivity can be checked in parallel logarithmic complexity instead of the linear raster scan commonly utilized in the state-of-the-art algorithms [8].

1.1   Notations and Definitions

An image P can be represented using a 4-adjacent neighborhood graph $G = (V, E)$ where V corresponds to pixels of P and E relates neighboring pixels. 8-Adjacency could be used only if the image is well-formed [11], which is not satisfied in general cases. The gray-value of a pixel $g(p)$ becomes an attribute of the corresponding vertex v, $g(v) = g(p)$ and the $contrast(e) = |g(u) - g(v)|$ becomes an attribute of an edge $e(u, v)$ where $u, v \in V$. In the neighborhood graph of the binary image, the edges have only two values: zero and one. We call them accordingly: **zero-edge** and **one-edge** [1]. Furthermore we denote the set of all zero-edges as E_0 and the set of all one-edges as E_1. In this way, the edges of the graph are partitioned into $E = E_0 \cup E_1$.

Irregular Pyramid. [10] is a stack of successively reduced smaller graphs where each graph is built from the graph below by selecting a specific subset of vertices and edges. In each level of the pyramid, the vertices and edges disappearing in level above are called *non-surviving* and those appearing in the upper level *surviving* ones.

Definition 1 (Contraction Kernel (CK)). *A CK is a tree consisting of a surviving vertex as its root and some non-surviving neighbors with the constraint that every non-survivor can be part of only one CK.*

Two basic operations are used to construct the pyramid: **edge contraction** and **edge removal**. In the edge contraction, an edge $e = (v, w)$ is contracted while its two endpoints, v and w, are identified and the edge is removed. The edges that were incident to the joined vertices will be incident to the resulting vertex after the operation. An arrow over an edge is commonly used to indicate the direction of contraction, i.e., from non-survivor to survivor (cf. Fig. 2). Contracting an edge has the enormous advantage of preserving the connectivity of the graph.

During the edge removal, an edge is removed without changing the number of vertices or affecting the incidence relationships of other edges. Constraints are needed to make sure that edge removal does not disconnect the graph [6].

2   Distance Transform in a Graph

In a graph $G = (V, E)$ distances can be measured by the shortest length of paths. In this case the elements are the vertices V and neighbors $\mathcal{N}(v) = \{(v, w) \in E\}$ are related by edges. The distance between two vertices is the shortest path connecting the two vertices.

To compute the DT in a graph $G(V, E)$ with background $B \subset V$ and foreground $F \subset V$ vertices, the shortest distances of foreground vertices from the background should be computed. In this case the *seed* vertices $b \in B$ are initialized by $DT(b) = 0$. The foreground vertices $f \in F$ are initialized by $DT(f) = \infty$. Each one-edge $e = (b, f) \in E_1, b \in B, f \in F$ has two endpoints where $b \in B$ is a seed vertex with $DT(b) = 0$. The other vertex $f \in F$ belongs to the foreground and we initialize its distance by $DT(f) = 1$. The one-edges E_1 are frozen because they have no role in propagating the distances in the graph. Distances are propagated only through the E_0 edges of the foreground.

Using the total order on the foreground F proposed in [1,3], a spanning forest contains only edges E_0 spanning the foreground. The spanning forest is created in a single step with parallel constant complexity. Moreover, to propagate the distances we use the breath-first search (BFS) [4].

Proposition 1. *The parallel complexity of propagating DT is $\mathcal{O}(\delta(T))$ where $\delta(T)$ is the longest path in the spanning forest of the foreground.*

Proof. The complexity of propagating distances in a tree is $\mathcal{O}(|E|)$. Each connected component of the foreground is covered by a spanning tree which is processed independently [3]. Therefore, the longest path in the forest indicates the parallel complexity.   □

The propagation of the distances to the remaining vertices v of the foreground F follows:

$$D(v) = \min\{D(v), D(v_j) + 1|\ v_j \in \mathcal{N}(v)\}   v \in F \tag{1}$$

where the foreground neighbors $\mathcal{N}(v)$ are defined by:

$$\mathcal{N}(v) = \{v \in F\} \cup \{w \in F|e_0 = (v, w) \in E_0\} \tag{2}$$

The distances are propagated until there is no vertex $v \in F$ with $DT(v) = \infty$ (see Fig. 1). Algorithm 1 shows the steps of computing the DT in a graph.

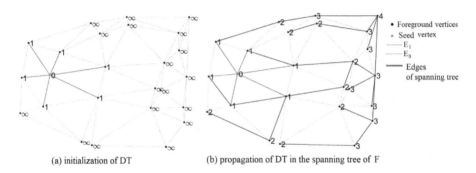

(a) initialization of DT   (b) propagation of DT in the spanning tree of F

Fig. 1. Computing the DT in a graph

Algorithm 1. Computing the DT in the Neighborhood Graph

1: **Input:** Neighborhood Graph: $G = (V, E) = (B \cup F, E_0 \cup E_1)$
2: Initialization: $DT(b) = 0 \; \forall b \in B, \quad DT(f) = \infty, \; \forall f \in F$
3: $DT(f) = 1 \; \forall (f, b) \in E_1, \; f \in F, \; b \in B$
4: **While** $\exists f \in F$ with $DT(f) = \infty$ **do**
5:     Propagate the distances by (1)
6: **end**

2.1 Geodesic Distance Transform

Geodesic DT (GDT) computes distances within the connected component of interest in a labeled image (or labeled neighborhood graph). The objects of interest are considered as the foreground objects and the remaining objects with different labels are considered as the background. A subset of points in the foreground are the seeds, $s \in S$, $S \subset F$, initialized by zero, $DT(s) = 0$. The aim is to compute the minimum distance of every point of the foreground to these seeds. The disjoint foreground objects keep the infinite distance if there is no seed in the connected component.

To compute the GDT we employ the irregular graph pyramid with logarithmic complexity. Each vertex receives a unique index and a total order is defined over the indices [1,3] that results in an efficient selection of contraction kernels (CKs). The CKs are only selected from E_0 edges which propagate the distances. The propagating distances are a set of power-of-two numbers. In Fig. 2 edges of CKs are shown by an arrow pointing towards the surviving vertex. The propagating distance i is shown by $\boxed{i}$ over an edge. By default all edges propagate distances by 1. Each surviving edge propagates the distance equal to 2^i into its adjacent unlabeled vertex. Next, to speed up the propagation of the distances with a power of two, the independent edges of a CK are identified by employing a logarithmic encoding. This logarithmic encoding indicates the priority of contractions through the construction of the pyramid. In Fig. 2a the numbers $1, 2, 3$ and $1', 2', 3'$ indicate the primary priorities that are different for each adjacent edge. The bottom-up construction of the pyramid (Fig. 2(a) to (d)) terminates when there is no edge remaining for the contraction. In top of the pyramid all surviving vertices have their distance values. At this stage the distances are propagated from top to down where the vertices with $DT(v) = \infty$ receive their distance from their adjacent vertices and adding the distance of an edge (Fig. 2(d) to (g)).

In order to correctly compute the GDT, each surviving vertex counts the number of contractions from its receptive field while this is not needed in computing the DT.

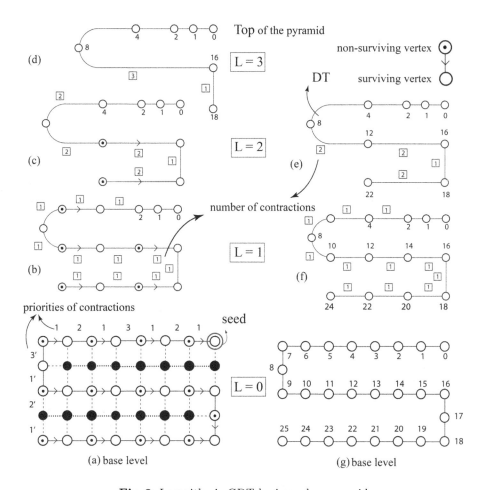

Fig. 2. Logarithmic GDT by irregular pyramid

Proposition 2. *Geodesic distance between two points in the higher dimension is always shorter or equal than in the lower dimension.*

Proof. Assume there is a distance between two points in the lower dimension that is shorter than distance between the same points in higher dimension. Since, every point in lower dimension is included in higher dimension, the shorter distance in lower dimension exists in the higher dimension as well which is in contradiction with the assumption.   □

3   Distance Transforms in n-Gmaps

An n-dimensional generalized map (n-Gmap) is a combinatorial data structure allowing to describe an n-dimensional orientable or non-orientable quasi-manifold with or without boundaries [12]. An n-Gmap is defined by a finite set of darts $\mathcal{D}$ on which act $n+1$ involutions[1] α_i, satisfying composition constraints of the following definition [7]:

Definition 2 (n-Gmap). *An n-dimensional generalized map, or n-Gmap, with $0 \leq n$ is an $(n+2)$-tuple $G = (\mathcal{D}, \alpha_0, ..., \alpha_n)$ where:*

1. *$\mathcal{D}$ is a finite set of darts,*
2. *$\forall i \in \{0, ..., n\}$: α_i is an involution on $\mathcal{D}$*
3. *$\forall i \in \{0, ..., n-2\}$, $\forall j \in \{i+2, ..., n\}$: $\alpha_i \circ \alpha_j$ is an involution.*

Let $(\mathcal{D}, \alpha_0, ..., \alpha_n)$ be an n-Gmap and let us consider its darts $d \in \mathcal{D}$ to be of a unit length. Similar to graphs, we first initialize the distance transform at any nonempty subset of *seed* darts $\mathcal{S} \subseteq \mathcal{D}$ as follows: $\delta(s) := 0 \ \forall s \in \mathcal{S}$ and $\delta(\bar{s}) := \infty$ $\forall \bar{s} \in \mathcal{D} \setminus \mathcal{S}$. Scenarios for the initialization (seeding) may include:

– single dart: $\mathcal{S} = \{d_0\}$,
– single i-cell: $\mathcal{S} = \{$all darts of the i-cell$\}$ (e.g., an edge), or
– any multi-combinations of the above, e.g., all edges (1-cells) connecting vertices of different labels resulting from segmentation or connected component labeling.

Similar to graphs, the distances are propagated from the seeds in the breath-first search. The difference to graphs, however, is that the propagation is more general and is driven along (some or all) *involutions* α_i rather than being restricted to the edges of the graph.

Figure 3b shows an example of a 2-Gmap – a 6×6 matrix of vertices (0-cells) of four labels A, B, C, and D where A and B have both two connected components. Edges $(d, \alpha_0(d), \alpha_2(d), \alpha_2(\alpha_0(d))$ connecting different labels[2] are initialized to 0 and distances are propagated following α_0, α_1, and α_2 involutions. Figure 3a illustrates the arrangement of darts around an implicit vertex (X).

The propagation of distances in Fig. 3b is performed equally in all dimensions, i.e., involving all involutions α_i. Excluding a fixed α_j, the propagation is constrained to manifolds of dimensions j. This makes the computation of the *geodesic* distance transforms on n-Gmaps viable.

---

[1] Self-inverse permutations.
[2] red separators in Fig. 3(b).

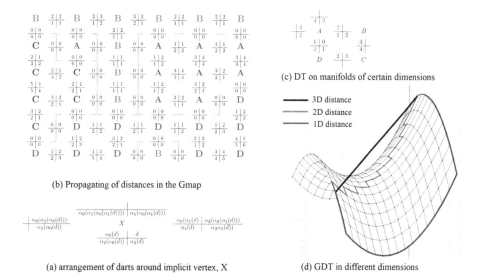

(b) Propagating of distances in the Gmap

(c) DT on manifolds of certain dimensions

— 3D distance
--- 2D distance
— 1D distance

(a) arrangement of darts around implicit vertex, X

(d) GDT in different dimensions

Fig. 3. DT in a Gmap (Color figure online)

We illustrate the effect by a simple 2D example (see Fig. 3c) where we initialize a single dart by zero and propagate distances only by pairs of involutions:

1. $\langle \alpha_0, \alpha_1 \rangle$ denotes the propagation[3] of the orbit $(\alpha_0^*, \alpha_1^*)^*(d_0)$ and identifies the (dual) 2-cell between A,B,C,D. α_2 does not propagate the distance.
2. $\langle \alpha_0, \alpha_2 \rangle$ denotes the propagation of $(\alpha_0^*, \alpha_2^*)^*(d_0)$ and identifies the 1-cell consisting of the four darts between A and D. In this case α_1 does not propagate the distance.
3. $\langle \alpha_1, \alpha_2 \rangle$ denotes the propagation of $(\alpha_1^*, \alpha_2^*)^*(d_0)$ and identifies the 0-cell (a point), the eight darts surrounding A. In this case α_0 does not propagate the distance.

Depending on the initialization and the choice of involutions, distances can thus be propagated along the *boundaries* of any i-cells, $i > 0$. For 3-Gmaps, in addition to 3-cells (volume elements), propagation of distances along their (2D) bounding surfaces or along (1D) curves bounding these surfaces becomes possible. Based on Proposition.2 the GDT in the higher dimension is shorter or equal than in lower dimension (Fig. 3d).

4   Results

As an example of the calculation of distance transforms on 2-Gmaps we refer to Fig. 4. The three black, zero-labeled pixels of the 4×5 image (Fig. 4a) are

---

[3] Blue distance values belong to the 2-cell, black distances to two types of cells (Fig. 3c).

used to seed the distance transform. Figure 4b represents the result of a graph-based distance transform where pixels correspond to vertices of the graph and its edges model the 4-connectivity of the image. The result of the 2-Gmap distance transform is displayed in Fig. 4c. Each pixel corresponds to eight darts which we choose to display by triangles colored by the minimum distance from the seeds. The axes-parallel and the diagonal lines between the triangles of one pixel correspond to α_0 and α_1, respectively. The axes-parallel *pixel-separating* lines correspond to α_2. The 3 seeds of Fig. 4a are represented by total of 24 black, zero-labeled triangles in Fig. 4c. It can be observed that in the 2-Gmaps the distances are propagated in a smoother and a more detailed way.

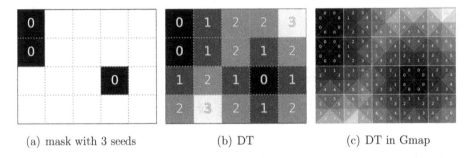

(a) mask with 3 seeds   (b) DT   (c) DT in Gmap

Fig. 4. Comparison of a graph-based (b) and Gmap-based (c) distance transforms of a binary image (a). Best viewed in color and magnified.

To exploit the advantage of the proposed method in a real application, several geodesic distance transforms (GDTs) are computed through a labeled 2D cross slice of a leaf scan (Fig. 5). The input image (Fig. 5a) has six different labels illustrating different regions inside the leaf. In this figure, the stomas act as gates to control the amount of CO_2 that is entering the leaf. The CO_2 propagates through the airspace to reach the cells and by combining with water and heat the photosynthesis takes place. To model various aspects of the photosynthesis, GDTs may aid in several ways. First, since we are interested in simulations of gas exchange in the leaf [13], we compute the GDT from the stomata through the airspace (Fig. 5b). This is intended to approximate how long it takes to reach the necessary CO_2 concentration. Second, bottlenecks of the airspace supposedly slow down the diffusion processes. We therefore compute the widths of the bottlenecks by the GDT inside the airspace seeded at ist boundary (Fig. 5c). Finally, the GDT from the stomata along the boundary of airspace is calculated (Fig. 5d). This is motivated by the observation that a longer boundary accommodates more cells to perform photosynthesis.

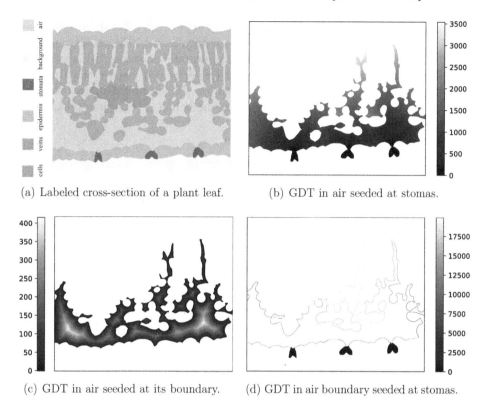

(a) Labeled cross-section of a plant leaf.   (b) GDT in air seeded at stomas.

(c) GDT in air seeded at its boundary.   (d) GDT in air boundary seeded at stomas.

Fig. 5. Computing GDT in a leaf.

It should be noted that the parallel logarithmic complexity of computing the GDT in the proposed method makes it useful for processing the big data. In our data-set each dimension of the 3D input image (leaf) is more than 2000 pixels. Therefore, fast computation of the DT with low complexity is required as shown in [2,3].

5   Conclusions

The paper presents a new algorithm to propagate distances in a graph which is based on a spanning forest of the foreground. The spanning forest is produced in parallel constant complexity and it reduces the linear search space to the length of the longest path in the spanning forest. By preserving the connectivity of connected components (CCs) and the topological information between CCs the proposed algorithm performs the connected component labeling (CCL) and the distance transform (DT) simultaneously. Using the hierarchical structure of the irregular pyramid the new method computes the geodesic distance transform (GDT) with parallel logarithmic complexity that makes it useful for processing of the big data.

We additionally introduce distance transforms for the generalized combinatorial maps (n-Gmaps). We show how they naturally result in a smoother and a higher resolution distance fields. More importantly, however, we show how geodesic distance transforms can efficiently be performed just by omitting relevant involutions from the distance propagation. Finally, we demonstrate how computing GDTs in n-Gmaps may support modelling of the gas exchange in plant leaves.

References

1. Banaeyan, M., Batavia, D., Kropatsch, W.G.: Removing redundancies in binary images. In: International Conference on Intelligent Systems and Patterns Recognition (ISPR), Hammamet, Tunisia, 24–25 March 2022. pp. 221–233. Springer, Cham (2022). https://doi.org/10.1007/978-3-031-08277-1_19

2. Banaeyan, M., Kropatsch, W.G.: Pyramidal connected component labeling by irregular graph pyramid. In: 2021 5th International Conference on Pattern Recognition and Image Analysis (IPRIA), pp. 1–5 (2021)

3. Banaeyan, M., Kropatsch, W.G.: Parallel $\mathcal{O}(log(n))$ computation of the adjacency of connected components. In: International Conference on Pattern Recognition and Artificial Intelligence (ICPRAI), Paris, France, June 1–3, 2022. pp. 102–113. Springer, Cham (2022).https://doi.org/10.1007/978-3-031-09282-4_9

4. Beamer, S., Asanovic, K., Patterson, D.: Direction-optimizing breadth-first search. In: SC '12: Proceedings of the International Conference on High Performance Computing, Networking, Storage and Analysis. pp. 1–10 (2012). https://doi.org/10.1109/SC.2012.50

5. Borgefors, G.: Distance transformations in arbitrary dimensions. Comput. Vis., Graphics image Process. **27**(3), 321–345 (1984)

6. Brun, L., Kropatsch, W.G.: Hierarchical graph encodings. In: Lézoray, O., Grady, L. (eds.) Image Processing and Analysis with Graphs: Theory and Practice, pp. 305–349. CRC Press (2012)

7. Damiand, G., Lienhardt, P.: Combinatorial Maps: Efficient Data Structures for Computer Graphics and Image Processing. CRC Press (2014)

8. Fabbri, R., Costa, L.D.F., Torelli, J.C., Bruno, O.M.: 2D Euclidean distance transform algorithms: a comparative survey. ACM Comput. Surv. **40**(1), 1–44 (2008)

9. Haxhimusa, Y.: The Structurally Optimal Dual Graph Pyramid and Its Application in Image Partitioning. DISKI, Berlin (2007)

10. Kropatsch, W.G.: Building irregular pyramids by dual graph contraction. IEE-Proc. Visi. Image Signal Process. **142**(No. 6), 366–374 (1995)

11. Latecki, L., Eckhardt, U., Rosenfeld, A.: Well-composed sets. Comput. Image Underst. **61**(1), 70–83 (1995)

12. Lienhardt, P.: Topological models for boundary representation: a comparison with n-dimensional generalized maps. Comput. Aided Des. **23**(1), 59–82 (1991)

13. Momayyezi, M., et al.: Desiccation of the leaf mesophyll and its implications for CO_2 diffusion and light processing. Plant, Cell Enviro. **45**(5), 1362–1381 (2022)

14. Pizlo, Z., Stefanov, E.: Solving large problems with a small working memory. J. Probll. Solv. **6**(1), 5 (2013)

15. Rosenfeld, A., Pfaltz, J.L.: Sequential operations in digital picture processing. Assoc. Comput. Macch. **13**(4), 471–494 (1966)

Retargeted Regression Methods for Multi-label Learning

Keigo Kimura[1][✉][iD], Jiaqi Bao[1][iD], Mineichi Kudo[1][iD], and Lu Sun[2][iD]

[1] Division of Computer Science and Information Technology, Graduate School of Information Science and Technology, Hokkaido University, Sapporo 060-0814, Japan
{kimura5,bao,mine}@ist.hokudai.ac.jp
[2] School of Information Science and Technology, ShanghaiTech University, Shanghai 201210, China
sunlu1@shanghaitech.edu.cn

Abstract. In Multi-Label Classification, utilizing label relationship is a key to improve classification accuracy. Label Space Dimension Reduction or Classifier Chains utilizes the relationship explicitly however those utilization are still limited. In this paper, we propose Retargeted Regression methods for Multi-Label classification by extending Retargeted Linear Least Squares originally proposed for Multi-Class Classification. Retargeted methods not only learn classifiers but also modify targets with margin constraints. Since in Multi-Label Classification, an instance may have more than one label, large margin constraints between all pairs of positive labels and negative labels are introduced. This enables to utilize the label relationship with taking ranking of labels for each instance into consideration. We also propose a simple heuristic to determine a threshold parameter for each instance to earn zero-one classification. On nine benchmark datasets, the proposed method outperformed conventional methods in the sense of instance-wise ranking. In best cases, classification accuracy was improved at 7% on AUC metric.

1 Introduction

Multi-Label Classification is a classification problem where an instance can belong to more than one class at the same time. This problem setting is suitable for several real applications such as document classifications or image annotations [1,3,9,11]. The key challenge on Multi-Label Classification is how to incorporate label relationships to improve the classification accuracy [8,10,11]. Label Space Dimension Reduction (LSDR) methods [2,4,6] are methods to utilize label relationship explicitly. LSDR assumes low-rank structures on the label matrix and utilizes that through a low-rank decomposition. LSDR methods can improve the classification accuracy and many decomposition strategies have been investigated, however, the improvement is limited since the assumption of low-rank structure on the label matrix is too strong [1]. Classifier chain (CC) proposed by Read is another method to utilize the label relationship directly. CC constructs a Directional Acyclic Graph (DAG) over labels and then learns classifiers sequentially according to the DAG. CC incorporates previous classification results on parent labels into feature vectors in turn. Thus, CC can directly utilize the

A. Krzyzak et al. (Eds.): S+SSPR 2022, LNCS 13813, pp. 203–212, 2022.
https://doi.org/10.1007/978-3-031-23028-8_21

label relationship. However, CC cannot utilize mutual interactions between classifiers since it uses a DAG and thus the utilization is still limited.

In this paper, we extend Retargeted Least Squares Regression (ReLSR) proposed by Zhang *et al.* [13] so as be applicable to Multi-Label Classification problems by utilizing label relationships and proposing Retargeted Multilabel Least Squares Regression (ReMLSR). ReMLSR learns not only linear classifiers but also targets, modified label vectors, at the same time as ReLSR does. In Multi-Class Classification, an instance has only one label, and thus, one-vs-all margin constraints have been used to guarantee correct classification. On the other hand, in Multi-Label Classification, an instance may have more than one label thus we extend its margin constraint into *positive label set vs. negative label set* margin constraint. Specifically, ReMLSR requires a large margin between all positive labels and all negative labels to guarantee correct classification. Those margin constraints and the target modification allows to take relative scores of labels into consideration and learn classifiers flexibly. In addition, we also propose a simple heuristic to determine a threshold for each test instance to obtain a zero-one vector prediction. It is necessary because ReMLSR modifies targets and a constant value of threshold such as 0.5 is not the midpoint of positive labels and negative labels anymore.

2 Proposal: Retargeted Multi-label Least Square Regression

2.1 Notations

We use $\mathbf{X}$ for a matrix, $\boldsymbol{x}$ for a vector and x for a scalar. In a matrix $\mathbf{X}_{ij}$ stands for ijth element of $\mathbf{X}$. In a vector, $\boldsymbol{x}_i$ stands for the ith element of $\boldsymbol{x}$ as well. The task of Multi-Label Classification is learning a relationship between F-dimensional feature vector $\boldsymbol{x} \in \mathbb{R}^F$ and an L-dimensional vector $\boldsymbol{y} \in \{0,1\}^L$. In this paper, we includ an intercept into Fth element of the feature vector $\boldsymbol{x}_F = 1$. We have N pairs of feature and label vectors $D = \{(\boldsymbol{x}^{(1)}, \boldsymbol{y}^{(1)}), (\boldsymbol{x}^{(2)}, \boldsymbol{y}^{(2)}), \ldots, (\boldsymbol{x}^{(N)}, \boldsymbol{y}^{(N)})\}$ as a training dataset and we use $\mathbf{X} = [\boldsymbol{x}^{(1)}, \boldsymbol{x}^{(2)}, \ldots, \boldsymbol{x}^{(N)}]^T \in \mathbb{R}^{N \times F}$ called feature matrix and $\mathbf{Y} = [\boldsymbol{y}^{(1)}, \boldsymbol{y}^{(2)}, \ldots, \boldsymbol{y}^{(N)}]^T \in \{0,1\}^{N \times L}$ called label matrix for simplicity. More specifically say, in this paper, the task of multi-label classification is a learning regression matrix $\mathbf{W} \in \mathbb{R}^{F \times L}$ for prediction of a test instance as $\hat{\boldsymbol{y}}^{(test)} = \mathbf{W}^T \boldsymbol{x}^{(test)}$.

2.2 Brief Review of ReLSR

Retargeted Least Square Regression (ReLSR) has been proposed for Multi-Class Classification by Zhang *et al.* [13]. The core idea of this method is not only to learn a weight matrix of the regression but also to modify the target matrix at the same time. For the correct classification with a consistent class decision $\hat{c} = argmax_i \hat{y}_i$, ReLSR introduces an instance-wise large margin constraint between the correct class and other classes. Thus, the optimization problem of ReLSR is:

$$\min_{\mathbf{W},\mathbf{T}} \|\mathbf{T} - \mathbf{XW}\|_F^2 + \lambda\|\mathbf{W}\|_F^2, \tag{1}$$

$$\text{subject to} \quad \mathbf{T}_{i,c^{(i)}} - \max_{j \neq c^{(i)}} \mathbf{T}_{i,j} \geq 1, \ i = 1, \ldots, N, \tag{2}$$

where $c^{(i)}$ is the correct class index of the ith instance. Note that ReLSR modifies Y to be the retarget matrix $\mathbf{T}$ with the margin constraints (2). This constraint and target matrix modification brings flexibility into the class separability for each instance.

2.3 Problem Formulation

We extend ReLSR to Retargeted Multi-Label Least Square Regression (ReMLSR) that is able to deal with Multi-Label Classification problems. Since each instance has only one label in Multi-Class Classification, the constraint of ReLSR only need to consider one-vs-all margin (2). However, in Multi-Label Classification, an instance can have more than one labels, and thus, it is necessary to consider all pairs of positive classes and negative classes. The objective function of ReMLSR can be written as follows:

$$\min_{\mathbf{W},\mathbf{T}} \|\mathbf{T} - \mathbf{X}\mathbf{W}\|_F^2 + \lambda\|\mathbf{W}\|_F^2, \tag{3}$$

$$\text{subject to} \quad \min_{j:y_j^{(i)}=1} \mathbf{T}_{ij} - \max_{k:y_k^{(i)}=0} \mathbf{T}_{ik} \geq 1 - \xi,\ i = 1,\dots,N, \tag{4}$$

where ξ is a hyperparameter we introduced to add flexibility. This margin constraint with ξ is slightly different from the soft margin constraint on SVM [5]. In ReMLSR, its margin width is fixed as $(1 - \xi)$ for all instances and this ξ is a hyperparameter. We observed a large margin is sometimes too strict and harms the classification accuracy. We will report the effects of this hyperparameter on the experiments.

It is easy to see that ReLSR is a special case of ReMLSR when each instance has only one label. However, this ReMLSR has another problem to be solved, a threshold to separate positive and negative labels in the prediction $\hat{\boldsymbol{y}}^{(test)} = \mathbf{W}^T\boldsymbol{x}_{test}$. In Multi-Class Classification, the label that earned the largest score can be considered as a prediction result, however, in Multi-Label Classification, the number of positive labels is not given. In a simple linear regression on multi-label classification, a constant value threshold such as 0.5 is introduced to earn a zero-one classification prediction. However, this does not work with ReMLSR because the target $\mathbf{T}$ is modified. To avoid this problem we propose a simple heuristic to determine the value of threshold and describe the detail later.

2.4 Optimization

Since the problem (3) is not convex for $\mathbf{W}$ and $\mathbf{T}$ at the same time, therefore, we optimizes the problem (3) by alternating optimization which conducts following steps until convergence:

1. Optimize the regression matrix $\mathbf{W}$ (with fixing the target matrix $\mathbf{T}$),
2. Optimize the target matrix $\mathbf{T}$ (with fixing the target matrix $\mathbf{W}$).

Here, we describe each optimization, respectively.

Optimize the Regression Matrix W. When the target matrix $\mathbf{T}$ is fixed, the constraint is kept satisfied and ignorable, thus, the optimization of $\mathbf{W}$ is the same to that of the original regression method. In our formulation, we use Ridge Regression for the regression. The optimal solution of $\mathbf{W}$ can be calculated as:

$$\mathbf{W}^* = (\mathbf{X}^T\mathbf{X} + \lambda\mathbf{I})^{-1}\mathbf{X}^T\mathbf{T}. \tag{5}$$

Optimize the Target Matrix T. In this optimization, regression matrix $\mathbf{W}$ is fixed and thus we use $\mathbf{R} = \mathbf{W}\mathbf{X}$ for simplicity. Since constraint (4) is row-wisely (instance-wisely) independent, we can decompose the optimization problem into a set of row-wise optimization problems. With row vectors $t^{(i)} = [\mathbf{T}_{i,1}, \mathbf{T}_{i,2}, \ldots, \mathbf{T}_{i,L}]$ and $r^{(i)} = [\mathbf{R}_{i,1}, \mathbf{R}_{i,2}, \ldots, \mathbf{R}_{i,L}]$, the optimization problem for ith row can be re-written as:

$$\min_{t^{(i)}} \|t^{(i)} - r^{(i)}\|_F^2 = \sum_l (t_l^{(i)} - r_l^{(i)})^2 \quad \text{s.t.} \quad \min_{j:y_j^{(i)}=1} t_j - \max_{k:y_k^{(i)}=0} t_k \geq 1 - \xi. \tag{6}$$

Hereafter, we omit the index i for r, t and y for simplicity.

The objective function (6) seems to need to take all constraints on all positive and negative label pair combinations into consideration, however, this can be easily solved by introducing a margin midpoint α defined as:

$$\alpha = (\min(t_{j:y_j=1}) + \max(t_{k:y_k=0}))/2. \tag{7}$$

The margin constraint in (6) requires a $(1-\xi)$-width margin between targets for all positive labels and those for all negative labels. With the margin midpoint α, this constraint can be considered to ask targets for all positive labels to be at least $(1-\xi)/2$ larger than α and those for all negative labels at least $(1-\xi)/2$ smaller than α. Therefore, with the midpoint α, it suffices to set targets for all positive labels as follows:

$$t_{j:y_j=1} = \begin{cases} r_j & r_j > \alpha + \frac{1-\xi}{2}, \\ \alpha + \frac{1-\xi}{2} & r_j \leq \alpha + \frac{1-\xi}{2}. \end{cases} \tag{8}$$

If r_j is larger than the top of margin $(\alpha + \frac{1-\xi}{2})$, the optimal solution of (5) is $t_j = r_j$. On the other hand, if r_j is smaller than that, then t_j must be $\alpha + \frac{1-\xi}{2}$ to satisfy the margin constraint. Similarly, targets for all negative labels are set as:

$$t_{k:y_k=0} = \begin{cases} \alpha - \frac{1-\xi}{2} & r_k \geq \alpha - \frac{1-\xi}{2}, \\ t_k & r_k < \alpha - \frac{1-\xi}{2}. \end{cases} \tag{9}$$

From (8) and (9), it is obvious that all values of the target vector t depend on the margin midpoint α. Therefore, it is sufficient to learn α to minimize (6). We can re-write objective function (6) with (8) and (9) as an objective function of the margin midpoint α:

$$\min_{\alpha} \sum_{j:y_j=1} \max(0, (\alpha + \frac{1-\xi}{2}) - r_j)^2 + \sum_{k:y_k=0} \max(0, r_k - (\alpha - \frac{1-\xi}{2}))^2. \tag{10}$$

Algorithm 1 Retargeted Multi-Label Least Square Regression

1: **Input:** Label matrix $\mathbf{Y}$, Feature matrix $\mathbf{X}$ and Parameters λ and ξ;
2: **Output:** Regression matrix $\mathbf{W}$, Target matrix $\mathbf{T}$;
3: Initialize $\mathbf{T} = \mathbf{Y}$
4: **while** until convergence **do**
5:    Learn $\mathbf{W}$ with fixed $\mathbf{T}$ as the label matrix by an LSDR method.
6:    Learn $\mathbf{T}$ row-wisely with fixed $\mathbf{W}$ by Algorithm 2.
7: **end while**

Algorithm 2 Learning Target Vector

1: **Input:** Reconstruction vector r, Target vector t, Parameter ξ;
2: **Output:** Target vector t;
3: Initialize $\alpha = (\min(\mathbf{r}_{j:\mathbf{y}_j=1}) + \max(\mathbf{r}_{k:\mathbf{y}_k=0}))/2$;
4: Learn α by minimizing (10) with a gradient decent;
5: Update t with learned α as $t_j = \begin{cases} \max(\mathbf{r}_j, \alpha + \frac{(1-\xi)}{2}) & \text{if } \mathbf{y}_j = 1, \\ \min(\mathbf{r}_j, \alpha - \frac{(1-\xi)}{2}) & \text{otherwise.} \end{cases}$

This objective function with α is convex and thus we can adopt a gradient descent method to find an optimal value of α. Once we learned α, optimal targets t are calculated by (8) and (9) for positive labels and negative labels, respectively.

In each iteration, the value of objective function of ReMLSR (3) is equal to the sum of row-wise objective function (10). Therefore, ReMLSR is minimizing squared hinge loss formed in (10). Note that ReMLSR is not equivalent to regressions with the squared hinge loss function. This is because this squared hinge loss function is across labels, not across instances. This is the difference between ReMLSR and other conventional regression-based methods.

In the sense of considering orders of labels in the consrtaint (4), the proposed ReMLSR is close to ranking SVM proposed by Elisseeff and Weston [5]. Ranking SVM optimizes instance-wise ranking with margin constraints of all pairs of positive and negative labels. Their constraint and our constraint have the same form. However, ReMLSR is slightly different from ranking SVM since ReMLSR uses a fixed width margin and does not maximize the margin. In addition, ReMLSR can use any regression method including kernel regressions or LSDR for multi-label classification [4]. On the other hand, ranking SVM with LSDR or other regression methods is not trivial.

The pseudo-code of ReMLSR algorithm is described in Algorithm 1 and Algorithm 2. The proposed algorithm converges to a local minimum since each step in alternating optimization minimizes the value of the objective function. We omit the proof due to the space limitation.

2.5 Computational Complexity

The complexity of learning regression matrix $\mathbf{W}$ depends on the regression to be used. For example, with ridge regression, $O(N^3)$ is required due to the inverse matrix calculation (5). On the other hand, the complexity of learning target matrix requires $O(L)$ for each instance and thus $O(NL)$ is required. As we can see, the complexity of learning

Table 1. Datasets used in experiments

Dataset	# Instances	#Labels	#Features	Cardinality	Distinct
bibtex	7395	159	1836	2.402	2856
CAL500	502	174	68	26.044	502
corel5k	5000	374	499	3.522	3175
emotions	593	6	72	1.869	27
enron	1702	53	1001	3.378	753
medical	978	45	1449	1.245	94
scene	2407	6	294	1.074	15
tmc2007	28596	22	500	1.074	1341
yeast	2417	14	103	4.237	198

target matrix is not large compared to that of learning of regression matrix. In total, the proposed method requires $O((N^3 + NL)\ell)$ with the number of loop ℓ. Unfortunately, the proposed ReMLSR requires to learn the regression matrix ℓ times and this is relatively large. However, some regression methods can reduce the cost while keeping some statistics in the loop. For example, on ridge regression, $(\mathbf{X}^T\mathbf{X} + \lambda\mathbf{I})^{-1}\mathbf{X}^T$ in (5) is fixed in the loop and thus if we keep this matrix, the cost of learning can be reduced.

2.6 Learning Threshold

Since the number of labels is not given for each test instance, we cannot determine which labels are positive or negative without a threshold. In traditional regression methods, this threshold parameter can be fixed to 0.5 since the target matrix is a zero-one binary matrix. However, as we mentioned before, ReMLSR is not the case because the target matrix is optimized and not a zero-one matrix anymore.

In this paper, to address with this issue, we propose a simple heuristic that uses k-Nearest Neighbors. For training instances, the midpoint of margin α (learned for each instance) is the best threshold to determine positive labels and negative labels. We utilize this information and learn $s^{(test)}$ for a test instance $x^{(test)}$ by k-Nearest Neighbors:

$$s^{(test)} = \frac{1}{K} \sum_{i \in \mathcal{K}^{(test)}} \alpha^{(i)}, \qquad (11)$$

where $\mathcal{K}^{(test)}$ is the set of k nearest neighbors of $x^{(test)}$ in training instances. Note that, here, we use $\alpha^{(i)}$ for ith instance is learned by Algorithm 2.

3 Experiments

3.1 Dataset and Evaluation Measurement

We conducted experiments on nine benchmark datasets summarized in Table 1. All datasets can be downloaded from Mulan [9]. To evaluate classification performances, we used Instance-AUC, Macro-F1, Micro-F1 and Hamming-score defined as follows:

$$\text{Instance-AUC} = \frac{1}{N}\sum_{i=1}^{N}\frac{\sum_{j:\mathbf{Y}_{ij}=1}\sum_{k:\mathbf{Y}_{ik}=0}\left[\!\left[\hat{\mathbf{Y}}_{ij} \geq \hat{\mathbf{Y}}_{ik}\right]\!\right]}{\sum_{j=1}^{L}[\![\mathbf{Y}_{ij}=1]\!]\cdot\sum_{k=1}^{L}[\![\mathbf{Y}_{ik}=0]\!]},$$

$$\text{Macro-F1} = \frac{1}{L}\sum_{j=1}^{L}\frac{2\sum_{i=1}^{N}\hat{\mathbf{Y}}_{ij}\cdot\mathbf{Y}_{ij}}{\sum_{i=1}^{N}\hat{\mathbf{Y}}_{ij}+\sum_{i=1}^{N}\mathbf{Y}_{ij}},$$

$$\text{Micro-F1} = \frac{2\sum_{j=1}^{L}\sum_{i=1}^{N}\hat{\mathbf{Y}}_{ij}\cdot\mathbf{Y}_{ij}}{\sum_{j=1}^{L}\sum_{i=1}^{N}\hat{\mathbf{Y}}_{ij}+\sum_{j=1}^{L}\sum_{i=1}^{N}\mathbf{Y}_{ij}},$$

$$\text{Hamming-score} = \frac{1}{NL}\sum_{j=1}^{L}\sum_{i=1}^{N}\left[\!\left[\hat{\mathbf{Y}}_{ij}=\mathbf{Y}_{ij}\right]\!\right],$$

where N is the number of instances, L is the number of labels, $\hat{\mathbf{Y}}$ is the predicted label matrix, $\mathbf{Y}$ is the ground-truth label matrix and $[\![\cdot]\!]$ is an indicator function that returns 1 if the statement in $[\![\cdot]\!]$ is true, otherwise 0. On all measurements, the higher score is, the better classification performance is.

3.2 Settings

We compared the following conventional linear classifiers:

- Binary Relevance (**BR**) [12]
- Conditional Principle Label Space Transformation (**CPLST**) [4]
- Classifier Chain (**CC**) [7]
- Retargeted Multi-Label Least Squares Regression (**ReMLSR**)

We used Ridge Regression as base learners for all compared methods. We conducted a five-fold cross validation and reported the average of them.

In parameter setting, the regularization parameter for ridge regression λ was tuned in the range of $[0.01, 0.1, 1, 10]$. For **CPLST**, the number of low-rank dimension is fixed as $0.8L$ for each dataset. We used a random single-path classifier chain for **CC**. In **ReMLSR**, there are two parameters, the margin constraint parameter ξ, the number k of nearest neighbors to determine the threshold to earn a zero-one classification prediction. We tuned the margin constraint parameter ξ from 0 to 0.9. We fixed the number of iterations as 30 according to a pre-experiment. The number of neighbors in the proposed heuristic was set to $k = 3$. We will report parameter sensitivity analyses later.

3.3 Results

First, we evaluated the instance-wise ranking. Table 2 shows that the results of Instance-AUC. The proposed **ReMLSR** outperformed conventional methods on six datasets. Even on other three datasets, **ReMLSR** was competitive. This shows how the proposed **ReMLSR** succeeded to learn the instance-wise ranking correctly. Especially, on *corel5k* and *bibtex*, significant improvements have been observed (more than 7%).

Table 2. Results: ROC-AUC

	bibtex	CAL500	corel5k	emotions	enron	medical	scene	tmc2007	yeast
BR	0.9008	0.8069	0.8171	0.8447	0.8754	0.9845	0.9249	0.9548	0.8359
CPLST	0.8891	0.7756	0.7981	0.8405	0.7566	0.9565	0.9065	0.9535	0.8294
CC	0.9017	**0.8074**	0.8032	0.8462	0.8737	0.9847	**0.9276**	0.9550	**0.8360**
ReMLSR	**0.9337**	0.8061	**0.8768**	**0.8472**	**0.9075**	**0.9849**	0.9237	**0.9580**	0.8353

Table 3. Results: Macro-F1

	bibtex	CAL500	corel5k	emotions	enron	medical	scene	tmc2007	yeast
BR	0.1806	0.0671	0.0099	0.6165	0.1635	0.2807	0.6317	0.4725	0.3261
CPLST	0.2040	**0.0940**	0.0157	0.6262	0.1616	**0.3672**	0.6428	0.4637	**0.3547**
CC	0.1784	0.0669	0.0085	0.6133	0.1619	0.2808	0.6172	0.4720	0.3253
ReMLSR	**0.3525**	0.0687	**0.0281**	**0.6329**	**0.2236**	0.2853	**0.6920**	**0.6142**	0.3404

Table 4. Results: Micro-F1

	bibtex	CAL500	corel5k	emotions	enron	medical	scene	tmc2007	yeast
BR	0.3728	0.3330	0.0738	0.6360	0.5440	0.7958	0.6323	0.6642	0.6273
CPLST	0.3926	**0.3554**	0.1072	0.6438	0.4121	0.7510	0.6393	0.6636	**0.6357**
CC	0.3700	0.3339	0.0579	0.6340	0.5429	0.7958	0.6191	0.6645	0.6271
ReMLSR	**0.3961**	0.3363	**0.1677**	**0.6497**	**0.5664**	**0.8017**	**0.6915**	**0.7190**	0.6314

Table 5. Results: Hamming-score

	bibtex	CAL500	corel5k	emotions	enron	medical	scene	tmc2007	yeast
BR	**0.9878**	0.8622	0.9906	0.7951	0.9493	0.9893	0.8940	0.9401	0.8005
CPLST	0.9877	0.8562	0.9906	0.7954	0.9097	0.9855	0.8881	0.9400	0.7998
CC	0.9877	**0.8624**	0.9906	0.7959	**0.9497**	0.9893	0.8931	0.9401	0.8006
ReMLSR	0.9720	0.8620	**0.9903**	**0.7963**	0.9346	**0.9898**	**0.8970**	**0.9423**	**0.8008**

Next, we evaluated the zero-one classification accuracy. Table 3, 4 and 5 shows the results of Macro-F1, Micro-F1 and Hamming-Score, respectively. The proposed **ReMLSR** also outperformed on six datasets in MacroF1, on seven datasets in Micro-F1, on six datasets in Hamming-score. Those results show that the proposed simple heuristic is enough for determining threshold parameters. It is noteworthy that **ReMLSR** outperformed other methods on four datasets in all three metrics.

We conducted parameter sensitivity analyses on **ReMLSR**. Figure 1 (a) shows that the result where the margin parameter ξ was varied from 0 to 0.9. As we can see, on some datasets, AUC performance was better with a smaller margin $(1 - \xi)$. This is probably because the large margin constraint is too strict on those datasets. On the other hand, on some datasets, a smaller margin damaged AUC performance. This is also because **ReMLSR** loses the generalization ability with the smaller margin and thus

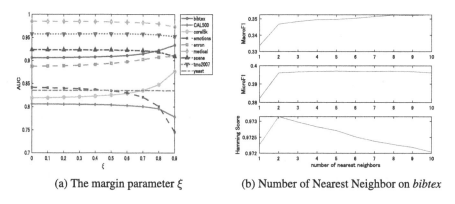

(a) The margin parameter ξ (b) Number of Nearest Neighbor on *bibtex*

Fig. 1. Parameter analysis

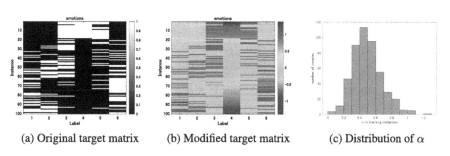

(a) Original target matrix   (b) Modified target matrix   (c) Distribution of α

Fig. 2. Example of how **ReMLSR** modifies the taget matrix on *emotions*. We sampled a hundred instances and sorted rows by modified values of the fourth label.

failed the classification. From those observations, **ReMLSR** requires careful tuning for that parameter. Figure 1 (b) shows the results where the number of nearest neighbors on the proposed heuristic from 1 to 10 on *bibtex* dataset. We can see that that parameter does not affect the result significantly. On Hamming-score, the difference is about 0.001.

Figure 2 (a) and (b) show an example of the original target matrix and the modified target matrix on *emotions* by **ReMLSR**, respectively. We sampled a hundred instances and sorted them by the modified values of the fourth label for both matrices to see the result clearly. **ReMLSR** modifies the target values for each instance with the margin constraint. Thus, the modified values were not uniform. This makes the regression enables to learn instance-wisely accurate classifiers. Figure 2 (c) shows the midpoint of the margin α learned by Algorithm 2. This α can be seen as a decision threshold to separate positive labels and negative labels. On Ridge regression without retargeting, the value of α is 0.5 for all instances. On the other hand, on the proposed **ReMLSR**, this is midpoint α is also modified while learning for each instance. This brings the flexibility to learn class separation and ranking of labels for each instance.

4 Conclusion

In this paper, we extended Retargeted Least Squares method (ReLSR) and proposed Retargeted Multi-label Least Squares method (ReMLSR) for Multi-Label Classification. ReMLSR optimizes not only the regression matrix but also the target matrix with a large margin constraints between all positive labels and negative labels for correct classifications. This enables to utilize label relationship and learn classifier flexibly with class separability for each instance. Since ReMLSR modifies the target matrix thus there is no explicit threshold values to earn zero-one prediction. To avoid this, we also proposed a simple heuristic to determine a threshold for each instance. Experiments performed on nine benchmark datasets demonstrated that the proposed ReMLSR has advantages in the sense of ranking compared to conventional methods.

Acknowledgment. This work was partially supported by JSPS KAKENHI (Grant Number 19H04128).

References

1. Bhatia, K., Jain, H., Kar, P., Varma, M., Jain, P.: Sparse local embeddings for extreme multi-label classification. In: Advances in Neural Information Processing Systems 28 (2015)
2. Bi, W., Kwok, J.: Efficient multi-label classification with many labels. In: International Conference on Machine Learning, pp. 405–413. PMLR (2013)
3. Boutell, M.R., Luo, J., Shen, X., Brown, C.M.: Learning multi-label scene classification. Pattern Recogn. **37**, 1757–1771 (2004)
4. Chen, Y.N., Lin, H.T.: Feature-aware label space dimension reduction for multi-label classification. In: Advances in Neural Information Processing Systems 25 (2012)
5. Elisseeff, A., Weston, J.: A kernel method for multi-labelled classification. In: Advances in Neural Information Processing Systems 14 (2001)
6. Lin, Z., Ding, G., Hu, M., Wang, J.: Multi-label classification via feature-aware implicit label space encoding. In: International Conference on Machine Learning, pp. 325–333. PMLR (2014)
7. Read, J., Pfahringer, B., Holmes, G., Frank, E.: Classifier chains for multi-label classification. Mach. Learn. **85**, 333–359 (2011)
8. Sorower, M.S.: A literature survey on algorithms for multi-label learning. Oregon State University, Corvallis, vol. 18, pp. 1–25 (2010)
9. Tsoumakas, G., Spyromitros-Xioufis, E., Vilcek, J., Vlahavas, I.: Mulan: a java library for multi-label learning. J. Mach. Learn. Res. **12**, 2411–2414 (2011)
10. Tai, F., Lin, H.T.: Multilabel classification with principal label space transformation. Neural Comput. **24**, 2508–2542 (2012)
11. Tsoumakas, G., Katakis, I.: Multi-label classification: an overview. Int. J. Data Warehous. Min. (IJDWM) **3**, 1–13 (2007)
12. Tsoumakas, G., Katakis, I., Vlahavas, I.: Random k-labelsets for multilabel classification. IEEE Trans. Knowl. Data Eng. **23**, 1079–1089 (2010)
13. Zhang, X.Y., Wang, L., Xiang, S., Liu, C.L.: Retargeted least squares regression algorithm. IEEE Trans. Neural Netw. Learn. Syst. **26**(9), 2206–2213 (2014)

Transformer with Spatio-Temporal Representation for Video Anomaly Detection

Xiaohu Sun⬧, Jinyi Chen⬧, Xulin Shen⬧, and Hongjun Li$^{(\boxtimes)}$⬧

School of Information Science and Technology, Nantong University, Nantong 226019, China
lihongjun@ntu.edu.cn

Abstract. With the popularity of smart surveillance devices and the increase of people's security awareness, video anomaly detection has become an important task. However, learning rich multi-scale spatio-temporal information from high-dimensional videos to predict anomalous behaviors is a challenging task due to the large local redundancy and complex global dependencies among video frames. Although Convolutional Neural Network (CNN) has extraordinary bias induction capabilities, their inherent localization limitations lead to their lack of ability to capture long-term spatio-temporal features. Therefore, we propose a Transformer with spatio-temporal representation for video anomaly detection. The network combines the convolution operation with the Transformer operation, and uses the convolution operation to extract shallow spatial features to facilitate the recovery of sampled images. At the same time, Transformer operation is used to encode patches and efficiently capture remote dependencies through a self-attention mechanism, and to reduce the limitations in local redundancy. Experimental results on the UCSD Ped2, CUHK Avenue and ShanghaiTech datasets demonstrate the effectiveness of the proposed network.

Keywords: Video anomaly detection · Transformer · Convolutional Neural Network

1 Introduction

With the development of intelligent video surveillance [1], the research on capturing sequences of interest from video and performing automatic detection has been greatly facilitated. Among the many research objectives, video anomaly detection [2, 3] plays an important role in monitoring video and activity recognition, among others. The high dimensionality of video data, the rarity, ambiguity, and high contextual relevance of unusual events lead to extremely challenging modeling of video anomaly detection.

With the great success of CNN in video anomaly detection. U-Net [4] based encoder-decoder networks are widely used for video anomaly detection and achieve better prediction results. Despite the extraordinary representational capabilities of CNN, each of their convolutional kernels focuses only on feature information about itself and its boundaries, lacking a large range of feature fusion, especially for target structures that exhibit large differences in texture, shape, and size, an inherent limitation that leads to a

© The Author(s), under exclusive license to Springer Nature Switzerland AG 2022
A. Krzyzak et al. (Eds.): S+SSPR 2022, LNCS 13813, pp. 213–222, 2022.
https://doi.org/10.1007/978-3-031-23028-8_22

lack of poor performance in modeling remote relationships. To overcome this drawback, existing studies have proposed some self-attention mechanisms based on CNN methods [5, 6], but these methods ignore the feature dependencies between temporal sequences.

In recent years, Transformer [7] has demonstrated powerful modeling capabilities for sequential data, and there has been a lot of work applying Transformer to the field of computer vision. Unlike CNN-based approaches, Transformer is not only powerful in global context modeling, but also shows superior transferability to downstream tasks under large-scale pre-training. For example, Dosovitskiy et al. [8] proposed the Vision Transformer technique for image recognition tasks.

In this paper, we propose the Transformer architecture with spatio-temporal representation. It combines the learning ability of U-Net network for different scale features and Transformer architecture for long time series features and applies it to video anomaly detection. The network uses convolutional operations to obtain feature maps, and then encodes patches of high-resolution images using Transformer, which strengthens the global connection between the encoding parts, and Transformer's self-attention mechanism is more conducive to detailed feature extraction. Finally, the encoding part is combined with the decoding part to achieve the prediction of future video frames.

2 Related Work

2.1 Video Anomaly Detection

With the development of computer vision and machine learning, many methods [9, 10] for video anomaly detection have made remarkable progress. Currently, video anomaly detection methods are divided into two categories: reconstruction-based methods [11, 12] and prediction-based methods [13, 14]. Among them, reconstruction-based methods assume that models learned only on normal data cannot accurately reconstruct anomalies, and thus anomalous samples will have large reconstruction errors based on this distribution. For example, Luo et al. [11] combined Convolutional Long Short-Term Memory (ConvLSTM) with a convolutional autoencoder to reconstruct the input sequence for better detection performance. In addition, Ravanbakhsh et al. [12] used the network structure of U-Net to achieve cross-modal reconstruction of video frames and optical flow maps. Unlike the reconstruction-based methods, the prediction-based methods focus on the prediction of the next frame by using the content of the previous frame and the current frame. Liu et al. [13] used a U-Net based approach to predict future frames and used the U-shaped structure of U-Net to obtain rich multi-scale feature information. Villegas et al. [14] used an LSTM-based motion encoder for all historical motion encoding to make predictions about future frames. Since anomaly detection is the prediction of events that do not match expectations, prediction-based methods is more natural.

2.2 Transformer

Transformer was first proposed by [7] for solving Natural Language Processing (NLP) tasks. Due to its ability to model long-term features of sequences, it can make up for some of the shortcomings of CNN in terms of global features. To make the Transformer applicable to computer vision tasks as well, many studies have started to combine Transformer structures with convolutional operations to obtain better performance. For example, Dosovitskiy et al. [8] implemented the classification of ImageNet images using Transformer with global self-attentiveness. Arnab et al. [15] explored the application of Video Vision Transformer in video classification. Considering Transformer's excellent learning ability on global temporal features and U-Net's advantage of bias induction on different scale feature maps. Therefore, we combine Transformer with U-Net to make it suitable for video anomaly prediction tasks.

3   Methodology

The overall architecture of our proposed Transformer with spatio-temporal representation for video anomaly detection is shown in Fig. 1. We input successive t frames sequentially into the encoder for feature extraction and into the context module to achieve multi-scale feature extraction as well as modeling of temporal information. Compared with U-Net, this network combines the Transformer mechanism with convolution in the encoding part, which retains the local bias induction ability of convolution and introduces the powerful modeling ability of Transformer in global information, and achieves better temporal feature learning using the self-attentive mechanism. During decoding, the encoded feature maps and decoded feature maps at the same feature scale are combined using skip connections, which preserves the detailed information, and finally outputs a frame as the prediction result of frame $t + 1$. The real frame $t + 1$ is used as Ground Truth to complete the prediction task.

Image Encoding. In order to preserve the texture, shape and other features of the image, we perform convolutional coding before Transformer coding. Suppose the input image $\mathbf{I}_{in} \in \mathbf{R}^{H \times W \times C}$ with spatial resolution $H \times W$ and the number of channels C. For the convolutional encoder, the image features after convolutional encoding can be expressed as $\mathbf{I}_{fea}$, as shown in Eq. (1).

$$\mathbf{I}_{fea} = \sigma(\text{BN}(\text{Conv}(\mathbf{I}_{in}))) \tag{1}$$

where σ denotes the ReLu function, BN denotes the batch normalization, and Conv denotes the convolution function. For each convolution, a maximum pooling operation is used to reduce the feature map size by a factor of two to obtain $\mathbf{I}_{max}$.

$$\mathbf{I}_{max} = \text{MaxPool}(\mathbf{I}_{fea}) \tag{2}$$

where MaxPool denotes the maximum pooling function. The purpose of the convolution operation is to preserve the low-level feature maps so that the shallow features after convolutional encoding can be effectively combined with the decoded feature maps after skip connection.

Linear Projection. We first reshape the feature map $\mathbf{I}_{max}$ into 2D patches $\left\{ \mathbf{x}_p^i \in \mathbf{R}^{P^2 \cdot C'} \middle| i = 1, \dots, N' \right\}$ of size $P \times P$ without overlapping each other, where $N' = H'W'/P^2$. Then we join the N' reshaped patch vectors together to obtain the $N' \times P^2$ matrix, and vectorize the patches $\mathbf{x}_p$ into a potentially d-dimensional space using a trainable linear projection.

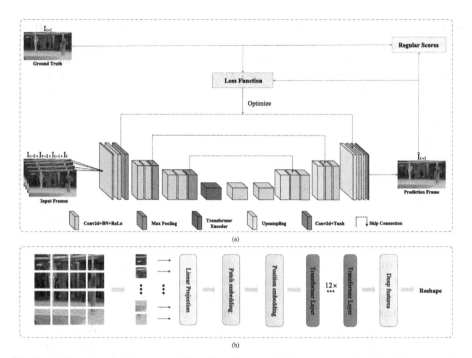

(a)

(b)

Fig. 1. (a) Overview of the proposed multi-path attention temporal method for video anomaly detection. (b) The architecture of the transformer encoder.

Position Embedding. To encode the patch space information, we embed the learned location-specific information into the patch embedding as shown in Eq. (3).

$$z_0 = \left[x_p^1 E; x_p^2 E; \cdots ; x_p^{N'} E \right] + E_{pos} \tag{3}$$

where $E \in R^{(P^2 \cdot C') \times D}$ is the patch embedding projection and $E_{pos} \in R^{N' \times D}$ denotes the location embedding.

Transformer Layer. The structure of Transformer encoder is similar to that of NLP, which consists of L Multihead Self-Attention (MSA) and Multi-Layer Perceptron (MLP) blocks. For the first layer of the encoder, assuming that the input is z_{l-1} and the output is z_l, the results of the output of the *l-th* layer can be expressed as follows, respectively.

$$z_l' = MSA(LN(z_{l-1})) + z_{l-1}, l = 1, \ldots, L \tag{4}$$

$$z_l = MLP(LN(z_l')) + z_l', l = 1, \ldots, L \tag{5}$$

where $LN(\cdot)$ denotes the layer normalization and z_l denotes the encoded image.

MSA, as the core module in the Transformer block, uses a parallel approach to train multiple value vectors and then stitch them together to get the output. Specifically, given an input matrix, we compute multiple sets of V, K, and Q matrices based on different parameter matrices, then compute multiple weighted V matrices by multiple attention functions, and finally concatenate these matrices by a weight to get the final output matrix W^O as follows.

$$\text{Multihead(Q, K, V)} = \text{Concat(head}_1, \ldots, \text{head}_N)W^O \tag{6}$$

$$\text{head}_i = \text{Attention}\left(QW_i^Q, KW_i^K, VW_i^V\right) \tag{7}$$

where $W_i^Q \in R^{d \times d_Q}$, $W_i^K \in R^{d \times d_K}$, $W_i^V \in R^{d \times d_V}$, $W^O \in R^{hd_v \times d}$.

To identify anomalies in the video, we provide a regularity score for each frame, with higher scores indicating a higher probability of normal. We use an image quality assessment method to better Peak Signal-to-Noise Ratio (PSNR). We obtain the PSNR values of all frames in each test video and normalize them to [0, 1] by Eq. (8).

$$s(t) = \frac{\text{PSNR}\left(I_t, \hat{I}_t\right) - \min_t \text{PSNR}\left(I_t, \hat{I}_t\right)}{\max_t \text{PSNR}\left(I_t, \hat{I}_t\right) - \min_t \text{PSNR}\left(I_t, \hat{I}_t\right)} \tag{8}$$

where the regularity score s(t) corresponds to the degree of normality of each frame in the video. $\hat{I}_t$ denotes the predicted frame and I_t denotes the real frame.

4 Experiments

4.1 Datasets and Evaluation Metrics

We train and test our model on three well-known video anomaly detection datasets: UCSD Ped2 [16], CUHK Avenue [17] and ShanghaiTech [18]. Ped2 contains 16 training videos and 12 test videos with a pixel resolution of 360×240. CUHK Avenue contains 16 training videos and 21 test videos with a pixel resolution of 640×360 per frame. ShanghaiTech contains 330 training videos and 107 test videos with a pixel resolution of 480×856. We use the Receiver Operating Characteristic (ROC) curve to evaluate the performance of our method, we use the Area Under the receiver operating Characteristic (AUC) curve and the Equal Error Rate (EER) as evaluation metrics. A higher AUC and a lower EER indicate better video anomaly detection performance of the model.

4.2 Implementation Details

Our model is trained and evaluated on platforms powered by NVIDIA TiTan RTX GPU and implemented using Pytorch. Next, we train using the Adam optimizer with batch size of 4, learning rate of 2e−4, and gradual decay using cosine annealing. To train the network, we resized all video frames to 256×256 and set the training epochs to 60, 60, and 10 for the UCSD Ped2, CUHK Avenue, and ShanghaiTech datasets.

4.3 Comparison with Existing State-of-the-Arts

To evaluate the effectiveness of our proposed Transformer with spatio-temporal representation for video anomaly detection method, we compare it with 8 different methods. Table 1 shows the AUC and EER results of different methods on UCSD Ped2, CUHK Avenue and ShanghaiTech datasets.

Table 1. AUC of different methods on the UCSD Ped2, CUHK Avenue and ShanghaiTech datasets. Bold numbers indicate the best results and underlined ones the second best.

Methods	UCSD Ped2		CUHK Avenue		ShanghaiTech	
	AUC (%)	EER	AUC (%)	EER	AUC (%)	EER
AMDN [3]	90.80	N/A	N/A	N/A	N/A	N/A
3D-ConvAE [9]	91.20	0.167	77.10	0.244	N/A	N/A
LSTM	84.70	0.243	71.10	0.345	N/A	N/A
ConvLSTM [12]	88.10	N/A	77.00	N/A	55.00	N/A
GRU	85.90	0.212	72.50	0.338	N/A	N/A
Stacked RNN [19]	92.00	N/A	81.70	N/A	68.00	N/A
SNRR-AE [10]	92.21	N/A	83.48	N/A	69.63	N/A
MNAD [20]	95.45	N/A	**84.81**	N/A	69.36	N/A
Ours	**96.00**	**0.123**	84.45	**0.220**	**74.35**	**0.319**

From Table 1 we observe that SNRR-AE outperforms 3D-ConvAE in the encoder-decoder-based approach because it captures temporal information better. Our proposed method achieves the best performance on UCSD Ped2 and ShanghaiTech, and closes to the best performance of MNAD on CUHK. For the temporal methods, our method achieves 11.30%, 10.10% and 13.35%, 11.95% performance improvement on the UCSD Ped2 and CUHK Avenue test sets, respectively. Compared to the standard LSTM and the standard GRU. This is because our proposed model is a hybrid model combining Transformer and CNN components, and the features extracted by Transformer include not only contextual information but also long-range global dependencies as well as inductive bias features of CNN, which makes it easier to discriminate the recognized objects in motion prediction. Although RNN-based anomaly detection methods are powerful and effective for temporal data processing, their potential for gradient disappearance or gradient explosion. The recent memory enhancement method MNAD shows good performance and can be integrated with our proposed framework in future work to further improve the performance. In addition, our method achieves lower EER at frame level with 0.123, 0.220 and 0.319 on three datasets. we also obtain better performance compared to temporal anomaly detection methods such as LSTM, GRU and Stacked RNN. As can be seen from Fig. 2, the frame-level ROC of our method on the three datasets UCSD Ped2, CUHK Avenue and ShanghaiTech is significantly better than the other methods.

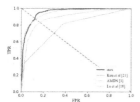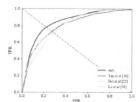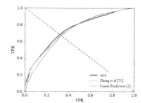

Fig. 2. The frame-level ROC curves of different methods on UCSD Ped2, CUHK Avenue and ShanghaiTech datasets.

We also qualitatively analyzed the anomaly detection performance of our method on three benchmark datasets and visualized it as shown in Fig. 3. In each subplot, a higher regularity score indicates a higher probability of normal. The cyan shaded part of the graph indicates anomalies in Ground Truth. It can be noticed that the regularity score significantly decreases when anomalies occur and increases when they disappear, which indicates that our method is able to detect the occurrence of anomalies.

4.4 Ablation Experiments

To evaluate the effectiveness of our method, we conducted ablation experiments on three datasets, UCSD Ped2, CUHK Avenue and ShanghaiTech, and reported the results of AUC-based anomaly detection, as shown in Table 2.

Table 2. Ablation study results on three datasets. The anomaly detection performance is reported in terms of AUC. Number in bold indicates the best result.

	UCSD Ped2		CUHK Avenue		ShanghaiTech	
U-Net	√	√	√	√	√	√
Transformer	×	√	×	√	×	√
AUC (%)	95.09	**96.00**	83.58	**84.45**	67.90	**74.35**

As shown in Table 2, the AUC of combining Transformer with U-Net are 0.91%, 0.87% and 6.45% improvement on the UCSD Ped2, CUHK Avenue and ShanghaiTech datasets, respectively. This indicates that our method can both preserve the detailed features of the image by convolution operation and preserve the global features of the image by using Transformer's effective learning ability of contextual information. On the other hand, it also reflects that the limitations of the convolution operation limit the ability of the model to extract global features of the image, while combining the convolution operation with the Transformer operation can better capture the spatio-temporal dynamics and thus allow for more accurate detection. At the same time, we found that the boosting effect on ShanghaiTech is much higher than the other two datasets, which is because Transformer is better at handling complex data, while the ShanghaiTech dataset has more scenarios and is more complex compared to the other two datasets. This further illustrates that combining CNN with Transformer is more suitable for practical and complex scenarios. Figure 4 shows the comparison results of AUC for U-Net and U-Net + Transformer on the three benchmark datasets. Our Transformer + U-Net approach predicts higher AUC values compared to the traditional U-Net approach, which indicates that the combination of Transformer and CNN has an advantage over other approaches in video dissimilarity prediction.

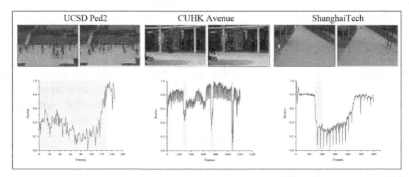

Fig. 3. Example of video anomaly detection.

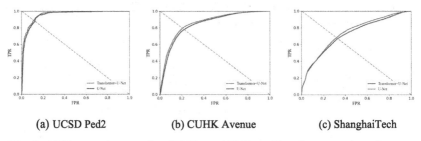

(a) UCSD Ped2   (b) CUHK Avenue   (c) ShanghaiTech

Fig. 4. AUC comparison results of U-Net and U-Net + Transformer on three datasets.

4.5   Running Time

With an NVIDIA TiTan RTX, we achieved an average speed of 25 fps on three datasets to detect anomalies in images of size 256×256, and reach the average or even better level in the field, such as 25 fps in [2], 10 fps in [21].

5   Conclusion

In this paper, we propose a Transformer with spatio-temporal representation for video anomaly detection. The network combines the bias induction capability of CNN and the powerful self-attention mechanism of Transformer for future frame prediction, which not only can effectively obtain global contextual information through the MSA of Transformer to compensate for the deficiency of convolutional operation in this aspect, but also can take advantage of the pyramidal structure of U-Net to make good use of the shallow CNN of feature map features to achieve better results. The experimental results on three benchmark video anomaly detection datasets show that our proposed method can better detect anomalous frames. In our future work, we will further explore how Transformer and CNN can be more perfectly combined to achieve its more powerful application in the field of video anomaly detection.

Acknowledgment. This work is supported in part by National Natural Science Foundation of China under Grant 61871241, Grant 61971245 and Grant 61976120, in part by Nantong Science and Technology Program JC2021131 and in part by Postgraduate Research and Practice Innovation Program of Jiangsu Province KYCX21_3084.

References

1. Balasundaram, A., Chellappan, C.: An intelligent video analytics model for abnormal event detection in online surveillance video. J. Real-Time Image Proc. **17**(4), 915–930 (2020)
2. Li, C.B., Li, H.J., Zhang, G.A.: Future frame prediction based on generative assistant discriminative network for anomaly detection. Appl. Intell. (2022).https://doi.org/10.1007/s10 489-022-03488-2
3. Xu, D., Yan, Y., Ricci, E., Sebe, N.: Detecting anomalous events in videos by learning deep representations of appearance and motion. Comput. Vis. Image Underst. **156**, 117–127 (2017)
4. D'Afflisio, En., Braca, P., Millefiori, L.M., Willett, P.: Detecting anomalous deviations from standard maritime routes using the Ornstein-Uhlenbeck process. IEEE Trans. Signal Process. **66**(24), 6474–6487 (2018)
5. Ronneberger, O., Fischer, P., Brox, T.: U-Net: convolutional networks for biomedical Image segmentation. In: Proceedings of the IEEE Conference on Computer Vision and Pattern Recognition, Boston, USA, pp. 234–241 (2015)
6. Schlemper, J., et al.: Attention gated networks: learning to leverage salient regions in medical images. Med. Image Anal. **53**, 197–207 (2019)
7. Wang, X., Girshick, R., Gupta, A., He, K.: Non-local neural networks. In: Proceedings of the IEEE Conference on Computer Vision and Pattern Recognition, Salt Lake City, USA, pp. 7794–7803 (2018)
8. Vaswani, A., et al.: Attention is all you need. In: Advances in Neural Information Processing Systems, pp. 5998–6008 (2017)
9. Zhao, Y., Deng, B., Shen, C., Liu, Y., Lu, H., Hua, X.S.: Spatiotemporal autoEncoder for video anomaly detection. In: Processing of the 25th ACM Multimedia Conference, pp. 1933–1941 (2017)
10. Yan, S.Y., Smith, J.S., Lu, W.J., Zhang, B.L.: Abnormal event detection from videos using a two-stream recurrent variational autoencoder. IEEE Trans. Cognit. Dev. Syst. **12**(1), 30–42 (2020)
11. Parmar, N., et al.: Image transformer. In: Proceedings of the 35th International Conference on Machine Learning, Stockholm, Sweden, pp. 4055–4064 (2018)
12. Luo, W., Liu, W., Gao, S.: Remembering history with convolutional LSTM for anomaly detection. In: Processing of the IEEE International Conference on Multimedia and Expo, pp. 439–444 (2017)
13. Ravanbakhsh, M., Sangineto, E., Nabi, M., Sebe, N.: Training adversarial discriminators for cross-channel abnormal event detection in crowds. In: Proceedings of the 19th IEEE Workshop on Application of Computer Vision, Waikoloa Village, USA, pp. 1896–1904 (2019)

14. Liu, W., Luo, W.X., Lian, D.Z., Gao, S.H.: Future frame prediction for anomaly detection – a new baseline. In: Processing of the IEEE Conference on Computer Vision and Pattern Recognition, Salt Lake City, USA, pp. 6536–6545 (2018)
15. Villegas, R., Yang, J., Hong, S., Lin X., Lee, H.: Decomposing motion and content for natural video sequence prediction. In: Processing of the International Conference on Learning Representations, Toulon, France, pp. 1–22 (2017)
16. Arnab, A., Dehghani, M., Heigold, G., Sun, C. Lučić, M., Schmid, C.: ViViT: a video vision transformer (2021). http://arxiv.org/abs/2103.15691
17. Li, W.X., Mahadevan, V., Vasconcelos, N.: Anomaly detection and localization in crowded scenes. IEEE Trans. Pattern Anal. Mach. Intell. **36**(1), 18–32 (2014)
18. Lu, C., Shi, J., Jia, J.: Abnormal event detection at 150 fps in MATLAB. In: Processing of the IEEE International Conference on Computer Vision, Sydney, pp. 2720–2727. IEEE (2013)
19. Luo, W., Liu, W., Gao, S.: A revisit of sparse coding based anomaly detection in stacked RNN framework. In: Processing of the IEEE International Conference on Computer Vision, Sydney, pp. 341–349. IEEE (2017)
20. Park, H., Noh, J., Ham, B.: Learning memory-guided normality for anomaly detection. In: Proceedings of the IEEE Conference on Computer Vision and Pattern Recognition, pp. 14360–14369 (2020)
21. Ye, M., Peng, X., Gan, W., Wu, W., Qiao, Y.: AnoPCN: video anomaly detection via deep predictive coding network. In: Processing of the 27th ACM International Conference on Multimedia, pp. 1805–1813 (2019)

Efficient Leave-One-Out Evaluation
of Kernelized Implicit Mappings

Mineichi Kudo[1][✉][ID], Keigo Kimura[1][ID], Shumpei Morishita[1][ID], and Lu Sun[2][ID]

[1] Division of Computer Science and Information Technology, Graduate School of
Information Science and Technology, Hokkaido University, Sapporo 060-0814, Japan
{mine,kimura5,morishita_s}@ist.hokudai.ac.jp
[2] School of Information Science and Technology, ShanghaiTech University,
Shanghai 201210, China
sunlu1@shanghaitech.edu.cn

Abstract. End-to-end learning is discussed in the framework of linear
combinations of reproducing kernels associated with training samples.
This paper shows that the leave-one-out (LOO) technique can be executed very efficiently in this framework. It is a simple extension of previous fast LOO algorithms for scalar-valued functions to vector-valued
functions, but opens the door for multiple analyses of the same data with
almost no cost. With a newly defined LOO matrix, we demonstrate the
effectiveness and the universality of this approach.

Keywords: Kernelized implicit mapping · Leave-one-out · LOO matrix

1 Introduction

Many machine learning algorithms aim at simulating an implicit mapping by an
explicit function on the basis of a finite number of input-output pairs. For example, linear regression models use linear combinations with basis functions, and
neural networks use a connected network with linear combinations and activation
functions. In this study, we put no assumption on neither probabilistic models
nor the network structure, but use a single reproducing kernel. Specifically, we
simulate an implicit mapping by linear combinations of kernel functions associated with individual training samples. Such trials have been already studied,
such as kernel ridge regression, but they are not simulators, that is, the training
error is larger than zero. Fast leave-one-out (LOO) evaluations have been also
proposed for those trials, but the output is a scalar. In this paper, we deal with
the case that the output is a vector. In this case, LOO evaluation compares the
true output vector and an estimated vector by LOO procedure. We show how
effective such LOO evaluation is for analyzing data.

2 Kernelized Implicit Mapping

We assume that n input-output pairs are given as

$$(\boldsymbol{x}_1, \boldsymbol{z}_1), (\boldsymbol{x}_2, \boldsymbol{z}_2), \ldots, (\boldsymbol{x}_n, \boldsymbol{z}_n), \quad \boldsymbol{x}_i \in \mathbb{R}^D, \boldsymbol{z}_i \in \mathbb{R}^d,$$

A. Krzyzak et al. (Eds.): S+SSPR 2022, LNCS 13813, pp. 223–232, 2022.
https://doi.org/10.1007/978-3-031-23028-8_23

but no explicit function is known to produce z_i's from x_i's. Therefore, the mapping is assumed to be implicit. Embeddings of high-dimensional data into a lower-dimensional space such as Laplacian eigenmaps are examples of implicit mappings. Let us express those pairs in two matrices separately:

$$X = (x_1, x_2, \ldots, x_n) \in \mathbb{R}^{D \times n}, \ Z = (z_1, z_2, \ldots, z_n) \in \mathbb{R}^{d \times n}.$$

We simulate the implicit vector-valued mapping by a set of linear combinations of kernels associated with training samples as

$$z = f(x) = C\,K(x) = \begin{pmatrix} c_1^T \\ c_2^T \\ \vdots \\ c_d^T \end{pmatrix} \begin{pmatrix} k(x, x_1) \\ k(x, x_2) \\ \vdots \\ k(x, x_n) \end{pmatrix} \in \mathbb{R}^d,$$

where c's are d coefficient vectors and $K(x)^T = (k(x, x_1), \ldots, k(x, x_n))$. This form is supported from the representer theorem [1]. Applying this mapping to all inputs X, we have

$$\hat{Z} = C\mathbb{G}, \quad \mathbb{G} = (k(x_i, x_j))_{n \times n}, \tag{1}$$

where $\mathbb{G}$ is called the *Gram matrix*. We determine the values of C from X and Z such that $\hat{Z} \simeq Z$ (end-to-end learning). If $\mathbb{G}$ is invertible, the optimal setting is given by

$$C = Z\mathbb{G}^{-1}. \tag{2}$$

From (1) and (2), it is evident that the implicit mapping is perfectly simulated on all the training data because $\hat{Z} = C\mathbb{G} = Z\mathbb{G}^{-1}\mathbb{G} = Z$. We call this approach a *Kernelized (nonlinear) Implicit Mapping* (shortly, KIM). This framework is proposed already in our previous paper [2] in the case that Z is the mapping result by Laplacian eigenmap of X. Note that this KIM simulation solves the so-called *out-of-sample problem* in which a new sample cannot be mapped in the same way as the training samples.

3   LOO Evaluation of KIM

Leave-one-out (LOO) evaluation is a useful tool for measuring the generalization performance of a mapping or a classifier. In KIM, the generalization ability is measured by the difference between two mappings over testing samples: one is the mapping by KIM designed from both training and testing samples and the other is the mapping by KIM designed from training samples only. In LOO procedure, one sample x_ℓ is excluded with z_ℓ in turn from X as a testing sample and is mapped by these two mappings (designed including or excluding (x_ℓ, z_ℓ)). We measure the difference of z_ℓ and $\hat{z}_\ell$ that is mapped from x_ℓ by KIM designed without (x_ℓ, z_ℓ).

In our framework, KIM is realized as

$$z = f(x) = Z\mathbb{G}^{-1}K(x). \tag{3}$$

Assume that the ℓth sample is excluded from X and Z and then it is mapped by KIM trained from all the other samples,

$$\hat{z}_\ell = Z_{-\ell} \mathbb{G}_{-\ell}^{-1} K_{-\ell}(x_\ell), \tag{4}$$

where the suffix $-\ell$ denotes the case that all the elements related to the ℓth sample are excluded. For example, $Z_{-\ell} = (z_1, z_2, \ldots, z_{\ell-1}, z_{\ell+1}, \ldots, z_n)$. In this paper, the normalized generalization error is measured by LOO procedure as

$$\epsilon_{LOO} = \frac{1}{n} \sum_{\ell=1}^{n} \|z_\ell - \hat{z}_\ell\|^2 / \|z_\ell\|^2. \tag{5}$$

To calculate (4), we propose an efficient way to calculate

$$\mathbb{G}_{-\ell}^{-1} K_{-\ell}(x_\ell), \quad \ell = 1, 2, \ldots, n.$$

It is known that a partitioned matrix of $((n-1)+1) \times ((n-1)+1)$

$$\mathcal{A} = \begin{pmatrix} A & a \\ a^T & a \end{pmatrix} \quad ('T' \text{denotes the transpose})$$

can be inverted as

$$\mathcal{B} = \begin{pmatrix} B & b \\ b^T & b \end{pmatrix} \stackrel{\text{def}}{=} \mathcal{A}^{-1} = \begin{pmatrix} A^{-1} + A^{-1}ac^{-1}a^T A^{-1} & -A^{-1}ac^{-1} \\ -c^{-1}a^T A^{-1} & c^{-1} \end{pmatrix},$$

where $c = a - a^T A^{-1} a$ [3].

Applying this to the gram matrix of n samples,

$$\mathbb{G} = \begin{pmatrix} \mathbb{G}_{-n} & g \\ g^T & g \end{pmatrix}, \quad \begin{pmatrix} g \\ g \end{pmatrix} = \begin{pmatrix} K_{-n}(x_n) \\ k(x_n, x_n) \end{pmatrix},$$

we have

$$\mathbb{H} = \begin{pmatrix} H & h \\ h^T & h \end{pmatrix} \stackrel{\text{def}}{=} \mathbb{G}^{-1} = \begin{pmatrix} \cdots & -\mathbb{G}_{-n}^{-1}gc^{-1} \\ \cdots & c^{-1} \end{pmatrix}$$

where $c = g - g^T \mathbb{G}_{-n}^{-1} g$. Hence, $-h/h = \mathbb{G}_{-n}^{-1} g$ where $g = K_{-n}(x_n)$. Therefore, without an explicit inverting of $\mathbb{G}_{-n}$, we can calculate $\mathbb{G}_{-n}^{-1} K_{-n}(x_n)$ by just scanning the last column of $\mathbb{G}^{-1}$.

This technique is applicable to any position $\ell(= 1, 2, \ldots, n-1)$ other than n. Let us consider a partial cyclic permutation matrix

$$C_{\ell n} = \begin{array}{c} \\ \\ \\ \\ \ell > \\ \\ \\ \\ n > \end{array} \begin{bmatrix} 1 & 0 & \cdots & & \cdots & & 0 \\ \vdots & \ddots & & & & & \vdots \\ 0 & \cdots & 1 & 0 & \cdots & 0 & 0 \\ 0 & \cdots & \cdots & 0 & \cdots & 0 & 1 \\ 0 & \cdots & \cdots & 1 & \ddots & 0 & 0 \\ 0 & \cdots & \cdots & 0 & \ddots & \ddots & 0 \\ 0 & \cdots & \cdots & 0 & \cdots & 1 & 0 \end{bmatrix}.$$

to move the ℓth row and the ℓth column to the last ones. Note that $C_{\ell n}^T C_{\ell n} = C_{\ell n} C_{\ell n}^T = I_n$. Then, we can see

$$\begin{aligned}
\mathbb{G}^{-1} &= C_{\ell n}(C_{\ell n}^T \mathbb{G}^{-1} C_{\ell n}) C_{\ell n}^T \\
&= C_{\ell n} \mathbb{H}_\ell^{-1} C_{\ell n}^T,
\end{aligned}$$

where $\mathbb{H}_\ell$ has the ℓth row and the ℓth column of $\mathbb{G}$ in the last ones. Therefore, we have

$$- h_{\ell,-\ell}/h_{\ell\ell} = \mathbb{G}_{-\ell}^{-1} K_{-\ell}(x_\ell), \tag{6}$$

where $h_{\ell,-\ell}$ is the ℓth column of $\mathbb{H} = (h_1, h_2, \ldots, h_n) = \mathbb{G}^{-1}$ excluding $h_{\ell\ell}$.

By gathering all the results for $\ell = 1, 2, \ldots, n$, the *LOO prediction* is given by

$$\hat{Z} = Z \mathbb{G}_{\mathrm{LOO}}^{-1}, \tag{7}$$

where

$$\mathbb{G}_{\mathrm{LOO}}^{-1} = \tilde{\mathbb{H}} = (\tilde{h}_1, \tilde{h}_2, \ldots, \tilde{h}_n),$$

$$\tilde{h}_\ell = (\tilde{h}_{i\ell}), \quad \tilde{h}_{i\ell} = \begin{cases} -h_{i\ell}/h_{\ell\ell} & (i \neq \ell) \\ 0 & (i = \ell) \end{cases}. \tag{8}$$

We call $\mathbb{G}_{\mathrm{LOO}}^{-1}$ the *LOO matrix*. $\mathbb{G}_{\mathrm{LOO}}^{-1}$ is able to be obtained by just scanning $\mathbb{G}^{-1}$ once, without calculating $\mathbb{G}_{-\ell}^{-1} (\ell = 1, 2, \ldots, n)$. The computational cost $O(n^2)$ is almost ignorable compared with the calculation cost $O(n^3)$ of the inverse of $\mathbb{G}$. Note that the same $\mathbb{G}_{\mathrm{LOO}}^{-1}$ is applicable for different Z's in (7).

An example is shown for

$$\mathbb{G} = \begin{pmatrix} 1 & -1 & 4 \\ -1 & 1 & 2 \\ 4 & 2 & 1 \end{pmatrix} \quad \text{and} \quad \mathbb{H} = (h_1, h_2, h_3) = \mathbb{G}^{-1} = \begin{pmatrix} 1/12 & -1/4 & 1/6 \\ -1/4 & 5/12 & 1/6 \\ 1/6 & 1/6 & 0 \end{pmatrix}.$$

For $\ell = 2$, from (6), we have

$$-h_{2,-2}/h_{22} = - \begin{pmatrix} -1/4 \\ 1/6 \end{pmatrix}/(5/12) = \begin{pmatrix} 3/5 \\ -2/5 \end{pmatrix}$$

which is equal to

$$\mathbb{G}_{-2}^{-1} h_{2,-2} = \begin{pmatrix} 1 & 4 \\ 4 & 1 \end{pmatrix}^{-1} \begin{pmatrix} -1 \\ 2 \end{pmatrix} = \begin{pmatrix} 3/5 \\ -2/5 \end{pmatrix}.$$

4   Positioning of this Study

In the scalar-valued version (regression or classification), KIM is one of the simplest transformations adopted in several approaches such as Kernel ridge regression and least square kernel machines [5–9]. In addition, the same or a similar efficient LOO evaluation is already proposed by the authors. We have just described it in a more intuitive way, but in a vector-valued form.

In the light of these researches, the contributions of this study are 1) LOO error estimation of a scalar-valued function is extended to LOO prediction of a vector-valued function, 2) The LOO prediction is decomposed into a target independent LOO matrix and a target matrix, so that LOO prediction to different targets can be executed with relatively no cost, and 3) It becomes possible to analyze the information of data X, through the Gram matrix, by investigating the nature of the LOO matrix.

A KIM is easily extended to an affine transformation

$$f(x) = CK(\boldsymbol{x}) + \boldsymbol{c}_0 \in \mathbb{R}^d.$$

In addition, we can relax the criterion from the (perfect) simulation to a regularized criterion to minimize

$$\|Z - \hat{Z}\|_F^2 + \mu\, C\mathbb{G}C^T \qquad (\mu \geq 0).$$

Then, we can show that the optimal $\boldsymbol{c}_0$ is given by $\boldsymbol{c}_0 = Z(\mathbb{G} + \mu I)^{-1}\mathbf{1}/\mathbf{1}^T(\mathbb{G} + \mu I)^{-1}\mathbf{1}$, where $\mathbf{1}$ is the vector of all one's. Hence, by modifying the output as $\tilde{Z} = Z - \boldsymbol{c}_0\mathbf{1}^T$, the affine transformation turns a linear transformation, a usual KIM form. Then, the LOO prediction is obtained by $(Z - \boldsymbol{c}_0\mathbf{1}^T)\mathbb{G}_{\text{LOO}}^{-1} + \boldsymbol{c}_0\mathbf{1}^T$ after replacing $\mathbb{G}$ with $\mathbb{G} + \mu I$.

It is also possible to execute k-fold cross validation instead of LOO at the expense of $O((n/k)^3)$ for the inversion of a block matrix of size $(n/k) \times (n/k)$ instead of the reciprocal of $h_{\ell\ell}$ in (8).

5   Kernels

In this paper, we consider the following three reproducing kernels:

1) RBF (Gaussian) kernel

$$k(\boldsymbol{x}, \boldsymbol{y}) = \exp\left(-\frac{\|\boldsymbol{x} - \boldsymbol{y}\|^2}{\sigma^2}\right),$$

2) Polynomial kernel

$$k(\boldsymbol{x}, \boldsymbol{y}) = \left(\frac{\boldsymbol{x}^T\boldsymbol{y}}{D} + 1\right)^g, \quad \text{and}$$

3) Sinc kernel

$$k(\boldsymbol{x}, \boldsymbol{y}) = \frac{\sin\sigma(\|\boldsymbol{x} - \boldsymbol{y}\|)}{\sigma\pi\|\boldsymbol{x} - \boldsymbol{y}\|}.$$

Here, D is the dimensionality of the original feature space. Note that each kernel has a single parameter. Among these three kernels, it is known that the Gram matrix of RBF (Gaussian) kernel is always nonsingular regardless of the value of the parameter σ^2, as long as the samples are all distinct [4]. In practice, we set the value of σ^2 by the way described in [2]. To make the other two nonsingular, we may use regularization with μ.

6   Multiple Applications of LOO Matrix

In what follows, we will show three applications (three Z's) of (the same) LOO matrix when RBF kernel is used. Remember that $\mathbb{G}^{-1}K(\boldsymbol{x})$ in (3) and $\mathbb{G}_{\mathrm{LOO}}^{-1}$ in (7) are separable from Z, respectively, as long as X is the same.

6.1   Visualization and Model Selection

First, we visualized `Digit` dataset by a supervised Laplacian eigenmap (SLE) [2] that is enhanced in the class separability with parameter $\lambda = 0.4$. In this case, Z is the mapped $2D$ vectors by SLE. We simulated the SLE by several kernels and parameters. The results are shown in Fig. 1. We see how well each method works with the LOO prediction $\hat{Z}$ as well as the LOO error. LOO prediction $\hat{Z}$ shows to what degree the KIM is stable and which part (class) is strongly or weakly simulated. This is a remarkable advance in model selection compared with the single measure of LOO error. The second and third rows show that parameter selection is more important than kernel selection. The last row visualizes the effect of regularization coefficient μ. For LOO prediction for several values of μ, Tanaka et al. [9] shows a more efficient algorithm using a spectral decomposition. In this study, for the choice of the value of μ, we use the accumulated contribution ratio k_μ. That is, μ is chosen as the eigenvalue when k_μ is attained by accumulating eigenvalues in the decreasing order. We can observe that the samples spread from the original positions keeping the form of the class regions. This might be the reason why the regularization brings better classification performance.

6.2   Nonlinear Classification

Second, we tried classification that is an important implicit mapping. Typically we do not know by ourselves how to label, but only numbers of labeled samples are given. By replacing Z with $Y \in \{0, 1\}^{L \times n}$ (L is the number of labels), we have the kernelized nonlinear classifier $\phi(\boldsymbol{x})$:

$$\hat{\boldsymbol{y}} = \phi(\boldsymbol{x}) = Y\mathbb{G}^{-1}K(\boldsymbol{x}) \in \mathbb{R}^L.$$

Note that Y can express multiple labels for multi-label problems. Since ϕ outputs an approximate one-hot vector in single-label problems, we take the largest value of the prediction $\hat{\boldsymbol{y}}$ for classification. It is clear that the training error is zero as $\phi(X) = Y\mathbb{G}^{-1}\mathbb{G} = Y$. The LOO prediction $\hat{Y}$ is given by $\hat{Y} = Y\mathbb{G}_{\mathrm{LOO}}^{-1}$.

We have examined the performance of the KIM classifier in `CIFAR10` and `FashionMNIST` (both from [10]), varying the number of training data. The dataset `CIFAR10` consists of 60,000 32×32 color images (50,000 for training and 10,000 for testing) in 10 classes, with 6000 images per class. The classes in the dataset are $\{airplane, automobile, bird, cat, deer, dog, frog, horse, ship, truck\}$. The dataset `FashionMNIST` is a collection of labeled fashion images. Each sample is a 28×28 gray-scale image associated with a label from 10 classes. This dataset contains 70,000 samples (60,000 for training and 10,000 for testing). The results are

Parameters and ϵ_{LOO} of individual figures

Z SLE ($\lambda = 0.4$)	This table	
RBF $1/\sigma^2 = 0.001$ $\epsilon = 0.046$	**Sinc** $\sigma = 0.01$ $\epsilon = 0.116$	**Polynomial** $g = 3$ $\epsilon = 0.063$
RBF $1/\sigma^2 = 0.00001$ $\epsilon = 0.088$	**RBF** $1/\sigma^2 = 0.0001$ $\epsilon = 0.053$	**RBF** $1/\sigma^2 = 0.01$ $\epsilon = 0.59$
RBF $1/\sigma^2 = 0.001$ $k_\mu = 0.9$ $\epsilon = 0.046$	**RBF** $1/\sigma^2 = 0.001$ $k_\mu = 0.8$ $\epsilon = 0.065$	**RBF** $1/\sigma^2 = 0.001$ $k_\mu = 0.7$ $\epsilon = 0.085$

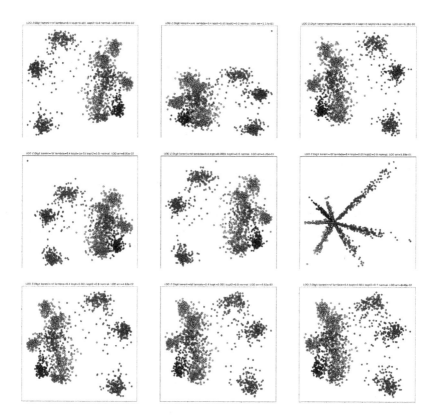

Fig. 1. Model selection on `Digit` dataset (1797 handwritten digits of 10 classes). Top left is Z produced by a class-separability enhanced Laplacian eigenmap and the others are LOO predictions $\hat{Z}$ by three kernels with several values of the parameters. The parameter values and the LOO errors are shown in the top right table. The 2nd row shows $\hat{Z}$'s by three kernels, the 3rd row those by rbf kernel with different parameters, and the last row those by different degree of regularization (the smaller of k_μ, the larger of the effect).

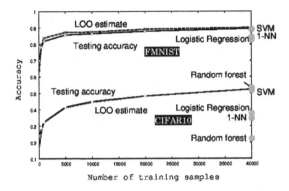

Fig. 2. Accuracy of the learned classifiers in `FashionMNIST` and `CIFAR10`. The solid lines are the accuracies for testing samples and the broken lines are their LOO estimates. The accuracies of the other four classifiers are plotted in 40,000 training samples only.

shown in Fig. 2 where the accuracies of the other four classifiers are also shown. For the four classifiers, we used *scikit-learn* python library [11] with the default parameter values. From Fig. 2, The KIM classifier is almost best among compared classifiers and the accuracy reaches 89.4% for `FashionMNIST` and 52.2% for `CIFAR10`, although it does not reach the accuracies of fully tuned deep neural networks. We can use the same LOO matrix $\mathbb{G}_{\mathrm{LOO}}^{-1}$ both for visualization with Z and for classifier with Y. For example, we might visualize it first, to confirm the state of the separability among classes, and then construct ϕ.

6.3   Recovery

Last, we tried a recovery of an image. We set Z to X itself. Then the LOO prediction $\hat{Z}$ shows how well each image can be recovered by the other images in KIM situation. An example is shown in Fig. 3 with the corresponding weights $\tilde{h}$ in (8) that show the contribution ratios of the other samples to recovery.

7   Analysis of LOO Matrix

Here we analyze LOO matrices $\mathbb{G}_{\mathrm{LOO}}^{-1}$. Its ith column $\tilde{h}_i$ in (8) plays a role of weights to linearly combine the samples other than ith sample as

$$\hat{z}_i = \tilde{h}_{1i}z_1 + \cdots + \tilde{h}_{(i-1)i}z_{i-1} + \tilde{h}_{(i+1)i}z_{i+1} + \cdots + \tilde{h}_{ni}z_n.$$

Since the same LOO matrix is applicable to arbitrary target Z, it is expected that the elements of $\tilde{h}_i$ sum to one. In other words, the contributions of individual elements should be relative. Indeed, as shown in Fig. 4, "sum to one" rule holds for almost all columns. This is also validated by considering Y of one-hot vectors as Z. Since all the columns of Y are one-hot vectors, the elements of $\tilde{h}_i$ corresponding to a class should sum to almost one, and the other elements should sum to almost zero, if the LOO approximation is good enough.

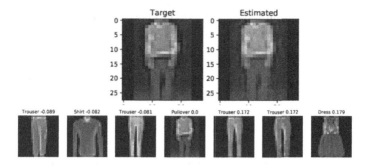

Target picture #1 and the pictures with the lowest and highest coefficients

Fig. 3. An example of a target picture and the estimated picture by a weighted linear combination of the other pictures in `FashionMNIST`. The lower pictures are those with smallest negative weights, the target, and those with largest positive weights.

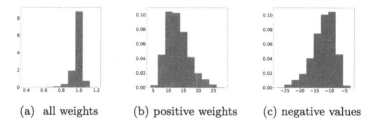

(a) all weights   (b) positive weights   (c) negative values

Fig. 4. Histograms of sum of weights in each column of $\mathbb{G}_{\text{LOO}}^{-1}$.

8 Discussion

As already stated, the fast LOO technique (8) shown in this paper is not new. Even so, (7) is valuable because it is more intuitive and suggestive compared with the formulation of [5–9]. For example, the LOO approximation $\hat{Z} = Z\mathbb{G}_{\text{LOO}}^{-1}$ can be seen as a variant of the exact recovery $Z = Z\mathbb{G}^{-1}\mathbb{G}$. Indeed, $\mathbb{G}_{\text{LOO}}^{-1}$ can be rewritten as

$$\mathbb{G}_{\text{LOO}}^{-1} = \left(\mathbb{G}^{-1}\left(-\text{diag}(1/h_{11}, \ldots, 1/h_{nn})\right)\right) \odot (\mathbf{1}\mathbf{1}^T - I_n),$$

where h_{ii} is the (i, i)th element of $\mathbb{G}^{-1}$ and $\odot$ is the element-wise product (Hadamard product) and $\mathbf{1}$ is the vector of all one's. The last operator forbids returning the target value itself.

9 Conclusion

In the framework of simulating an implicit vector-valued function by a linear transformation of kernels associated with training samples, we have shown an efficient way of the leave-one-out (LOO) prediction with LOO accuracy estimation. The LOO prediction requires almost no cost relative to the calculation cost of the inverse of the Gram matrix. We have shown some applications of the same LOO matrix. In addition, we analyzed the LOO matrix to understand the universality.

Acknowledgment. This work was partially supported by JSPS KAKENHI (Grant Number 19H04128).

References

1. Schölkopf, B., Smola, A.J.: Learning with Kernels: Support Vector Machines, Regularization, Optimization, and Beyond. The MIT Press, Cambridge (2002)
2. Tai, M., et al.: Kernelized supervised Laplacian eigenmap for visualization and classification of multi-label data. Pattern Recogn. **123**, 108399 (2022)
3. Magnus, J.R., Neudecker, H.: Matrix Differential Calculus with Applications in Statistics and Econometrics (Revised Version), p. 11. Wiley, Chichester (1999)
4. Knaf, H.: Kernel fisher discriminant functions - a concise and rigorous introduction. Technical report 117, Fraunhofer (ITWM) (2007)
5. Cawley, G.C., Talbot, N.L.C.: Fast exact leave-one-out cross-validation of sparse least-squares support vector machines. Neural Netw. **17**, 1467–1475 (2004)
6. Cawley, G.C.: Leave-one-out cross-validation based model selection criteria for weighted LS-SVMs. In: Proceedings of the 2006 IEEE International Joint Conference on Neural Networks, Vancouver, pp. 1661–1668 (2006)
7. Cawley, G.C., Talbot, N.L.C.: Efficient approximate leave-one-out cross-validation for kernel logistic regression. Mach. Learn. **71**, 243–264 (2008)
8. An, S., Liu, W., Venkatesh, S.: Fast cross-validation algorithms for least squares support vector machine and kernel ridge regression. Pattern Recogn. **40-8**, 2154–2162 (2007)
9. Tanaka, A., Imai, H.: A fast cross-validation algorithm for kernel ridge regression by eigenvalue decomposition. IEICE Trans. Fundam. Electron. Commun. Comput. Sci. **9**, 1317–1320 (2019)
10. François, C., et al.: Keras (2015). https://keras.io
11. Pedregosa, F., et al.: Scikit-learn: machine learning in Python. J. Mach. Learn. Res. **12**, 2825–2830 (2011)

Graph Similarity Using Tree Edit Distance

Shri Prakash Dwivedi[1]([✉]) [iD], Vishal Srivastava[2] [iD], and Umesh Gupta[2] [iD]

[1] Department of Information Technology, G.B. Pant University
of Agriculture and Technology, Pantnagar, India
`shriprakashdwivedi@gbpuat-tech.ac.in`
[2] School of Computer Science Engineering and Technology,
Bennett University, Greater Noida, India
{`vishal.srivastava,umesh.gupta`}`@bennett.edu.in`

Abstract. Graph similarity is the process of finding similarity between two graphs. Graph edit distance is one of the key techniques to find the similarity between two graphs. The main disadvantage of graph edit distance is that it is computationally expensive and in order to do exhaustive search, it has to perform exponential computation. Whereas for tree, which is a special kind of graph, we have relatively efficient method known as tree edit distance, which finds the minimum number of modifications required to convert one tree to another. In this paper we use tree edit distance to find the similarity between two graphs. To find the similarity between two graph we convert them to spanning trees, find the tree edit distance between these two spanning tree, and then augment the tree edit distance with additional edit operations to perform the graph matching.

Keywords: Graph similarity · Tree edit distance · Graph edit distance

1 Introduction

The graph is an essential and universal discrete mathematical structure in engineering and computer science. Its power lies in the flexibility to depict itself as a structural model in various science and engineering disciplines. Graph Edit Distance (GED) is the least number of modifications needed to convert one graph into another. The common edit operations are addition, deletion, and substitution of nodes and edges. The edit operations were initially used as string edit operations for converting one string to another using the lowest count of edit operations, like addition, deletion, or substituting of alphabets. Later on the concept of string edit distance was applied to tree edit distance and further generalized to GED. Because of its strength, the edit distance-based method is powerful, and it can be applied to diverse problems.

© The Author(s), under exclusive license to Springer Nature Switzerland AG 2022
A. Krzyzak et al. (Eds.): S+SSPR 2022, LNCS 13813, pp. 233–241, 2022.
https://doi.org/10.1007/978-3-031-23028-8_24

One of the substantial applications of GED is Graph Matching (GM). In GM, we evaluate the similarity between two objects depicted in the form of graphs. It is primarily organized into two classes, exact and inexact GM. For Exact GM (EGM), strict correspondence must be between the nodes and edges of the two graphs. In the Error-Tolerant GM (ETGM), also known as Inexact GM (IGM), some flexibility or tolerance is allowed to match the two graphs. The limitation of exact GM is that it can only be used to find a strict matching between the two graphs, and therefore it can not take into account distortion incurred to the graph due to the existence of noise in the process of matching. ETGM offers flexibility during the process of GM [3]. There are many approaches to ETGM but GED, being very adaptable, is the crucial technique for the GM problem. A comprehensive review of diverse GM methods is described in [4] and [14].

In [15], the authors introduce a technique to show patterns by trees instead of strings. Trees are more general high-dimensional structures than strings, so their use will lead to an efficient characterization of a higher-dimensional pattern. Then tree system representation of patterns is used for syntactic pattern recognition. A distance measure for attributed relational graphs [29] utilizing the calculation of the minimum number of modifications is given in [27]. This paper also considers the costs of identifying the vertices in the distance computation. In [2], the authors presented, using heuristic information, the IGM of attributed graphs derived from a state-space search. The matching procedure is generalized to random graphs, and the edit cost functions are constructed so that the GED satisfies the properties of metrics in some circumstances.

GED is suitable for a wide spectrum of applications as it allows precise edit cost functions to be designed for diverse applications. A considerable constraint of GED is that its computation becomes too expensive. It employs exponentially high execution time to compute GED concerning the number of nodes in the input graph. The paper in [30] shows the GED problem to be NP-hard. Since a fast deterministic algorithm is inaccessible for this problem, many approximate and suboptimal algorithms have been proposed recently.

The paper [18] offered an algorithm for GED computation of attributed planar graphs by iteratively matching small subgraphs to enhance structural resemblance. Then it applied the above strategy to the fingerprint classification problem. In [17], the authors present the fast suboptimal algorithm for the GED by diminishing the space necessary to compute the GED utilizing A^* search technique [16]. They explain the various variants of A^*, such as A^*-beamsearch and A^*-pathlength, to decrease the search space that may not be relevant to certain classification tasks. The computation of approximate GED by considering only local rather than global edge structures through the optimizing process is provided in [23]. In [28], the authors describe the GED as a basis in the label space and use it to describe a class of GED cost. They also present the various characteristics and uses of this GED cost. A refinement over the above method by manipulating the initial assignments of the approximate algorithm so that the assignment is ordered based on individual confidence [13]. Evaluating exact GED assuming lower and upper bounds of bipartite approximation using regression

analysis is given in [25]. Various search techniques for enhancing the approximation of bipartite GED computation utilizing the beam search, iterative search
and greedy search, etc., are provided by [24]. The book [21] provides a detailed
exposition of structural pattern recognition and represents different algorithms
for structural pattern recognition utilizing GED.

A promising class of structural pattern recognition using GED is recently
presented in [5], which diminishes the graph size, reducing search space by skip-
ping the less relevant vertices using some measure of importance. In [8], the
authors described homeomorphic GED for topologically equivalent graphs, and
utilized it to perform GM by measuring the structural similarity between two
graphs. The presented technique utilizes the path contraction to remove the vertices having degree two to construct simple paths of input graphs in which every
node except first and last have degree two. An extension to GED utilizing the
notion of node contraction in which a graph is changed into another by contracting the lesser degree vertices is given in [9]. In [12], the authors proposed
centrality GED to perform IGM using the various centrality measures [20] for
removing the vertices having least centrality value in the graph. Some other
recent works are given in [6,7,10,11,26].

This paper describes the technique of using TED algorithm to find the similarity between two graphs. The idea is to first convert the graphs to spanning
trees, apply TED between the spanning trees and finally augment the TED
with additional edit operations over edges to find the similarity between the two
graphs.

This paper is outlined as follows. Section 2 describes basic concepts and motivation for the proposed work. Section 3 presents algorithm to compute the similarity between two graphs using the concept of TED. Section 4 demonstrates
some experiments and results, and at last, Sect. 5 includes the conclusion.

2   Motivation and Basic Concepts

GED is the generalization of Tree Edit Distance (TED). As discussed above,
one of the disadvantage of GED is its high computational complexity due to its
exhaustive nature of search for optimal sequence of edit operations. Whereas for
the ordered trees the efficient algorithms for finding the TED are available. For
the sparse graphs, it may be useful to apply TED. For example, if by removing
only few edges the graph becomes a tree. Then it can be more advantageous to
compute the tree edit distance between such sparse graphs. After that augment
the tree edit distance with the required edit operations only on edges as all the
nodes are already added to the spanning tree. One technique can be to perform
LSAP of remaining edges from G_1 to G_2 which require minimum edge edit
operations consisting of insertion, deletion and substitution of edges only.

Now, we describe the basic concepts and definitions related to GED and TED
[1,19]. We define a *graph* G as a tuple $G = (V, E, \mu, \nu)$; here, V and E are defined
as the set of vertices and edges respectively, $\mu : V \to L_V$, and $\nu : E \to L_E$. Here,
μ is a function which assign each vertex $v \in V$, a unique label $l_v \in L_V$. Similarly,
ν is a mapping that assign each vertex $e \in E$, a unique label $l_e \in L_E$.

A graph may be directed or undirected based on its edges; if $\nu(u,v) = \nu(v,u)$, then the graph is undirected since from both directions, edges have the same value, whereas for directed graphs $\nu(u,v) \neq \nu(v,u)$. When $L_V = L_E = \epsilon$, i.e., vertex label as well as edge label sets are empty, the graph G is called an unlabeled graph.

A graph can be converted into another graph using a set of edit operations. A set of edit operations is inserting, deleting, and substituting nodes and edges. A chain of edit operations that convert an input graph to the output one is defined as an *edit path* from the first graph to the second one. To insert a node u, we denote $\epsilon \to u$, to delete a node u we represent $u \to \epsilon$, to substitute the vertex u by vertex v we represent $u \to v$. Likewise, to insert an edge e we denote by $\epsilon \to e$, $e \to \epsilon$ represents deletion of the edge e, and $e \to f$ defines substitution of the edge e by edge f.

Definition 1. *The GED from G_1 to G_2 is defined by*

$$GED(G_1, G_2) = min_{(e_1,\ldots,e_k) \in \varphi(G_1,G_2)} \sum_{i=1}^{k} c(e_i)$$

here $c(e_i)$ represents the costs of corresponding edit operations of e_i and $\varphi(G_1, G_2)$ denotes the sequence of edit path converting G_1 to G_2.

A *tree* T is an acyclic connected graph. Tree is an special kind of graph in which there is no loop and all nodes are connected. *Forest* is a collection of trees. Forest can also be defined as an acyclic graph. The degree of a node u in a tree is the number of children of u. If a node u has no children then it is called as leaf nodes and otherwise they are called as internal nodes.

Similar to GED, we can define the TED as follows.

Definition 2. *The TED from T_1 to T_2 is defined by*

$$TED(T_1, T_2) = min_{(e_1,\ldots,e_k) \in \varphi(T_1,T_2)} \sum_{i=1}^{k} c(e_i)$$

here $c(e_i)$ represents the costs of corresponding edit operations of e_i and $\varphi(T_1, T_2)$ denotes the sequence of edit path converting T_1 to T_2.

3   Algorithm

To find the similarity between two graphs, firstly we convert these graphs to corresponding spanning tree. Apply the TED between these spanning trees and update the edit operations for edges to transform spanning tree to graphs. We denote this modified TED to Graph to Tree Edit Distance (GTED).

Definition 3. *The GTED from G_1 to G_2 is defined by*

$$GTED(G_1, G_2) = TED'(ST(G_1), ST(G_2))$$

where TED' represents the modified TED which includes edit operations over remaining edges to transform $ST(G_1)$ to G_1 and $ST(G_2)$ to G_2.

In the above definition $ST(G_1)$ represents the spanning tree obtained from the graph G_1 and $ST(G_2)$ represents the spanning tree obtained from the graph G_2.

Algorithm 1 describes the steps to perform the Graph to Tree Edit Distance (GTED) computation. The input to the GTED algorithm is two graphs $G_1 = (V_1, E_1, \mu_1, \nu_1)$, $G_2 = (V_2, E_2, \mu_2, \nu_2)$. The output to the algorithm is the minimum cost GTED between G_1 and G_2. Lines 1–2 of the algorithm transforms the graph G_1 and G_2 to corresponding minimum spanning trees T_1 and T_2 respectively. For the graph datasets in which each node has a coordinate point, the Minimum Spanning Tree (MST) of the graphs can easily be constructed by considering the edge weight of every edge to be its corresponding length. MST procedure in lines 9–18, selects the minimum spanning tree of the graph G. It initializes an empty set A in line 10, then sort the edges in non-decreasing order in line 11. During the for loop of lines 12–17, it selects the safe edge which does not form the loop and append it to set A. Final set of edges is returned in line 15. Line 3 calls the subroutine TED to compute tree edit distance between T_1 and T_2. TED procedure in lines 19–36 uses dynamic programming to store the intermediate results. During each iteration it checks whether the result is there in the table in line 20, if it is the case it return the results in line 21. Otherwise it recursively computes the TED in lines 32–34 and return the minimum cost edit path to transform T_1 to T_2 in line 35. After getting the values of TED the algorithm in for loop of lines 4–8, finds the minimum cost edge edit distances to transform T_1 and T_2 to G_1 and G_2 respectively.

Proposition 1. *The GTED algorithm executes in $O((n \log n)(m \log m))$ time.*

The MST procedure takes $O(n \log n)$ time for G_1 and $O(m \log m)$ time for G_2. The TED procedure takes $O((n \log n)(m \log m))$ time for n and m nodes in T_1 and T_2 respectively. Therefore overall complexity of GTED is $O((n \log n)(m \log m))$.

4   Experimental Results

To compare graph similarity computed using GTED algorithm and GED, we use Letter datasets of the IAM graph database [22]. Letter datasets consist of graphs depicting capital letters of alphabets, created using straight lines only. Modifications of three different levels are applied to prototype graphs to produce three classes of Letter datasets, which are high, medium and low. Letter graphs in the high class are more distorted than that of a graph in the medium or low class. The results in this section are computed using the system having 16 GB of memory running the i7 processor on the Linux operating system. Table 1 shows the comparison of GTED with GED computed between the first graph and the following ten graphs of each three classes of Letter dataset. $GTED_{HIGH}$, $GTED_{MED}$ and $GTED_{LOW}$ in this table represents GTED computed for graphs of high, medium and low classes respectively. Similarly, GED_{HIGH}, GED_{MED} and GED_{LOW} denotes GED computed for graphs of high, medium and low classes respectively.

Algorithm 1 : GTED (G_1, G_2)

Require: Two Graphs G_1, G_2, where $V_1 = \{u_1, ..., u_n\}$ and $V_2 = \{v_1, ..., v_m\}$
Ensure: A minimum cost GTED between G_1 and G_2
1: $T_1 \leftarrow MST(G_1)$
2: $T_2 \leftarrow MST(G_2)$
3: $TED(T_1, T_2)$
4: **for** each $(e_i \in G_1 \land e_i \notin T_1)$ **do**
5:      **for** each $(e_j \in G_2 \land e_j \notin T_2)$ **do**
6:          **return** $min(e_i \rightarrow \epsilon, e_i \rightarrow e_j, \epsilon \rightarrow e_j)$
7:      **end for**
8: **end for**
9: **procedure MST**(G)
10:     $A \leftarrow \phi$
11:     Sort the edges $E(G)$ in non-decreasing order $(e_1, ...e_n)$
12:     **for** each $(e_i \in G)$ **do**
13:         **if** $(e_i$ is safe) **then**
14:             $A \leftarrow A \cup e_i$
15:             **return** A
16:         **end if**
17:     **end for**
18: **end procedure**
19: **procedure TED**(T_1, T_2)
20:     **if** (results (T_1, T_2) is known) **then**
21:         **return** results (T_1, T_2)
22:     **end if**
23:     **if** $(T_1 = \phi \land T_2 = \phi)$ **then**
24:         **return** 0
25:         **else if** $(T_1 \neq \phi \land T_2 = \phi)$
26:         $v_1 \leftarrow$ rightmost node of T_1
27:         **return** $TED(T_1 - v_1, \phi) + c(v_1 \rightarrow \epsilon)$
28:         **else if** $(T_1 = \phi \land T_2 \neq \phi)$
29:         $v_2 \leftarrow$ rightmost node of T_2
30:         **return** $TED(\phi, T_2 - v_2) + c(\epsilon \rightarrow v_2)$
31:     **end if**
32:     $ted_{del} \leftarrow TED(T_1 - v_1, T_2) + c(v_1 \rightarrow \epsilon)$
33:     $ted_{ins} \leftarrow TED(T_1, T_2 - v_2) + c(\epsilon \rightarrow v_2)$
34:     $ted_{sub} \leftarrow TED(T_1 - v_1, T_2 - v_2) + c(v_1 \rightarrow v_2)$
35:     **return** results $(T_1, T_2) \leftarrow min(ted_{del}, ted_{ins}, ted_{sub})$
36: **end procedure**

In this table, we observe that largest GTED under $GTED_{HIGH}$ also corresponds to largest GED under GED_{HIGH}. One of the advantage of using GTED over GED is that it reduces the search space required to find the similarity between two graphs using their tree representation.

Table 1. GTED versus GED

$GTED_{HIGH}$	GED_{HIGH}	$GTED_{MED}$	GED_{MED}	$GTED_{LOW}$	GED_{LOW}
6.151	3.152	6.432	2.307	2.432	1.285
5.457	3.050	8.532	3.056	5.453	2.293
4.871	2.111	6.453	3.433	3.128	1.387
6.532	3.092	9.123	2.843	2.783	1.358
7.342	3.067	8.235	4.061	4.452	2.458
9.435	4.148	7.453	2.371	3.752	1.317
4.235	2.808	6.365	2.402	3.146	1.339
4.342	2.342	5.246	3.830	4.127	2.336
5.237	2.318	5.146	3.528	3.138	1.036
5.213	2.238	4.782	2.025	2.347	1.778

5 Conclusion

In this paper, we described an approach to measure the similarity between two graphs utilizing the concept of TED. This technique works by transforming the input graphs to the spanning trees consisting of all the nodes but some of the edges may be missing. It performs the TED between the converted trees and then augment the TED with the required minimum edit distances for the remaining edges between the two input graphs. This method might be useful for the sparse graphs, which becomes tree after removing few edges.

References

1. Aggarwal, C.C., Wang, H.: Managing and Mining Graph Data. Advances in Database Systems, Springer, New York (2010). https://doi.org/10.1007/978-1-4419-6045-0
2. Bunke, H., Allerman, G.: Inexact graph matching for structural pattern recognition. Pattern Recogn. Lett. **1**, 245–253 (1983)
3. Bunke, H.: Error-tolerant graph matching: a formal framework and algorithms. In: Amin, A., Dori, D., Pudil, P., Freeman, H. (eds.) SSPR/SPR 1998. LNCS, vol. 1451, pp. 1–14. Springer, Heidelberg (1998). https://doi.org/10.1007/BFb0033223
4. Conte, D., Foggia, P., Sansone, C., Vento, M.: Thirty years of graph matching in pattern recognition. Int. J. Pattern Recogn. Artif. Intell. **18**(3), 265–298 (2004)
5. Dwivedi, S.P.: Some algorithms on exact, approximate and error-tolerant graph matching. Ph.D. thesis, Indian Institute of Technology (BHU), Varanasi. arXiv:2012.15279 (2019)
6. Dwivedi, S.P.: Inexact graph matching using centrality measures. arXiv:2201.04563 (2021)
7. Dwivedi, S.P.: Approximate bipartite graph matching by modifying cost matrix. In: Sanyal, G., Travieso-González, C.M., Awasthi, S., Pinto, C.M.A., Purushothama, B.R. (eds.) International Conference on Artificial Intelligence and Sustainable Engineering. LNEE, vol. 837, pp. 415–422. Springer, Singapore (2022). https://doi.org/10.1007/978-981-16-8546-0_34

8. Dwivedi, S.P., Singh, R.S.: Error-tolerant graph matching using homeomorphism. In: International Conference on Advances in Computing, Communication and Informatics (ICACCI), pp. 1762–1766 (2017)

9. Dwivedi, S.P., Singh, R.S.: Error-tolerant graph matching using node contraction. Pattern Recogn. Lett. **116**, 58–64 (2018)

10. Dwivedi, S.P., Singh, R.S.: Error-tolerant geometric graph similarity. In: Bai, X., Hancock, E.R., Ho, T.K., Wilson, R.C., Biggio, B., Robles-Kelly, A. (eds.) S+SSPR 2018. LNCS, vol. 11004, pp. 337–344. Springer, Cham (2018). https://doi.org/10.1007/978-3-319-97785-0_32

11. Dwivedi, S.P., Singh, R.S.: Error-tolerant geometric graph similarity and matching. Pattern Recogn. Lett. **125**, 625–631 (2019)

12. Dwivedi, S.P., Singh, R.S.: Error-tolerant approximate graph matching utilizing node centrality information. Pattern Recogn. Lett. **133**, 313–319 (2020)

13. Ferrer, M., Serratosa, F., Riesen, K.: Improving bipartite graph matching by assessing the assignment confidence. Pattern Recogn. Lett. **65**, 29–36 (2015)

14. Foggia, P., Percannella, G., Vento, M.: Graph matching and learning in pattern recognition in the last 10 years. Int. J. Pattern Recogn. Artif. Intell. **28**, 1450001.1-1450001.40 (2014)

15. Fu, K.S., Bhargava, B.K.: Tree systems for syntactic pattern recognition. IEEE Trans. Comput. **22**, 1087–1099 (1973)

16. Hart, P.E., Nilson, N.J., Raphael, B.: A formal basis for heuristic determination of minimum cost paths. IEEE Trans. Sys. Sci. Cybern. **4**, 100–107 (1968)

17. Neuhaus, M., Riesen, K., Bunke, H.: Fast suboptimal algorithms for the computation of graph edit distance. In: Yeung, D.-Y., Kwok, J.T., Fred, A., Roli, F., de Ridder, D. (eds.) SSPR/SPR 2006. LNCS, vol. 4109, pp. 163–172. Springer, Heidelberg (2006). https://doi.org/10.1007/11815921_17

18. Neuhaus, M., Bunke, H.: An error-tolerant approximate matching algorithm for attributed planar graphs and its application to fingerprint classification. In: Fred, A., Caelli, T.M., Duin, R.P.W., Campilho, A.C., de Ridder, D. (eds.) SSPR/SPR 2004. LNCS, vol. 3138, pp. 180–189. Springer, Heidelberg (2004). https://doi.org/10.1007/978-3-540-27868-9_18

19. Neuhaus, M., Bunke, H.: Bridging the Gap Between Graph Edit Distance and Kernel Machines. World Scientific, Singapore (2007)

20. Newman, M.E.J.: Networks-An Introduction. Oxford University Press, Oxford (2010)

21. Riesen, K.: Appendix B: data sets. In: Riesen, K., et al. (eds.) Structural Pattern Recognition with Graph Edit Distance. ACVPR, pp. 149–156. Springer, Cham (2015). https://doi.org/10.1007/978-3-319-27252-8_9

22. Riesen, K., Bunke, H.: IAM graph database repository for graph based pattern recognition and machine learning. In: da Vitoria Lobo, N., Kasparis, T., Roli, F., Kwok, J.T., Georgiopoulos, M., Anagnostopoulos, G.C., Loog, M. (eds.) SSPR/SPR 2008. LNCS, vol. 5342, pp. 287–297. Springer, Heidelberg (2008). https://doi.org/10.1007/978-3-540-89689-0_33

23. Riesen, K., Bunke, H.: Approximate graph edit distance computation by means of bipartite graph matching. Image Vis. Comput. **27**(4), 950–959 (2009)

24. Riesen, K., Bunke, H.: Improving bipartite graph edit distance approximation using various search strategies. Pattern Recogn. **48**(4), 1349–1363 (2015)

25. Riesen, K., Fischer, A., Bunke, H. (2015) Estimating graph edit distance using lower and upper bounds of bipartite approximations. Int. J. Pattern Recogn. Artif. Intell. **29**(2), 1550011 (2015)

26. Robles-Kelly, A., Hancock, E.: Graph edit distance from spectral seriation. IEEE Trans. Pattern Anal. Mach. Intell. **27**(3), 365–378 (2005)
27. Sanfeliu, A., Fu, K.S.: A distance measure between attributed relational graphs for pattern recognition. IEEE Trans. Syst. Man Cybern. **13**(3), 353–363 (1983)
28. Sole-Ribalta, A., Serratosa, F., Sanfeliu, A.: On the graph edit distance cost: properties and applications. Int. J. Pattern Recogn. Artif. Intell. **26**(5), 1260004.1-1260004.21 (2012)
29. Tsai, W.H., Fu, K.S.: Error-correcting isomorphisms of attributed relational graphs for pattern analysis. IEEE Trans. Syst. Man Cybern. **9**, 757–768 (1979)
30. Zeng, Z., Tung, A.K.H., Wang, J., Feng, J., Zhou, L.: Comparing stars: on approximating graph edit distance. PVLDB **2**, 25–36 (2009)

Data Augmentation on Graphs for Table Type Classification

Davide Del Bimbo(ID), Andrea Gemelli(✉)(ID), and Simone Marinai(✉)(ID)

AI Lab, DINFO, University of Florence, Florence, Italy
davide.delbimbo@stud.unifi.it, {andrea.gemelli,simone.marinai}@unifi.it

Abstract. Tables are widely used in documents because of their compact and structured representation of information. In particular, in scientific papers, tables can sum up novel discoveries and summarize experimental results, making the research comparable and easily understandable by scholars. Since the layout of tables is highly variable, it would be useful to interpret their content and classify them into categories. This could be helpful to directly extract information from scientific papers, for instance comparing performance of some models given their paper result tables. In this work, we address the classification of tables using a Graph Neural Network, exploiting the table structure for the message passing algorithm in use. We evaluate our model on a subset of the Tab2Know dataset. Since it contains few examples manually annotated, we propose data augmentation techniques directly on the table graph structures. We achieve promising preliminary results, proposing a data augmentation method suitable for graph-based table representation.

Keywords: Graph Neural Network · Data augmentation · Table classification

1 Introduction

Tables within scientific documents represent an essential source of knowledge. Their use is necessary for the intelligibility of a document as they provide useful information in a structured and well-organized form, allowing the reader to understand the data through visual content. In particular, in scientific documents, the tables can summarize data from experiments, observations and much more, providing essential information to reconstruct the state of the art of different fields of research [6]. Since different users write different documents, tables usually present different layouts: sometimes, they can be irregular or contain unique abbreviations that are difficult to disambiguate automatically. It would be helpful if their contents were interpreted and transcribed into a Knowledge Base (KB), a database in which tables are translated using a single standard vocabulary. The use of the KB could be helpful to those who need to make use of the information and data contained in the tables without having to access the documents directly [9]. In this scenario, it appears necessary to define a way to classify tables into entities that share common features.

© The Author(s), under exclusive license to Springer Nature Switzerland AG 2022
A. Krzyzak et al. (Eds.): S+SSPR 2022, LNCS 13813, pp. 242–252, 2022.
https://doi.org/10.1007/978-3-031-23028-8_25

In this work we present a model to classify scientific tables given their content and structure. The label of a table is related to its purpose within the paper and, as proposed in [9], we try to classify them into four different types: *Observation*, *Input*, *Example* and *Other*. This classification is useful in areas such as the automatic comprehension of an article or the summarization of information in a document. To address the task just described, we make use of Graph Neural Networks (GNNs), which have been widely considered recently in Document Analysis and Table Understanding. This choice is motivated by their ability to consider the structural information. In addition, we propose some data augmentation techniques working directly on the graph representation of tables, which led to promising preliminary results.

This work is organized as follows. In Sect. 2 works that mostly inspired our paper are explored, focusing on the most significant ones. The proposed approach[1] is discussed in Sect. 3 including the preprocessing of the tables of scientific papers, the data augmentation techniques and the implementation of the GNN model. Experimental results on the Tab2Know dataset are presented in Sect. 4, while conclusions are drawn in Sect. 5.

2 Related Works

In this section we summarize previous work related to the proposed approach.

Table Related Tasks. Usually, to extract information from tables in documents two steps are used: first tables are detected, then their structure is described in terms of rows and columns. As shown in [4] different techniques have been used in the past to tackle these tasks, making use of both computer vision and natural language processing techniques. Recently, two new approaches have beaten the state-of-the-art: combining vision, semantic and relations for layout analysis and table detection [15] and applying a soft pyramid mask learning mechanism in both the local and global feature maps for complicated table structure recognition [11]. In addition to Table Detection (TD) and Table Structure Recognition (TSS), the authors who released PubTables-1M [14] proposed to perform Functional Analysis (FA) to distinguish table headers from table cells. In [3] we proposed a Graph Neural Network method to perform FA along with TD, TSS and document layout analysis to enrich the information of extracted tables with a context.

Information Extraction from Scientific Literature. Automatic extraction of table information can help scholars in several disciplines. In addition to values in table cells it is also useful to classify tables according to their type. To track progresses in scientific research, authors of [6] propose an automatic machine learning pipeline for extracting results from papers. The pipeline is split into three steps, the first one being table type classification. Since the focus is on results extraction, result and ablation tables are identified. The extracted information is summarized into a leaderboard, sorted by the best scores given certain

---
[1] Code available here (https://github.com/AILab-UniFI/DA-GraphTab).

metrics. Another work, Tab2Know [9], proposes to classify tables in four types and recognize table headers and columns. The aim is to extract and link tables into a knowledge base to answer user queries trying to identify relevant information over years of research in a given field. Table classification is referred by the authors as table type detection.

Data Augmentation Techniques. The Tab2Know dataset can be used for performing table classification. However, the manually labeled subset is small and therefore we need to implement suitable Data Augmentation (DA) techniques. DA is widely used in machine learning in order to make models better generalize on unseen samples and unbalanced datasets. In object detection, DA techniques involve color operations (contrast, brightness), geometric operations (translations, rotations), and bounding box operations [17]. None of these can be used in our case since we are considering graphs to represent the tables and augmentation operations commonly used in vision and language have no analogs for graphs [16]. Similarly to what we did for trees [1] and inspired by [7], we applied some of their augmentations on table examples directly in the graph structure (Sect. 3.2). Operations that can be performed on tables are random deletion of rows, row replication, column deletion and column replication. Instead of working directly on images, we therefore extract the table structure (Sect. 3.1) and then apply DA on their graph representation, by means of node deletion, edge deletion and inversion of node contents (Sect. 3.2).

3   Method

In this section we present the main steps of the proposed approach.

3.1   Preprocessing

The first step to apply a GNN for table classification is the conversion of tables in PDF papers in graphs. To this purpose, we use PyMuPDF, a toolkit for viewing and rendering PDF and XPS files [10]. The library is used to extract words and their bounding boxes; by using the positions of the tables in the annotations, only the words within them can be considered (Fig. 1). One graph for each table is built, where words correspond to nodes. A feature vector is associated to each word and contains information about its position and the embedding of the textual content, extracted using spaCy language models [5]. Edges represent the mutual position of bounding boxes and are identified by a visibility graph, like the one described in [13] (e.g. see Fig. 1). Each node is connected to its nearest visible nodes when their bounding boxes intersect horizontally or vertically. Each graph, representing a table, is associated with the annotation corresponding to its type.

Tables without annotation and those of which a graph cannot be built are discarded. At the end we obtain 320 graphs split into four classes: Observation

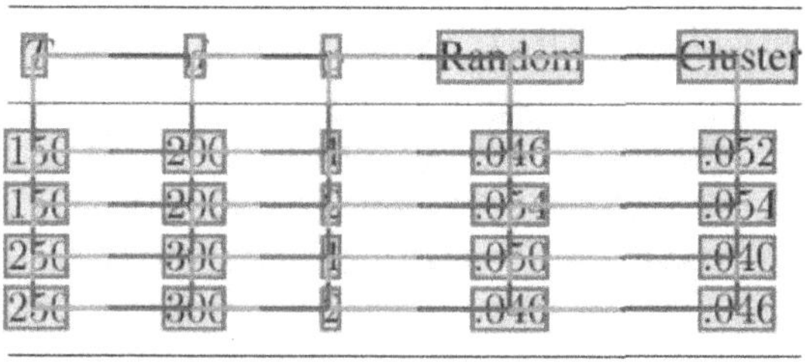

Fig. 1. Words and bounding boxes extracted from one PDF paper using PyMuPDF. Nodes are connected through a visibility graph.

(235), Input (43), Example (13), and Other (29). Examples of classes can be seen in Fig. 2. Each node in the graph corresponds to a feature vector. In addition to the geometric features of the nodes, such as position and size, textual content embeddings are added using spaCy. In particular, two spaCy models are used and compared: *en_core_web_lg* and *en_core_sci_lg*. The first one is the largest english vocabulary which associates each word with a numerical vector of 300 values; the other model, trained on a biomedical corpus, associates each word a numerical vector of 200 values. The results obtained using the two models are compared in the experiments.

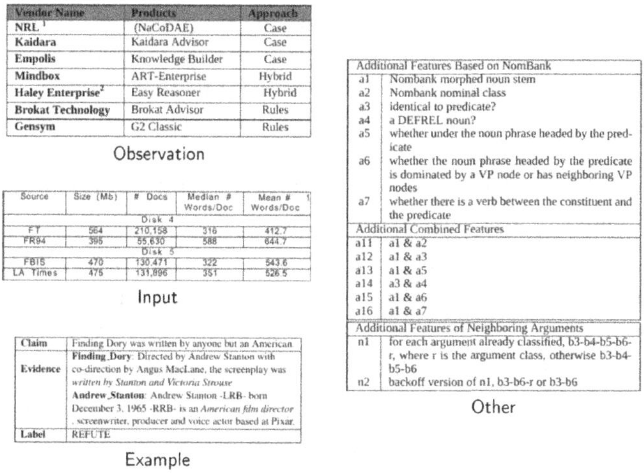

Fig. 2. Different types of tables with their classes.

312 313 314 315 316 317 318 319 320 321 322 323 324 325 326 327 328
0 0 1 1 1 0 1 1 1 1 0 1 1 1 1 0 0
POS
Dev. Acc.
Test Acc.
BERT
97.56
97.91
NN
97.33
97.64

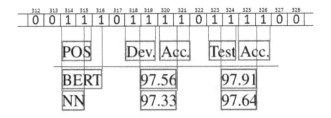

Fig. 3. Recognition of columns. Group of 1s in the projected vector indicate different columns.

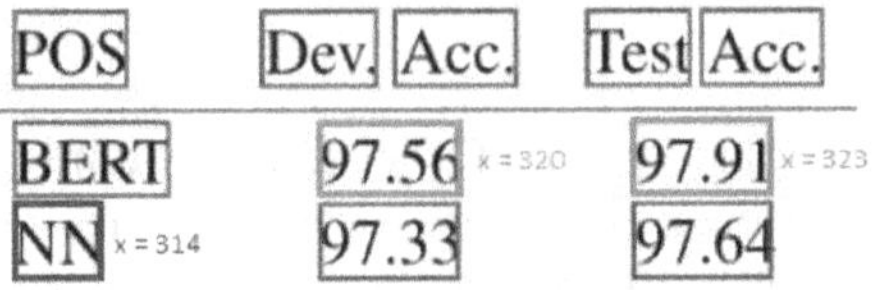

POS
Dev. Acc.
Test Acc.
BERT
97.56
x = 320
97.91
x = 323
NN
x = 314
97.33
97.64

Fig. 4. Recognition of rows. The blue bounding box is detected belonging to a new row since its x coordinate is lower than the previous green block. (Color figure online)

3.2 Data Augmentation

Since the dataset (Sect. 4.1) is unbalanced, a Data Augmentation strategy has been implemented to generate new training data from the available ones. For DA, new graphs can be obtained by modifying their structure and the information associated with the nodes and edges. Since the embedding for each table is evaluated through a message passing algorithm that strongly relies on the table structure and content, removing elements of the graph and changing node features helps to generate more variability of examples for each class. This not only improves the generalization capability of the model, but can help to reduce the class imbalance.

Random Removal of Nodes and Edges. In these operations, a random sample of nodes or arcs within the table is removed from the graph. By doing so, it is possible to generate a new graph similar to the initial one, but with different information.

- nodes removal: a random subset of node indexes is removed. The size of the sample depends on the number of nodes in the graph, a random number between 1% and 20% of the total number of nodes.
- edges removal: a random subset of edge indexes is removed. The size of the sample depends on the number of edges in the graph, a random number between 1% and 20% of the total number of edges.

The amount of randomly removed nodes/edges is an arbitrary choice. We did not want to: (i) discard too much information and (ii) introduce any bias in the decision.

Inversion of Rows and Columns. The row and column inversion technique is more complex, due to the fact that the internal structure of the tables is not known. Therefore, it is necessary to define an approach to approximate this structure. Once identified, rows or columns can be inverted, by means of swapping their node features.

– Column inversion: table columns identification is made with a projection-profile based approach which defines a vector of size equal to the width of the table region. Each element of the vector is initialized to 0. Then, for each word, the coordinates x_1 and x_2 of the corresponding bounding box are extracted and projected, setting to 1 the vector values whose indices correspond to these coordinates. The obtained result is shown in Fig. 3: adjacent 0s should identify column boundaries, while adjacent 1s the coordinates of each column. Thus, two columns can be inverted by swapping their contents, that is, the features of the nodes whose center of the bounding box belongs to those columns. The limitation of this technique is visible whenever there is a space between words belonging to the same column.
– Rows inversion: To reverse rows, it is necessary to compare the positions of "successive" bounding boxes. PyMuPDF reads and orders the content from left to right and from top to bottom. So, when a bounding box appears positioned ahead of the next one, it means that the latter is on a new row. In Fig. 4 the orange, green and blue bounding boxes are successive ones: the last one is on a new row since its x coordinate is lower than the green one. Once the structure of the rows has been identified, they can be reversed by swapping the features of the nodes belonging to them. The limitation of this technique is visible in the case of multi-row tables.

3.3 Model

Our baseline model uses two Graph Convolutional layers [8] and a Linear output one. Each node embedding is updated through the message passing algorithm: (i) firstly each node collects the embeddings of connected nodes; in our case study, each cell of the tables collects the embeddings of the visible ones; (ii) then a weighted sum is applied to aggregate the collected information and to update each node vector. (iii) At the end, a fully connected layer is used to learn the new node representations between the layers of the network We applied this procedure twice and then all the nodes of each graph are aggregated using a *redout* function. Every graph in the data may have its unique structure, as well as its node and edge features. In order to make a single prediction per each table, we aggregated and summarized over all the node information. Given a graph, the average node feature readout we use is

$$h_g = \frac{1}{|V|} \sum_{v \in V} h_v$$

where h_g is the representation of graph g, V is the set of nodes in g, h_n is the feature of node n.

4   Experiments

In this section, we present the dataset used in the experiments that are subsequently discussed.

4.1   The Tab2Know Dataset

The Tab2Know dataset, proposed in [9], contains information regarding tables extracted from scientific papers in the Semantic Scholar Open Research Corpus. Tables are extracted using PDFFigures [2], a tool that finds figures, tables, and captions within PDF documents, and Tabula[2], that outputs a CSV per each table reflecting its structure and content. After the conversion, each table is saved as an RDF triple addressable by an unique URI. Each CSV is then analyzed to recognize headers, type of table and columns type. The authors define an ontology of 27 different classes, 4 of which are defined as "root" ones (Example, Input, Observation and Other): the others are given depending on the type of columns found inside each table (e.g. Recall is a subclass of Metric that is a subclass of Observation). Their training corpus is composed of 73k tables, labeled using Snorkel [12] and starting from a small pre-labeled set of tables obtained through human supervision using SPARQL queries. Human annotators then looked at 400 of them, checking their labeling correctness and, after resolving their conflicts when disagreeing, used this subset as the test set.

4.2   Using the Dataset

To extract and group information on tables from Tab2Know, we built a conversion system to derive a JSON object for each available table. The information is the table numbering in the document, the page number where the table is located, the number of rows that make up the header, the document URL, the table class definition, and the caption text. We also added some information not represented in the RDF graph, such as the position of the table and the location of the caption (the latter information is obtained using PDFFigures and Tabula). Then we downloaded tables corresponding PDF papers, accessed from the Semantic Scholar Open Research Corpus. From each paper, the pages containing the tables are extracted. Unfortunately it is not possible to use the whole Tab2Know dataset. For instance, some papers are no longer available or an updated version do not match anymore the annotations provided. From the total, only the data whose annotations match are used, discarding the others. We obtained a subset containing 33,069 tables extracted from 11,800 scientific documents (45% of the original one). In addition, this dataset is very unbalanced (80% Observation, 10% Input, 7% Other, 3% Example) and it contains several missing or wrong annotations (55% of column classes have been labeled as 'others', across 22 different classes). For these reasons, we only use in this preliminary work the test set that was manually classified and corrected by humans. Specifically, this dataset contains 361

---

[2] https://github.com/tabulapdf/tabula.

Table 1. Results without data augmentation (*No Aug.*); Data Augmentation with Rows and Columns inversion (*R/C*); Data Augmentation with Rows and Columns inversion and random removal of nodes and edges (*All*). *P*, *R* and *F*1 correspond to Precision, Recall and F1 score.

	No Aug.					
	train size: 63					
	web_lg			sci_lg		
Classes (#)	P	R	F1	P	R	F1
Observation (185)	0.85	0.84	0.84	0.87	0.92	0.89
Input (35)	0.37	0.49	0.42	0.47	0.54	0.51
Example (10)	0.42	0.5	0.45	0.67	0.2	0.31
Other (23)	0.00	0.00	0.00	0.43	0.26	0.32
All (253)	0.41	0.46	0.43	0.61	0.48	**0.51**

	R/C											
	train size: 200						train size: 400					
	web_lg			sci_lg			web_lg			sci_lg		
Classes (#)	P	R	F1	P	R	F1	P	R	F1	P	R	F1
Observation (185)	0.82	0.84	0.83	0.85	0.92	0.89	0.82	0.84	0.84	0.84	0.94	0.88
Input (35)	0.37	0.43	0.39	0.52	0.46	0.48	0.34	0.34	0.34	0.52	0.40	0.45
Example (10)	0.80	0.40	0.53	0.60	0.30	0.40	0.57	0.40	0.47	0.50	0.30	0.37
Other (23)	0.06	0.04	0.05	0.41	0.30	0.35	0.05	0.04	0.05	0.50	0.30	0.38
All (253)	0.51	0.43	0.45	0.60	0.50	**0.53**	0.45	0.41	0.42	0.59	0.48	0.52

	All											
	train size: 200						train size: 400					
	web_lg			sci_lg			web_lg			sci_lg		
Classes (#)	P	R	F1	P	R	F1	P	R	F1	P	R	F1
Observation (185)	0.81	0.83	0.82	0.85	0.94	0.89	0.81	0.82	0.81	0.84	0.93	0.88
Input (35)	0.32	0.37	0.34	0.48	0.40	0.44	0.33	0.40	0.36	0.52	0.43	0.47
Example (10)	0.80	0.40	0.53	0.67	0.40	0.50	0.60	0.30	0.40	0.50	0.30	0.37
Other (23)	0.06	0.04	0.05	0.57	0.35	0.43	0.05	0.04	0.05	0.36	0.22	0.27
All (253)	0.50	0.41	0.44	0.64	0.52	**0.56**	0.45	0.39	**0.41**	0.55	0.47	0.50

tables extracted from 253 scientific papers. The distribution of tables according to the class is as follows: 235 *Observation*, 43 *Input*, 13 *Example*, 29 *Other* (41 were 'unclassified', and we do not consider them during training). We retain 20% of this subset as training (randomly sampled keeping the same class occurrences) and, through the data augmentation techniques described before, we evaluated the generalization capabilities of the proposed model.

4.3   Results

The main experiments performed are summarized in Table 1 that compares results obtained by applying different Data Augmentation techniques and spaCy models en_core_web_lg and en_core_sci_lg with baseline results. In bold we

Table 2. Summary of F1 score for different data augmentation approaches. *No Aug.* indicates no Data Augmentation technique was applied, *R/C* indicates row and column inversion technique and *All* indicates row and column inversion technique and random removal of nodes and arcs.

	No Aug.	R/C		All	
Train size	64	200	400	200	400
en_core_web_lg	0.43	**0.45**	0.44	**0.49**	0.41
en_core_sci_lg	0.52	**0.53**	0.52	**0.56**	0.51

highlight the most significant results of the F1 score for each technique applied. These values are also summarized in Table 2 to discuss the outcomes of the experiment. Table 2 summarizes the best F1 score values obtained considering some DA combinations. We can observe that the models appear rather inaccurate. This is mainly caused by the dataset itself, that is unbalanced toward the Observation class and small in size. It can also be seen that models using the en_core_sci_lg embedding show better results than those using en_core_web_lg, since the first one is a biomedical-based embedding that is most likely capable of appropriately characterizing and recognizing terms present in the tables extracted from scientific documents. In particular, models that exploit en_core_web_lg and do not use data augmentation techniques turn out to be less accurate and fail to recognize any table of class Other. In general, models that employ data augmentation result in higher F1 score values. Furthermore, observing Table 2, it can be seen that better values are obtained for the models in which data augmentation techniques are applied: particularly among these, the one obtained by alternating the inversion of rows and columns with random removal of nodes and arcs is preferable.

5   Conclusions

In this work we presented a GNN model for classifying tables in scientific articles applying Data Augmentation techniques. The results achieved are promising but still show limitations. In particular, the imbalance in the available data and a very low number of examples demonstrate that it is difficult to achieve good generalization. However, the use of Data Augmentation techniques made it possible to improve the results obtained by an increase in the F1-Score measure in the ablation studies presented. The proposed solution has some aspects that could be deepened or improved to further develop the work started. First, it might be useful to test the implemented Data Augmentation techniques on other datasets to analyze their potential and efficiency. In addition, other Data Augmentation techniques could be implemented, such as adding or removing rows and columns to a table, or improving the techniques already implemented. For example, the row and column recognition techniques could be improved, especially in the case of multi-row tables. We conclude by noting how such Data Augmentation techniques applied directly to graphs could prove to be an interesting clue for the application of Graph Neural Networks in the presence of resource-limited datasets, a very common situation in many application domains.

References

1. Baldi, S., Marinai, S., Soda, G.: Using tree-grammars for training set expansion in page classification. In: Proceedings of the Seventh International Conference on Document Analysis and Recognition, pp. 829–833 (2003)
2. Clark, C., Divvala, S.: PDFFigures 2.0: mining figures from research papers. In: Proceedings of the 16th Joint Conference on Digital Libraries, JCDL 2016, pp. 143–152. ACM (2016)
3. Gemelli, A., Vivoli, E., Marinai, S.: Graph neural networks and representation embedding for table extraction in PDF documents. In: 26th International Conference on Pattern Recognition, ICPR 2022, Montreal, QC, Canada, 21–25 August 2022, pp. 1719–1726. IEEE (2022). https://doi.org/10.1109/ICPR56361.2022.9956590
4. Hashmi, K.A., Liwicki, M., Stricker, D., Afzal, M.A., Afzal, M.A., Afzal, M.Z.: Current status and performance analysis of table recognition in document images with deep neural networks. IEEE Access 9, 87663–87685 (2021)
5. Honnibal, M., Montani, I.: Natural language understanding with bloom embeddings, convolutional neural networks and incremental parsing. Unpublished software application (2017). https://spacy.io
6. Kardas, M., et al.: AxCell: automatic extraction of results from machine learning papers. arXiv preprint arXiv:2004.14356 (2020)
7. Khan, U., Zahid, S., Ali, M.A., Ul-Hasan, A., Shafait, F.: TabAug: data driven augmentation for enhanced table structure recognition. In: Lladós, J., Lopresti, D., Uchida, S. (eds.) ICDAR 2021. LNCS, vol. 12822, pp. 585–601. Springer, Cham (2021). https://doi.org/10.1007/978-3-030-86331-9_38
8. Kipf, T.N., Welling, M.: Semi-supervised classification with graph convolutional networks. CoRR abs/1609.02907 (2016). http://arxiv.org/abs/1609.02907
9. Kruit, B., He, H., Urbani, J.: Tab2Know: building a Knowledge Base from tables in scientific papers. arXiv abs/2107.13306 (2020)
10. McKie, J.X.: PyMuPDF documentation. github (2022)
11. Qiao, L., et al.: LGPMA: complicated table structure recognition with local and global pyramid mask alignment. In: Lladós, J., Lopresti, D., Uchida, S. (eds.) ICDAR 2021. LNCS, vol. 12821, pp. 99–114. Springer, Cham (2021). https://doi.org/10.1007/978-3-030-86549-8_7
12. Ratner, A., Bach, S.H., Ehrenberg, H., Fries, J., Wu, S., Ré, C.: Snorkel: rapid training data creation with weak supervision. In: Proceedings of the VLDB Endowment. International Conference on Very Large Data Bases, vol. 11, p. 269. NIH Public Access (2017)
13. Riba, P., Dutta, A., Goldmann, L., Fornés, A., Ramos, O., Lladós, J.: Table detection in invoice documents by graph neural networks. In: 2019 International Conference on Document Analysis and Recognition (ICDAR), pp. 122–127. IEEE (2019)
14. Smock, B., Pesala, R., Abraham, R.: PubTables-1M: towards comprehensive table extraction from unstructured documents. CoRR abs/2110.00061 (2021). https://arxiv.org/abs/2110.00061

15. Zhang, P., et al.: VSR: a unified framework for document layout analysis combining vision, semantics and relations. In: ICDAR, vol. 12821, pp. 115–130 (2021)

16. Zhao, T., Liu, Y., Neves, L., Woodford, O.J., Jiang, M., Shah, N.: Data augmentation for graph neural networks. CoRR abs/2006.06830 (2020). https://arxiv.org/abs/2006.06830

17. Zoph, B., Cubuk, E.D., Ghiasi, G., Lin, T.-Y., Shlens, J., Le, Q.V.: Learning data augmentation strategies for object detection. In: Vedaldi, A., Bischof, H., Brox, T., Frahm, J.-M. (eds.) ECCV 2020. LNCS, vol. 12372, pp. 566–583. Springer, Cham (2020). https://doi.org/10.1007/978-3-030-58583-9_34

Improved Training for 3D Point Cloud Classification

Sneha Paul[✉][ID], Zachary Patterson[ID], and Nizar Bouguila[ID]

Concordia Institute for Information Systems Engineering (CIISE), Concordia University,
Montreal, Canada
sneha.paul@mail.concordia.ca,
{zachary.patterson,nizar.bouguila}@concordia.ca

Abstract. The point cloud is a 3D geometric data of irregular format. As a result, they are needed to be transformed into 3D voxels or a collection of images before being fed into models. This unnecessarily increases the volume of the data and the complexities of dealing with it. PointNet is a pioneering approach in this direction that feeds the 3D point cloud data directly to a model. This research work is developed on top of the existing PointNet architecture. The ModelNet10 dataset, a collection of 3D images with 10 class labels, has been used for this study. The goal of the study is to improve the accuracy of PointNet. To achieve this, a few variations of encoder models have been proposed along with improved training protocol, and transfer learning from larger datasets in this research work. Also, an extensive hyperparameter study has been done. The experiments in this research work achieve a 6.10% improvement over the baseline model. The code for this work is publicly available at https://github.com/snehaputul/ImprovedPointCloud.

Keywords: Point cloud classification · Transfer learning · 3D classification

1   Introduction

A point cloud is a set of data points in 3-dimensional space commonly collected by LiDAR or radar scan. This unstructured data consists of a set of points, which is unlike 2D image data (2D pixel arrays) or 3D voxel arrays. Each point in a point cloud contains x, y, and z coordinates and sometimes some additional information such as surface normal and intensity. These 3D geometric data are irregular in data format. Since convolution networks require pre-defined structured data formats, point clouds or meshes need

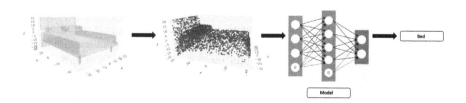

Fig. 1. Overview of 3D object classification method.

© The Author(s), under exclusive license to Springer Nature Switzerland AG 2022
A. Krzyzak et al. (Eds.): S+SSPR 2022, LNCS 13813, pp. 253–263, 2022.
https://doi.org/10.1007/978-3-031-23028-8_26

to be transformed into regular 3D voxel grids or collections of images. This transformation makes the data voluminous and reduces the natural variance of data. Feeding the data directly to the model reduces the issues arising from data transformation (Fig. 1).

In this study, we have worked on top of the existing PointNet [10] architecture that feeds the 3D point clouds directly to the model and predicts class labels. The Model-Net10 [12] dataset containing 10 classes has been used. Like PointNet, we have fed the point clouds directly to the proposed model as input. The output of the model consists of the class labels for the entire dataset. The goal of this study is to improve the accuracy of PointNet. The prediction accuracy of the existing PointNet architecture is 86.1% which we have taken as the baseline accuracy for this research work. Our objective in this research work is to improve the accuracy of PointNet by proposing a combination of a new encoder, loss function, and improved training protocol. We have proposed a few variations of the encoder model and conducted an extensive sensitivity study on the model. The experiments in this study achieve a **6.10%** improvement over the baseline model. The contributions made are summarized as follows:

- We have proposed a variant of 1D CNN and RNN models that improves the accuracy of the original PointNet on the ModelNet10 dataset.
- We have also introduced different loss functions to get better results for the model.
- Extensive hyper-parameter study has helped to find the best settings for training the model.

The rest of the paper is summarized as follows. First, we describe the related work relevant to this research. This is followed by the proposed method and the results from the proposed method. Finally, we present the concluding remarks and future research direction of the research.

2   Related Work

PointNet [10] is a pioneering work in the direction of classifying 3D geometry points. The main idea is to feed point clouds directly into models without any transformation. PointNet++ [11] is an upgraded version of PointNet. It is a hierarchical neural network that divides the overlapping points at local regions based on the distance metric of the underlying space. After that, local features, capturing fine-grained geometric features from the small neighborhood, are extracted from these points. These local points are then grouped to capture higher-level features. Due to the irregular format of point clouds. Typical CNN operations cannot be done directly on point clouds. PointCNN [7] uses a convolution operation on the X-transformed point clouds so that the input features are weighted and permuted in a conical order. POINTVIEW-GCN [9] uses a multi-level Graph Convolutional Network to aggregate single-view point clouds in hierarchical order and encode both the geometric clues and their relation with multiple views. This makes the global shape features more descriptive, which in turn, improves classification accuracy.

SO-Net [6] is a permutation invariant network that uses the spatial distribution of point clouds by transforming them into SOM (self-organized maps). The model extracts the hierarchical features from SOMs with adjustable overlaps of receptive fields. The

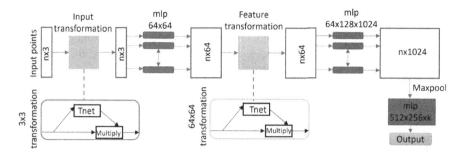

Fig. 2. Overview of the PointNet architecture.

proposed point clouded autoencoder has been shown to significantly improve training speed and performance accuracy. Brock et al. used a Variational Autoencoder (VAE) to learn the latent features of the voxels. This helps to find factors causing variation between different objects. For object classification, a deep CNN architecture has been used [1]. Yu et al. used bilinear pooling for aggregating the local convolutional features and harmonized the components found in the bilinear features to get a discriminative representation. Harmonized bilinear pooling is then used as a layer in the Multiview Harmonized Bilinear Network (MHBN) [13]. PointGrid [5] is a 3D convolutional neural network that uses a specific number of points at every grid cell to learn the high order local approximation and local geometric shape. Although previous works have tried to find different and specific solutions, there is still room for improvement over the original PointNet model. In this work, we propose a solution based on a few improvements over the existing architecture that results in better accuracy.

3   Method

In this section, we will discuss the preliminaries followed by the proposed method. The proposed method is composed of three components, proposed method improvement, proposed loss function, and proposed training protocol.

3.1   Preliminaries

The 3D point cloud dataset represents points in a 3D mesh, which requires a special form of the model to learn important features from the data. Convolutional neural networks (CNN) with 1×1 kernel are the primary building block of the PointNet [10] architecture. This architecture consists of two transformation blocks, namely Tnet and Transform, which consists of a few CNNs. Finally, the PointNet is just a composition of linear layers on top of the transformation blocks. An overview of the PointNet model is shown in Fig. 2.

3.2 Proposed Method

In this work, we explored 3 directions to improve the accuracy of point cloud classification as follows: (1) improved model, (2) improved loss function, and (3) improved training protocol.

Proposed Model Improvement. We have explored different variants and combinations of CNNs both for Tnet and Transformer and analyzed their effects on overall model accuracy. In particular, different kernel sizes and strides were tested with the CNN. For kernel size, 3 and 5 were used. For the stride, 2 and 3 were used for both the Tnet and Transformer blocks of the original PointNet architecture.

We have also explored variants of Tnet and Transform block with RNNs. More specifically, we used LSTM [3,4], GRU [2], and "vanilla" RNNs in place of the 1D CNN. For this, we have used an embedding dimension of 1024 with 2 hidden layers for all three models. Finally, a different number of linear layers are also explored in this experiment.

Proposed Loss Functions. The original PointNet proposed to use negative log-likelihood loss function in combination with a regularization loss applied on the intermediate output of the Tnet and Transform block. We experimented with different loss functions to train the model, for example, negative log-likelihood and Categorical cross-entropy functions are used.

The ModelNet10 dataset is not a balanced dataset with an equal amount of samples per class. So, we have proposed a popular loss function called Focal Loss [8] which is specialized in handling class imbalanced datasets. The formula for focal loss is represented as follows:

$$Loss_{focal} = -(1 - p_t)^\gamma log(p_t) \tag{1}$$

here, p_t is the output probability from the model, $\gamma \in [0, 5]$ is a hyper-parameter that controls the penalty. If a prediction with high probability p_t is wrong, then the loss factor will be very high.

3.3 Proposed Training Protocol

In this subsection, we describe the training protocol of the proposed method. They are: (a) Augmentations, (b) Transfer learning, and (c) Training setup.

Augmentations. An augmentation module generates a slight variant of the original image by applying a small transformation to the image. It is widely known for computer vision applications that applying augmentations can generate more data for training which in turn can reduce overfitting in the training. For point cloud data, there are not as many augmentation options available as the 2D image. In this work, we adopted the "rotation along z-axis" and "random noise" as the augmentations. Here, rotation along the z-axis changes the view of the angle for the object, which does not change the semantics of the data. The random noise augmentation adds some random noise to each of the points in the data point in point clouds.

Transfer Learning. Transfer learning is a widely used method for computer vision tasks, where a model pre-trained on a task is used for another similar task. Using a pre-trained model on a different but similar task is called fine-tuning. The pretrained model is used as a starting point for fine-tuning the model on a different task. This approach prevents the model from the random initialization of weights and the knowledge learned by the model in the pretraining stage reduces the overfitting tendency of models. Thus it increases the accuracy of the model.

Table 1. Summary of hyper-parameters study.

Experiment	Best parameter	Accuracy (%)
Dropout	0.3	86.1
Epoch	20	86.67
Weight decay	0.00	86.1
Batch size	32	86.1
Learning rate	0.0001	87.78

Training Setup. The baseline model is trained with the Adam optimizer and a constant learning rate of 0.001. The model is trained for 15 epochs with a batch size of 32. In the experiment, we have shown the impact of different hyper-parameters and chosen the best parameters where applicable.

4   Experimental Results

For this study, we used the PointNet [10] as the baseline model. First, we ran the model with the original PointNet on the ModelNet10 dataset and obtained an accuracy of 86.1% on PointNet. Considering this implementation of the PointNet on ModelNet10 as a baseline, this study aims to improve this accuracy. Below we present a short description of the dataset, followed by the main results.

Dataset. ModelNet10 [12] is a popular 3D point cloud dataset. It includes 10 categories of 3,991 and 908 samples for training and testing partitions. The data consists of sparse coordinates of points, where each point $x \in \mathcal{R}^3$. Here, the 3 dimensions are the x, y, and z coordinates of a point. Each sample in the dataset contains N points. So the overall dimension of 3D data is $N \times 3$, where N is different for each sample. However, the input of the model needs to be of the same size for each sample. So, we randomly sample n points from the set of N. As a result, the final input to the model is a vector $x \in \mathcal{R}^{n \times 3}$.

Since the dataset is already split into training and testing data, no further splitting is required. The data need to be normalized before using this as input. No other pre-processing is required. The data is sparse, although there is no missing data. The output of the model is the probability distribution over the number of classes, in this case, 10.

4.1   Hyper-parameter Sensitivity

At first, we started with the hyper-parameter sensitivity study on the baseline method. We have graphed the results in Fig. 3. The sensitivity study results are summarized in Table 1 which shows the effect of different settings of dropout, the number of epochs, weight decay, batch size, and learning rate.

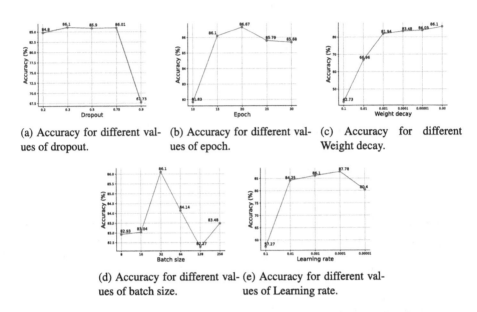

(a) Accuracy for different values of dropout.
(b) Accuracy for different values of epoch.
(c) Accuracy for different Weight decay.

(d) Accuracy for different values of batch size.
(e) Accuracy for different values of Learning rate.

Fig. 3. Sensitivity study on different parameters on the methods on all the datasets.

We also experimented with the effect of different dropout values on model accuracy. We have experimented dropout for 0.2, 0.5, 0.75 and 0.9. Among them, the model gives the highest accuracy which is 86.01% for a 0.75 dropout value. A trend can also be seen in the model for different values of dropout. Model accuracy increases as the dropout rate decreases to 0.75. After that, model accuracy falls to 67.73% for a 0.9 dropout rate. A higher value (up to a certain limit) of dropout is shown to be useful in the training when the size of the dataset is relatively low. As a result, the higher values of dropout up to 0.75 shows improvement in the accuracy.

For the experiment with training epochs, we find the highest accuracy of 86.67% for 20 epochs. This result is higher than the baseline accuracy of the model. The accuracy of the model decreases with an increase in the number of epochs higher than 20. This is because a very high number of training epochs have the tendency of causing the model to over-fit.

Weight decay is a regularization factor that penalizes by adding a penalty term to the loss function of the model that shrinks the weights during back-propagation and thus reduces over-fitting. An increasing trend in the model's accuracy is seen for decreasing values of weight decay. The highest accuracy observed for the model is 86.1% when no weight decay is imposed. So, for further experiments, weight decay can be avoided.

For the experiment with batch size, the highest accuracy is observed for 32. The model's accuracy increases as the batch size are increased up to 32. After that, the model's accuracy starts to decrease.

Table 2. Experiments on different model variants.

Experiment	Accuracy (%)
PointNet baseline	**86.1**
Baseline w/o dropout and normalization	85.90
Tnet; 1D Conv (k = 3, s = 2)	84.14
Tnet; 1D Conv (k = 3, s = 3)	86.12
Tnet; 1D Conv (k = 5, s = 3)	86.67
Tnet; Linear 2 layer	84.91
Transform; 1D Conv (k = 3, s = 2)	**87.00**
Transform; 1D Conv (k = 3, s = 3)	84.36
Transform; 1D Conv (k = 5, s = 3)	83.70
Tnet & Transform; 1D Conv (k = 3, s = 2)	69.82
Tnet & Transform; 1D Conv (k = 3, s = 3)	53.96
Tnet & Transform; 1D Conv (k = 5, s = 3)	49.23
LSTM	83.26
GRU	79.96
RNN	77.43

The learning rate controls the convergence rate of a model. Smaller learning rates result in smaller updates to the model's convergence, thus requiring a large number of epochs. A higher learning rate results in larger update steps to the model's convergence, which can cause overshooting and delay the model's convergence. An increasing trend in the model's accuracy can be seen by decreasing the learning rate to 0.0001. The highest accuracy of the model is 87.78% for the 0.0001 learning rate, which is higher than the baseline accuracy. After that, the model's accuracy began to decrease as the smaller learning rate makes the model's convergence speed slower.

4.2   Results for Proposed Model Variants

The results for different variants of the model are presented in Table 2. As we can see from the table, replacing the 1D CNN of kernel size 1 in the original Tnet block of PointNet with higher kernel and strides shows mixed results. For instance, k = 3 and s = 2 hurt the model and get worse results, but there is a slight improvement for the same k with s = 3. There is a good amount of improvement for k = 5 and s = 3 compared to the baseline result. However, the best improvement is achieved when the 1D CNN of the

'Transform' block is changed with k = 5 and s = 2. It shows a 0.9% improvement over the baseline. An interesting result is shown when both the Tnet and Transform blocks of the PointNet are changed to higher kernel size, in which case the results drop quickly. This indicates that the sparse point structure of the point cloud needs to be processed with kernel size 1 (no relation comes into consideration from the neighboring points) in at least one of the pre-processing blocks.

As mentioned in the method, we have replaced the 1D CNNs in the Tnet with RNNs, more specifically, LSTM, GRU, and vanilla RNN. The results on these variants of the model do not perform as well as the 1D CNNs. Here the best results are achieved by LSTM with an accuracy of 83.26% (Table 2).

Table 3. Summary of learning rate scheduler study.

Experiment	Accuracy (%)
Step LR sc. (step size = 4, gamma = 0.1)	88.11
Step LR sc. (step size = 6, gamma = 0.1)	87.33
Step LR sc. (step size = 4, gamma = 0.5)	**89.43**
Step LR sc. (step size = 6, gamma = 0.5)	88.33
Plateau sc.	87.67

4.3   Results for Explored Loss Functions

As mentioned in the Sect. 3.2, we have explored different variants of loss functions. The results for different loss functions are summarized in Table 4. At first, we have removed the regularization part of PointNet leaving the Negative log-likelihood (NLL) only, which shows about a 0.5% drop in accuracy. Next, we replaced the NLL with Categorical Cross-entropy loss, which shows

Table 4. Effect of different loss functions.

Experiment	Accuracy (%)
PointNet	86.1
Negative log-likelihood	85.68
Categorical cross-entropy	**87.11**
Focal loss	86.78

a 1% improvement in the result compared to the original PointNet. The focal loss also shows 0.7% improvement compared to the baseline.

4.4   Explored Training Protocol

The learning rate scheduler is used in models to adjust the learning rate between epochs as the training progresses. Here we have used two different scheduler methods to adjust the learning rate. One is learning rate scheduler and another is plateau scheduler. The plateau scheduler gets an accuracy of 87.67%, which is better than the contestant LR that is used in the baseline model (Table 3).

However, better results are achieved with Step LR. For the learning rate scheduler, four different combinations are used for two different parameters. They are step size and gamma. Step size is the number of epochs after which the learning rate will be updated. Gamma is the amount by which the learning rate will be changed. We have used the combination of step sizes 4 and 6 with gamma 0.1 and 0.5. The highest accuracy comes for step size 4 with gamma 0.5, which is **89.43**%.

Table 5. Effect of augmentation.

Experiment	Accuracy (%)
Baseline	86.1
Without random noise	86.78
Without z axis rotation	**91.52**
Without any of them	90.42

4.5 Effect of Augmentation

In this study, we have adopted two commonly known augmentation methods and considered that with the baseline. We also present ablation on augmentation by removing them one at a time. Surprisingly, all the results show better accuracy than the baseline. The best result here is achieved by removing rotation along the z-axis. This resulted in

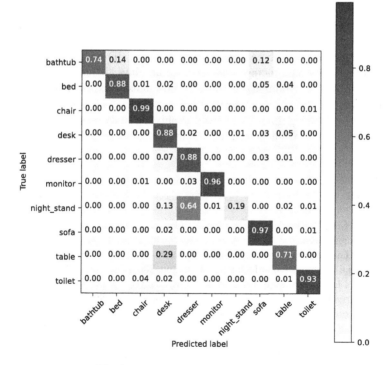

Fig. 4. Confusion matrix on the test data.

an accuracy of 91.52% (Table 5). We suspect that the use of augmentation might require a bigger model with some other training settings to get the best out of augmentation.

4.6   Transfer Learning

In the transfer learning setting, we first pre-train the model on the ModelNet40 dataset. The weights learned by the model are used to fine-tune the model on the ModelNet10 dataset. A learning rate of 0.0001 is used while fine-tuning. The accuracy of the model after fine-tuning is **92.20%** which is the highest accuracy achieved by the proposed model. This is a significant **6.10%** improvement over the PointNet baseline.

4.7   Confusion Matrix

Figure 4 represents a comparison between the true label and the predicted label of the model. As we can see from the figure, the model performs reasonably well for most of the classes. Some of the classes with higher error rates include bathtubs, nightstands, and tables. For 'bathtub', the highest amount of confusion is observed for 'bed', followed by 'sofa'. One other high miss-classification observed is for 'night stand' which is mostly confused with 'dresser'. The 'night stand' also misclassifies 'desk' sometimes. Class 'table' is often confused with the class 'desk'.

5   Conclusions

In this paper, we proposed three improvement areas for 3D point cloud classification. We have done this by introducing a new architecture, investigating different loss functions, and improving the training protocol. We did extensive experiments and the results show the impact of the proposed design choice of the model that showed significant improvement over the baseline. We also did an ablation study on the augmentations and loss functions. Overall we achieved a 6.1% improvement over the baseline results of PointNet.

References

1. Brock, A., Lim, T., Ritchie, J.M., Weston, N.: Generative and discriminative voxel modeling with convolutional neural networks. arXiv preprint arXiv:1608.04236 (2016)
2. Chung, J., Gulcehre, C., Cho, K., Bengio, Y.: Gated feedback recurrent neural networks. In: International Conference on Machine Learning, pp. 2067–2075. PMLR (2015)
3. Graves, A.: Long short-term memory. In: Graves, A. (ed.) Supervised Sequence Labelling with Recurrent Neural Networks. SCI, vol. 385, pp. 37–45. Springer, Heidelberg (2012). https://doi.org/10.1007/978-3-642-24797-2_4
4. Hochreiter, S., Schmidhuber, J.: Long short-term memory. Neural Comput. **9**(8), 1735–1780 (1997)
5. Le, T., Duan, Y.: PointGrid: a deep network for 3D shape understanding. In: Proceedings of the IEEE Conference on Computer Vision and Pattern Recognition, pp. 9204–9214 (2018)

6. Li, J., Chen, B.M., Lee, G.H.: SO-Net: self-organizing network for point cloud analysis. In: Proceedings of the IEEE Conference on Computer Vision and Pattern Recognition, pp. 9397–9406 (2018)
7. Li, Y., Bu, R., Sun, M., Wu, W., Di, X., Chen, B.: PointCNN: convolution on X-transformed points. In: Advances in Neural Information Processing Systems 31 (2018)
8. Lin, T.Y., Goyal, P., Girshick, R., He, K., Dollár, P.: Focal loss for dense object detection. In: Proceedings of the IEEE International Conference on Computer Vision, pp. 2980–2988 (2017)
9. Mohammadi Seyed, S., Wang, Y., Del Bue, A.: Pointview-GCN: 3D shape classification with multi-view point clouds (2021)
10. Qi, C.R., Su, H., Mo, K., Guibas, L.J.: PointNet: deep learning on point sets for 3D classification and segmentation. In: Proceedings of the IEEE Conference on Computer Vision and Pattern Recognition, pp. 652–660 (2017)
11. Qi, C.R., Yi, L., Su, H., Guibas, L.J.: PointNet++: deep hierarchical feature learning on point sets in a metric space. In: Advances in Neural Information Processing Systems 30 (2017)
12. Wu, Z., et al.: 3D ShapeNets: a deep representation for volumetric shapes. In: Proceedings of the IEEE Conference on Computer Vision and Pattern Recognition, pp. 1912–1920 (2015)
13. Yu, T., Meng, J., Yuan, J.: Multi-view harmonized bilinear network for 3D object recognition. In: Proceedings of the IEEE Conference on Computer Vision and Pattern Recognition, pp. 186–194 (2018)

On the Importance of Temporal Features in Domain Adaptation Methods for Action Recognition

Donatello Conte[1]([✉]) [iD], Giuliano Giovanni Fioretti[2] [iD], and Carlo Sansone[2] [iD]

[1] Université de Tours, Laboratoire d'Informatique Fondamentale et Appliquée de Tours (LIFAT - EA6300), 64 Avenue Jean Portalis, 37200 Tours, France
donatello.conte@univ-tours.fr
[2] Department of Electrical Engineering and Information Technology (DIETI), University of Naples Federico II, via Claudio 21, 80125 Naples, Italy
giul.fioretti@studenti.unina.it, carlo.sansone@unina.it

Abstract. One of the most common vision problems is Video based Action Recognition. Many public datasets, public contests, and so on, boosted the development of new methods to face the challenges posed by this problem. Deep Learning is by far the most used technique to address Video-based Action Recognition problem. The common issue for these methods is the well-known dependency from training data. Methods are effective when training and test data are extracted from the same distribution. However, in real situations, this is not always the case. When test data has a different distribution than training one, methods result in considerable drop in performances. A solution to this issue is the so-called Domain Adaptation technique, whose goal is to construct methods that adapt test data to the original distribution used in training phase in order to perform well on a different but related target domain. Inspired by some existing approaches in the scientific literature, we proposed a modification of a Domain Adaptation architecture, that is more efficient than existing ones, because it improves the temporal dynamics alignment between source and target data. Experiments show this performance improvement on public standard benchmarks for Action Recognition.

Keywords: Domain Adaptation · Action Recognition · Temporal Shift Model

1 Introduction

Human action recognition has become a challenging topic during the last years due to the impact that it represents in the comprehension of human behaviour (see [19,21,27] for some examples). Unsurprisingly, the action recognition field is extremely more difficult task than object recognition, for various reasons. In addition to the typical difficulties related to image recognition, such as scale

A. Krzyzak et al. (Eds.): S+SSPR 2022, LNCS 13813, pp. 264–273, 2022.
https://doi.org/10.1007/978-3-031-23028-8_27

variation, lighting or contrast, other obstacles must be taken into account to perform action recognition: the shooting point of view, the color scheme, the background clutter, and most importantly, the temporal dimension that further complicates the task.

Deep learning is by far the most used technique to address this problem (e.g. [7,16,24,29], and [28] for a good survey of these methods). While very effective, these techniques suffer from the problem of being too dependent on training data. The vast majority of supervised learning methods share a common prerequisite: training data and testing data are extracted from the same distribution [26]. However this may not always be the case. When this constraint is violated, the classifier trained on a dataset, which will be referred to as the *source domain*, exhibits a considerable drop in performance when tested on a different dataset, called the *target domain*.

A solution to this issue can be the use of the so-called Domain Adaptation. Domain adaptation refers to the goal of learning a concept from labeled data in a source domain that performs well on a different but related target domain. There are many domain adaptation proposals in the scientific literature, also in the context of Action Recognition (see Sect. 2).

The main goal of this paper is to propose a new architecture of Domain Adaptation for Action Recognition. The contribution is to propose integrating a different temporal module within an existing architecture, in order to improve the temporal adaptability between source and target domains, and consequently, improving the performances.

The rest of the paper is organized as follows: Sect. 2 illustrates a brief survey of the main domain adaptation techniques for Action Recognition. In Sect. 3 we recall the basic principles of an existing architecture for Domain Adaptation which served as our basis for proposing our modification described in Sect. 4. In Sect. 5, after describing the test protocol, we present some experimental results that prove the effectiveness of our approach. Some conclusions and perspectives are drawn in the last Sect. 6.

2   Related Works

Domain adaptation methods make the assumption that the tasks are the same and the differences are only caused by domain divergence. According to Wang et al. [5], considering the labeled data of the target domain, domain adaptation algorithms can be classified in three categories:

- Supervised (e.g. [17]): a small amount of labeled target data are present. The issue is that the labeled data are commonly not sufficient for tasks.
- Semi-supervised (e.g. [20]): there are limited labeled data and redundant unlabeled data in the target domain. This only allows the network to learn the structure information of the target domain.
- Unsupervised (e.g. [14]): there are labeled source data and only unlabeled target data available for training the network.

Supervised action recognition state-of-the-art methods all use deep learning algorithms, by leveraging CNNs for spatio-temporal information (see [2,25] for some examples). However, as we said in the Sect. 1, these supervised action recognition approaches are still limited by the dependency on annotated labels for each clip. There is no guarantee of robust performances if the algorithms are directly transferred to another domain, due to the presence of domain shift.

In this context, many Domain Adaptation techniques are proposed and applied to the Action Recognition problem. First works in this direction was focused geometric transformations of videos in the context of Supervised Domain Adaptation. Some works utilise supervisory signals such as skeleton or pose [12] and corresponding frames from multiple viewpoints [8,22].

On the contrary, Unsupervised Domain Adaptation (UDA) has been used for recovering and adapting changes, more general than only geometrical ones. At the first time, UDA for action recognition used shallow models to align source and target distributions of handcrafted features [1,4]. By the advent of Deep Learning, more proposals have been done for Domain Adaptation, but especially for image-based tasks. Authors of [13,15] proposed some Deep Networks to align the joint distributions by minimizing maximum mean discrepancy (MMD) or joint maximum mean discrepancy (JMMD) between source and target domains. Recently, some works deals with video domain adaptation. In [6] authors utilize an adversarial learning framework with 3D CNN to align source and target domains. TA3N [3] leverages a multi-level adversarial framework with temporal relation and attention mechanism to align the temporal dynamics of feature space for videos.

Inspired by this last work, and being convinced that temporal alignment remains the key feature in domain adaptation, our contribution can be stated as follow: we propose adapting an existing temporal alignment module to a Domain Adaptation Deep Network for improving the temporal dynamics correspondences between source and target videos.

3 Recalls Basics of an Architecture for Domain Adaptation

The architecture proposed by Chen et al. [3] consists of 3 main components: a spatial module, a temporal module, and a class predictor. The input data comes from frame-level feature vectors, which are extracted directly from the raw action videos via a ResNet convolutional network.

Spatial Module. This component uses fully-connected layers to translate feature vectors into features useful for the action recognition task.

Temporal Module. This module consists of a TRN that extracts the most representative frames of the videos and identifies the temporal relationships between them.

To capture temporal relations at different time scales, in the temporal relation module, multiple representations of relation features are generated, each of which leverages a different number of ordered frames.

In mathematical formula, considering a video input V composed by n ordered frames, the temporal relation between k frames can be expressed as:

$$T_k(V) = h_\phi \left(\sum_{i<j<...<k} g_\theta(f_i, f_j, \ldots, f_k) \right) \tag{1}$$

where f_i indicates the feature representation of the ith frame and the h and g functions aim to fuse together the features of multiple frames.

This formula can then be used to accumulate frame relations at multiple time scales to obtain multiscale temporal relations:

$$MT_N(V) = T_2(V) + T_3(V) + \cdots + T_N(V) \tag{2}$$

Then, an attention mechanism focused on domain discrepancy is used. This is made up of a series of domain attention blocks with relation discriminators for each representation of relation features. Finally, the outputs of the blocks are added together.

Class Predictor. Finally, this classifier is also a fully connected layer that converts features extracted from previous modules into the final class. For this purpose an attentive entropy loss is generated by domain entropy and class entropy.

In addition to these modules, there is a set of discriminators trained with the adversarial learning technique inspired by DANN [5]. The aim of these networks is to learn domain-invariant features and a symmetric mapping of source and target distributions in order to align them both spatially and temporally.

Spatial Discriminator. It applies an image-based domain adaptation to learn the spatial parameters by maximizing the spatial domain discrimination loss.

Temporal Discriminator. It applies a temporal-based domain adaptation to learn the features encoded in the temporal dynamics by maximizing the temporal domain discrimination loss.

Relation Discriminator. It generates a domain attention value, used to attend local temporal features, maximizing the relation domain loss.

4   The New Designed Architecture

The architecture described above focuses on aligning the features that contribute more than others to the overall domain shift, i.e. those that have a greater discrepancy between domains.

A larger domain gap, however, can be caused by frames that are not relevant to the action recognition task. Therefore, aligning those irrelevant frames can lead to suboptimal results in some cases.

Moreover, this module performs temporal modeling only after feature extraction, identifying the most descriptive frames of actions and attending on those.

Therefore, several pieces of information may be lost during the feature extraction process.

These considerations led to the idea of designing a new architecture, with a temporal module that overcomes these disadvantages.

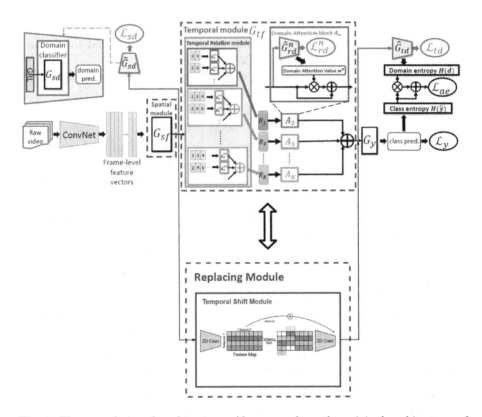

Fig. 1. The new designed architecture. Above we show the original architecture of domain adaptation. There are three main parts: on the left the feature extractor, in the middle the temporal module which aims to align the time instants of the sequences in the two domains, and on the right the classification part. The adaptation of the architecture that we propose, consists in replacing the temporal module (Temporal Relation Network, TRN) with the Temporal Shift Module (TSM) that is illustrated below. Note that, contrary to the original architecture, features of the video are directly fed in TSM, avoiding frames alignment as in TRN.

Specifically, the use of the Temporal Shift Module proposed by Lin et al. [11] in place of the Temporal Relation Network (TRN) was considered (as shown in Fig. 1). This module has the advantage of enabling all levels of temporal modeling, even during the feature extraction itself, just like methods based on 3D CNNs, but maintaining a reasonable computational time. Consequently, there is no need to attend frame relations. Furthermore, this module avoids the risk of aligning unnecessary frames.

As explained by Lin et al. [11], an activation in a video model can be represented as $A \in \mathbb{R}^{N \times C \times T \times H \times W}$, where N is the batch size, C is the number of channels, T is the time, H is the height and W is the width of the image.

TSM performs temporal modelling shifting by ± 1 along the temporal dimension a fraction of the features extracted from the frame, about $\frac{1}{8}$, to merge them with the features extracted from the frames immediately preceding and immediately following.

Consequently, after the shifting operation, the information of the current frame is fused together with the neighboring frames. An example of a tensor with C channels and T frames can be seen in Fig. 2, where different colors for each row indicate the features extracted from different frames.

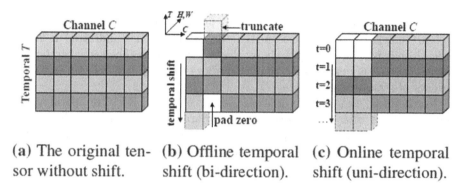

(a) The original tensor without shift.   (b) Offline temporal shift (bi-direction).   (c) Online temporal shift (uni-direction).

Fig. 2. Temporal modeling performed by TSM, obtained shifting a fraction of the features along the temporal dimension. The offline and online temporal shift are shown. (Figure from [11])

The new network formed by this temporal module was subsequently tested on the most common action datasets.

5   Experiments and Results

In this section we describe the used experimental protocol (Datasets and metrics) and we present the results of some results together with some comments on them.

5.1   Datasets and Metrics

The datasets considered for the experimental analysis are: UCF101 [23], Olympic Sports [18], and HMDB51 [10], all of which contain actions from real-world scenarios. UCF101 is an action recognition data set of realistic action videos, collected from YouTube, having 101 action categories. It contains $13,320$ videos from 101 action categories. The videos in 101 action categories are grouped into 25 groups, where each group can consist of 4–7 videos of an action. The videos from the same group may share some common features, such as similar background, similar viewpoint, etc. Olympic Sports dataset contains sport activities from YouTube sequences. It contains 16 sport classes, with 50 sequences per class. The HMDB51 dataset is a collection of realistic videos from various sources, including movies and web videos. The dataset is composed of 6,849 video clips from 51 action categories (such as "jump", "kiss" and "laugh"), with each category containing at least 101 clips.

For the domain adaptation test, only the subsets of overlapping categories across UCF101 and Olympic Sports datasets and the overlapping categories across UCF101 and HMDB51 datasets are used. Therefore, the experimental tests are conducted over two action datasets, namely $UCF-Olympic$ and $UCF-HMDB_{full}$, that are the most commonly used to benchmark the performance of domain adaptation and action recognition algorithms [9].

The labeled target data are used in the learning phase to set an upper bound on how well the model can perform. This experiment will be indicated as *Target-only*, since it is conducted entirely on the target domain. The same can be done with a *Source-only* experiment to obtain a lower bound, which is a result of the absence of adaptation.

Furthermore, the architecture that uses the original TRN as temporal module will be referred to as *TRN-Model*, while the architecture that uses the TSM as temporal module, that is our proposition, will be referred to as *TSM-Model*.

The Accuracy, that is the percentage of correct classified videos, will be used as the performance metric for method comparison.

5.2   Parameters Setting Details

The data used for this experiments consists of frame-level feature vectors pre-extracted from a ResNet-101 model pre-trained on ImageNet. The number of frame-level feature vectors sampled for each video is fixed to 5. Only RGB inputs are considered for the temporal alignment operation. The stochastic gradient descent (SGD) is utilized as the optimizer with momentum fixed to 0.9. The initial learning rate is 3×10^{-2}, and then it is decreased following the common strategy shown in DANN [5]. The weight decay is 10^{-4} and the batch size is scaled proportionally to the ratio between source and target datasets.

Parameters settings for TRN-Model and TSM-Model are the follows. For TRN-Model, The optimization was conducted as described by Chen et al. [3]: the optimized values of λ^r, λ^t and λ^s, have been found using a coarse-to fine grid search approach. This means that firstly was used a coarse grid with the geometric sequence $[0, 10^{-2}, 10^{-1}, 10^0, 10^1]$. Then, after finding the optimal range of values, being $[0, 1]$, a new fine-grid search was conducted in this range with the arithmetic sequence $[0, 0.25, 0.50, 0.75, 1]$. The final values found are: 1 for λ^r, 0.5 for λ^t and 0.75 for λ^s. A coarse search was also conducted on the γ value, whose best value is 0.3. For TSM-Model, it should be noted that in this case the λ^r representing the trade-off weighting for relational domain loss, has no influence, since the domain attention blocks have been removed. The other values of λ^t, λ^s and γ have been found similarly to the other architecture. The final values found are: 0.5 for λ^t, 0.75 for λ^s and 0.7 for γ.

5.3   Results and Comments

Table 1. Accuracy (%) for the $UCF - Olympics$ adaptation. Gain represents the absolute difference from the Source only accuracy.

Source → Target	U → O	Gain	O → U	Gain
Source only	90.74		85.83	
TRN-model	96.30	+5.56	90.42	+4.59
TSM-model	**96.30**	**+5.56**	**92.08**	**+6.25**
Target only	90.74		90.74	

Table 1 shows the accuracy for the $UCF - Olympic$ experiments. As we said before, we show the lower (training only on source data) and upper bound (training only on target data) and the performances of original architecture (TRN-Model) and the proposed one (TSM-Module). We tested the two configuration when UCF101 dataset was the source and target was the Olympic dataset and viceversa. It is clearly shown the gain when we perform a domain adaptation on data for classification and it is clear that our proposition outperform existing one.

Table 2. Accuracy (%) for the $UCF - HMDB_{full}$ adaptation. Gain represents the absolute difference from the source only accuracy.

Source → Target	U → H	Gain	H → U	Gain
Source only	72.5		85.83	
TRN-model	75.28	+2.78	80.56	+7.71
TSM-model	**77.22**	**+4.72**	**83.19**	**+10.34**
Target only	85.28		94.57	

Similarly, Table 2 show the results of the $UCF - HMDB_{full}$ experiment. The gain with TSM-Model is even more pronounced.

The results show the importance of the spatial and temporal alignment across domains with high discrepancy. The domain shift is evident from the performance gap between baselines trained exclusively on the source domain and the upper bound defined by the networks trained on target domain data. The new designed architecture, TSM-Model, successfully improves the results of previous works on this application. Overall, the temporal shift allows to better understand the temporal dynamics of the videos.

6   Conclusions

In this paper we propose adapting an existing temporal alignment module to a Domain Adaptation Deep Network for improving the temporal dynamics correspondences between source and target videos. Experimental results show the importance of learning temporal consistency between source and target domains, in order to improve data adaptation within a deep learning context for Action Recognition.

In the future we plan to analyse what is the mutual contribution of spatial and temporal alignment in domain adaptation and we plan to design some new architecture that exploit this knowledge in order to boost current performances.

References

1. Cao, L., Liu, Z., Huang, T.S.: Cross-dataset action detection. In: 2010 IEEE Computer Society Conference on Computer Vision and Pattern Recognition, pp. 1998–2005. IEEE (2010)
2. Carreira, J., Zisserman, A.: Quo vadis, action recognition? A new model and the kinetics dataset. In: proceedings of the IEEE Conference on Computer Vision and Pattern Recognition, pp. 6299–6308 (2017)
3. Chen, M.H., Kira, Z., AlRegib, G., Yoo, J., Chen, R., Zheng, J.: Temporal attentive alignment for large-scale video domain adaptation. In: Proceedings of the IEEE/CVF International Conference on Computer Vision, pp. 6321–6330 (2019)
4. Davar, N.F., de Campos, T., Windridge, D., Kittler, J., Christmas, W.: Domain adaptation in the context of sport video action recognition. In: Domain Adaptation Workshop, in conjunction with NIPS. University of Surrey (2011)
5. Ganin, Y., et al.: Domain-adversarial training of neural networks. J. Mach. Learn. Res. **17**(1), 2030–2096 (2016)
6. Jamal, A., Namboodiri, V.P., Deodhare, D., Venkatesh, K.: Deep domain adaptation in action space. In: BMVC, vol. 2–3, p. 5 (2018)
7. Karpathy, A., Toderici, G., Shetty, S., Leung, T., Sukthankar, R., Fei-Fei, L.: Large-scale video classification with convolutional neural networks. In: Proceedings of the IEEE Conference on Computer Vision and Pattern Recognition, pp. 1725–1732 (2014)
8. Kong, Y., Ding, Z., Li, J., Fu, Y.: Deeply learned view-invariant features for cross-view action recognition. IEEE Trans. Image Process. **26**(6), 3028–3037 (2017)
9. Kong, Y., Fu, Y.: Human action recognition and prediction: a survey. arXiv preprint arXiv:1806.11230 (2018)
10. Kuehne, H., Jhuang, H., Garrote, E., Poggio, T., Serre, T.: HMDB: a large video database for human motion recognition. In: 2011 International Conference on Computer Vision, pp. 2556–2563. IEEE (2011)
11. Lin, J., Gan, C., Han, S.: TSM: temporal shift module for efficient video understanding. In: Proceedings of the IEEE/CVF International Conference on Computer Vision, pp. 7083–7093 (2019)
12. Liu, M., Liu, H., Chen, C.: Enhanced skeleton visualization for view invariant human action recognition. Pattern Recogn. **68**, 346–362 (2017)
13. Long, M., Cao, Y., Wang, J., Jordan, M.: Learning transferable features with deep adaptation networks. In: International Conference on Machine Learning, pp. 97–105. PMLR (2015)

14. Long, M., Zhu, H., Wang, J., Jordan, M.I.: Unsupervised domain adaptation with residual transfer networks. In: Advances in Neural Information Processing Systems, vol. 29 (2016)
15. Long, M., Zhu, H., Wang, J., Jordan, M.I.: Deep transfer learning with joint adaptation networks. In: International Conference on Machine Learning, pp. 2208–2217. PMLR (2017)
16. Luvizon, D.C., Picard, D., Tabia, H.: 2d/3d pose estimation and action recognition using multitask deep learning. In: Proceedings of the IEEE Conference on Computer Vision and Pattern Recognition, pp. 5137–5146 (2018)
17. Motiian, S., Piccirilli, M., Adjeroh, D.A., Doretto, G.: Unified deep supervised domain adaptation and generalization. In: Proceedings of the IEEE International Conference on Computer Vision, pp. 5715–5725 (2017)
18. Niebles, J.C., Chen, C.-W., Fei-Fei, L.: Modeling temporal structure of decomposable motion segments for activity classification. In: Daniilidis, K., Maragos, P., Paragios, N. (eds.) ECCV 2010. LNCS, vol. 6312, pp. 392–405. Springer, Heidelberg (2010). https://doi.org/10.1007/978-3-642-15552-9_29
19. de Oliveira Silva, V., de Barros Vidal, F., Romariz, A.R.S.: Human action recognition based on a two-stream convolutional network classifier. In: 2017 16th IEEE International Conference on Machine Learning and Applications (ICMLA), pp. 774–778. IEEE (2017)
20. Saito, K., Kim, D., Sclaroff, S., Darrell, T., Saenko, K.: Semi-supervised domain adaptation via minimax entropy. In: Proceedings of the IEEE/CVF International Conference on Computer Vision, pp. 8050–8058 (2019)
21. Shu, N., Tang, Q., Liu, H.: A bio-inspired approach modeling spiking neural networks of visual cortex for human action recognition. In: 2014 international joint conference on neural networks (IJCNN), pp. 3450–3457. IEEE (2014)
22. Sigurdsson, G.A., Gupta, A., Schmid, C., Farhadi, A., Alahari, K.: Actor and observer: Joint modeling of first and third-person videos. In: Proceedings of the IEEE Conference on Computer Vision and Pattern Recognition, pp. 7396–7404 (2018)
23. Soomro, K., Zamir, A.R., Shah, M.: A dataset of 101 human action classes from videos in the wild. Center Res. Comput. Vision **2**(11), 1–7 (2012)
24. Wang, L., Qiao, Y., Tang, X.: Action recognition with trajectory-pooled deep-convolutional descriptors. In: Proceedings of the IEEE Conference on Computer Vision and Pattern Recognition, pp. 4305–4314 (2015)
25. Wang, L., et al.: Temporal segment networks: towards good practices for deep action recognition. In: Leibe, B., Matas, J., Sebe, N., Welling, M. (eds.) ECCV 2016. LNCS, vol. 9912, pp. 20–36. Springer, Cham (2016). https://doi.org/10.1007/978-3-319-46484-8_2
26. Wilson, G., Cook, D.J.: A survey of unsupervised deep domain adaptation. ACM Trans. Intell. Syst. Technol. (TIST) **11**(5), 1–46 (2020)
27. Yeffet, L., Wolf, L.: Local trinary patterns for human action recognition. In: 2009 IEEE 12th International Conference on Computer Vision, pp. 492–497. IEEE (2009)
28. Zhang, H.B., et al.: A comprehensive survey of vision-based human action recognition methods. Sensors **19**(5), 1005 (2019)
29. Zhao, Y., Xiong, Y., Lin, D.: Trajectory convolution for action recognition. In: Advances in Neural Information Processing Systems, vol. 31 (2018)

Zero-Error Digitisation and Contextualisation of Piping and Instrumentation Diagrams Using Node Classification and Sub-graph Search

Elena Rica[1], Susana Alvarez[1], Carlos Francisco Moreno-Garcia[2], and Francesc Serratosa[1(✉)]

[1] Universitat Rovira i Virgili, Tarragona, Spain
`francesc.serratosa@urv.cat`
[2] Robert Gordon University, Aberdeen, Scotland, UK

Abstract. Thousands of huge printed sheets depicting engineering drawings keep record of complex industrial structures from Oil & Gas facilities. Currently, there is a trend of digitising these drawings, having as final end the regeneration of the original computer-aided design (CAD) file, which can be better visualised and analysed through diverse computer applications. Most efforts in literature and commercial applications have focused on converting these sheets into CAD files in an automated way. Nonetheless, this needs to be a zero-error process; as the final CAD will always be verified by an engineer for integrity and inspection. In this paper, we present a method that, on the one hand, highlights which components in the CAD are most likely to have been incorrectly identified, and on the other hand, facilitates the engineer to search some groups of components in these huge assets. These techniques are based on graph embedding, computer neural networks and sub-graph matching.

Keywords: Piping and Instrumentation Diagram · Automatic validation · Sub-graph matching · Graph embedding

1 Introduction

Piping and Instrumentation Diagrams (P&IDs) are used to represent the structure and functionality of Oil & Gas facilities such as oil rigs and plants. P&IDs contain similar shapes to other complex engineering drawings such as circuit, architectural, mechanical, telephone manhole and chemical plant depictions. P&IDs are mostly generated by means of computer-aided design (CAD) tools and kept in an electronic record. However, in the past they were manually drawn on paper or using tools that are incompatible with modern software. Since these facilities are huge and composed of thousands of electric, electronic or mechanical components connected by a vast network of pipelines, printed handbooks

© The Author(s), under exclusive license to Springer Nature Switzerland AG 2022
A. Krzyzak et al. (Eds.): S+SSPR 2022, LNCS 13813, pp. 274–282, 2022.
https://doi.org/10.1007/978-3-031-23028-8_28

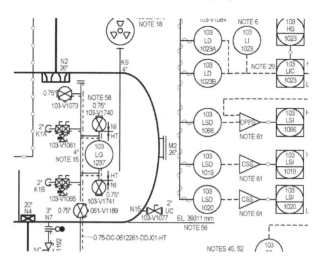

Fig. 1. An example of a P&ID.

composed of thousands of pages are required to depict them. Figure 1 shows a snippet of one sheet and portrays the complexity of a P&ID.

Analysing a facility using a P&ID handbook is an extremely complex process, due to the page quality and the variability of the electric, electronic and mechanic components. While several tools have been presented in recent years to generate a CAD file from these drawings in automated ways [8], the possibility of symbol miss-identification during the digitisation or that some properties have not been correctly associated to certain components becomes high [1,3]. Thus, it is expected that this process is not perfect and therefore, most systems enable human interaction to validate the symbol identification, connection and property association. In practice, the final CAD register is always verified by an engineer due to the need of being a zero-error process. Figure 2 shows a general flow diagram of the classical approach to extract a CAD given a sheet of P&ID. The automatic module is composed of two main steps: digitisation (converting the pulp and paper drawing into a digital register or parts count) and contextualisation (understanding the interaction between the digitised shapes, such as how symbols connect to each other and the text that describes each process, amongst others).

In this paper, we propose the integration of a couple of previously published tools [11,12] for engineers and risk analysts to reduce the amount of effort needed to validate the CAD model towards creating a zero-error digitisation and contextualisation process. The goal of the first tool, depicted in Fig. 3, is to aid in the validation of the automated digitisation process by ensuring that the engineer does not need to look at the whole diagram, but only at the highlighted components, which are the ones that have a chance of having incorrectly identified by the automatic method.

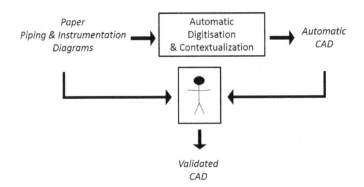

Fig. 2. The process of deducing the CAD document, in which a human is involved to validate the data.

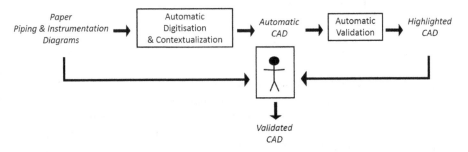

Fig. 3. Our model for automatic detection of possible incorrectly identified components and final human validation.

The aim of the second tool is to aid in the contextualisation by facilitating the search for a particular configuration of connected components. Figure 4 shows a P&ID sheet analysed by our proposed application. Note that CAD applications usually have enabled functionalities to search for specific components by their identity number and visualise them in their locations [2,5,7,9,10,14]. Nevertheless, while inspecting or analysing the gas facilities, engineers are usually interested in some structures, composed of a small set of connected components, instead of a specific type of component. For instance, they want to detect the appearances of structures that include a *valve check* connected to two *general valves* and a *butterfly valve*. Thus, engineers want to query a structure instead of a component and visualise it in the P&ID. Note that the aim of this example query is not to return the locations of the exact appearance of the specific structure, but the locations in structures similar to the one queried may appear. Thus, considering this example, it could be interesting to return the locations of structures composed of a *valve check* connected to one or three *general valves* and one or two *butterfly valves*.

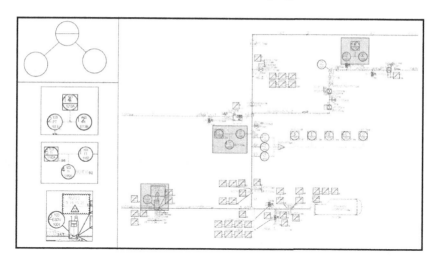

Fig. 4. Screenshot of the second tool proposed. Upper left corner: query structure. Lower left corner: returned structures ordered by increasing distance. Right: a portion of the P&ID in which three locations of the query have been detected.

Both tools fit with the topological challenge of P&ID contextualisation [6,8], whose objective is to understand the connectivity of the symbols. In comparison, we have been able to identify some commercial CAD applications[1] which work with P&IDs and that have online tools to quickly visualise the network of components. Moreover, we have presented some initial proof of concept tools that perform similar functions, such as NetVis or Netlist2CAD[2]. While these tools offer the possibility of component search, none of them are capable to highlight error-prone digitised symbols or sub-structures.

The paper is structured as follows. Section 2 explains the first tool, Sect. 3 explains the second one, and Sect. 4 concludes the paper.

2   Predicting Improperly Identified Components

The difference between the classical models (Fig. 2) and our model (Fig. 3) is the incorporation of the Automatic Validation module. The aim of this module is to deduce the identity of the components in the *Automatic CAD* and highlight the components that must be reviewed by an human expert, reducing in this way, the number of components that should be reviewed when any CAD document is generated by the Automatic Digitisation module. In the next two sub-sections, we detail the two main steps of this Automatic Validation module.

---

[1] https://www.geminivalve.com/best-piping-design-software/.
[2] http://cfmgcomputing.blogspot.com/p/software-demos.html.

2.1   Graph Representation and Data Embedding

The automatic validation method that we present is based on defining P&IDs as attributed graphs. In our graph, nodes represent components and edges represent pipelines that connect these components. Moreover, nodes have only one attribute, which is the component identity (valve, compressor,...) and edges are unattributed and unidirectional. In an attributed graph, a star (T_a) is defined as a local structure composed of a node, its connected edges and also the nodes that these edges connect (neighbour nodes). Our goal is to deduce the identity of each component given the set of pipelines connected to it and the components that connect these pipelines. For this reason we use the star, since by definition, this sub-structure contains this information.

Graphs have some limitations when they are applied to machine learning due to their intrinsic relational representation. This is because some trivial mathematical operations used in the traditional numeric machine learning representations have not an equivalence in the graph domain. Given an arbitrary set of graphs, a possible way to address this problem is to define an embedding function from the graph domain to a vector space [4]. Broadly speaking, an embedding function converts an attributed graph into a vector.

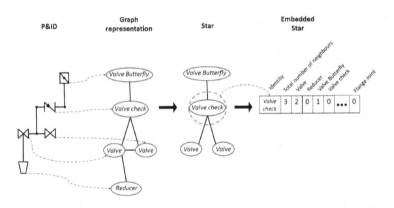

Fig. 5. An example of embedding a *Valve check* star into a vector.

Since we want to use classical machine learning techniques to deduce component identities, we embed stars into vectors. Thus, each star is embedded in a Euclidean space R^{n+2}, where n is the number of different component identities. The embedding of the ith node in the graph (or the ith component in the P&ID) is defined as a vector $E_i = (c_i, d_i, f_i^1, ..., f_i^n) \in R^{n+2}$ where c_i is the identity of the central node of the star; d_i is the number of edges in the star (or the number of connected pipelines to the central component); and f_i^p is the number of external nodes of the star that have the p identity, with $p = 1, ..., n$. A sample of a star embedding is presented in Fig. 5. The output of this step is the set of all embedded stars in the graph representation of the CAD model.

2.2   Machine Learning and Verification

The Machine learning and verification step performs the following tasks:

- Firstly, each component in the P&ID represented by an embedded vector is introduced into the machine learning algorithm that returns the predicted component identity.
- Secondly, the identities of the components returned by the machine learning algorithm are contrasted with identities obtained from the digitised and contextualised *netlist* of the *Automatic CAD*. Note these components have not been verified by the engineer. Thus, this task discerns whether the deduced identities by our machine learning algorithm are the same or they are different from the CAD model.
- Thirdly, it detects the components of the *Automatic CAD* whose identities are different from the identities obtained by the machine learning algorithm. These detected components are highlighted to be validated by the human expert.

The learning set is composed of a CAD model, validated by a human expert, which has been embedded using our graph representation and data embedding step (Sect. 2.1). This CAD model must include a representative number of components per identity in order to assure the proper learning of the data. Usually, the larger the learning set is, the better the prediction given by the machine learning algorithm.

3   Searching Groups of Components

Before explaining the algorithm that implements the second tool, we set the following three premises that condition the values of its input parameters:

- Only engineers know how similar are two components in the P&ID. This knowledge is introduced into the system through the cost of substituting components imposed by the engineers before doing the query. For instance, if the engineer queries a graph that has a *valve* and wants to visualise all the groups of components similar to this query that have this *valve*, then they have to impose the substitution cost between a *valve* and the rest of components to be infinite. Contrarily, if they know that two components are similar, then they can consider the substitution cost between them to be zero.
- In a similar way than the previous item, only engineers know how important is a component or a pipeline in the P&ID. This knowledge is considered in the node and edge deletion and insertion costs.
- Engineers want to visualise the locations where the query or similar queries appear in the P&ID. For this reason, it is desired that the method returns several connected components in the P&ID. These restrictions need to be handled by the search algorithm.

The rest of this section has been divided into two parts, in Sect. 3.1, the input and output parameters of our algorithm are defined and in Sect. 3.2 our algorithm is detailed.

3.1 Input and Output Parameters of Our Method

The **input** of our method is composed of:

- Q: A small graph that represents the query.
- G: A large graph that represents the P&ID.
- S_v: A square matrix of node substitution costs previously set by the engineer. Each cell is a non-negative real number that represents the cost of substituting two types of components. All the elements in the diagonal are zeros.
- D_v: A vector of node deletion costs previously set by the engineer. Each cell is a non-negative real number that depends on each specific component.
- D_e: Edge deletion cost. Since edges do not have attributes, the cost of deleting an edge is the same for all edges. It is a constant, D_e, previously set by the engineers.
- K: The number of compact sub-graphs the method has to return.

The **output** of the method is composed of:

- $\{f_1 : Q \to G, ..., f_K : Q \to G\}$. A list of K node-to-node mappings between the query graph Q and the P&ID graph G.
- $\{D(Q, G, f_1), ..., D(Q, G, f_K)\}$. A list of K distances, given the query graph, the P&ID and the above mappings f_p, $1 \leq p \leq K$. If we define G_p as the nodes in G reached by substitutions in f_p, then $D(Q, G, f_p)$ is the distance between Q and G_p.

The returned list of mappings $f_1,...,f_K$ hold the following four conditions:

- For each $1 \leq p \leq K$, the set of nodes $\{f_p(v) | v \in Q\} \subseteq G$ and their corresponding edges, defines a connected sub-graph.
- Components in P&ID can be reached by several mappings. Nevertheless, two mappings cannot be identical. Formally: $If\ f_p(v) = f_q(v), \forall v \in Q \Rightarrow p = q$.
- The mappings $f_1,...,f_K$ are listed in ascending order on their distances. Formally: $D(Q, G, f_1) \leq D(Q, G, f_2) \leq,..., \leq D(Q, G, f_K)$.
- The mappings $f_1,...,f_K$ might have to be the ones that return the minimum distance. Formally: $If\ f \notin \{f_1, ..., f_K\} \Rightarrow D(Q, G, f) \geq D(Q, G, f_p), \forall p = 1, ..., K$.

3.2 Algorithm

The algorithm uses the sub-optimal error-tolerant graph matching algorithm *Belief Propagation* [13]. Its computational cost is only linear with respect to the number of nodes. It needs some initial node-to-node mappings, which are called *Seeds*, which in some applications could be a drawback but in our case, it is going to be crucial to generate several solutions.

Our algorithm has three main steps (Matlab implementation in[3]). In the first one, a cost matrix C is computed with dimensions $m \times n$, where m and n are

---

[3] http://deim.urv.cat/francesc.serratosa/SW/.

the number of nodes of the query graph Q and the P&ID graph G, respectively. We assume, $m \leq n$. Each cell in C, $C(a, i)$, $1 \leq a \leq m$, $1 \leq i \leq n$, represents the cost of substituting the star T_a in Q by the star T_i in G.

In the second step, the K cells in the cost matrix that have the minimum value are selected. These substitution costs are used to set the K different *Seeds* that algorithm *Belief* is going to use in the K times it is run in the next step of our algorithm.

Finally, in the third step, the *Belief Propagation* algorithm [13] is executed K times. Each time, a different seed is used: $Seed_1$, ..., $Seed_K$. Using a different seed makes the algorithm to return a different mapping between the query Q and the P&ID G. This property could be considered a drawback in other methods but it becomes a must in our case.

Algorithm *Top-K-GED*
Input: Q, G, S_v, D_v, D_e, K
Output: f_1, ... , f_K, D_1, ... , D_K, being $D_p = D(Q, G, f_p)$, $p = 1, ..., K$
Begin Algorithm
$C = ComputeCostMatrix(Q, G, S_v, D_v, D_e)$
$(Seed_1, ..., Seed_K) = SelectLowerCostCells(C, K)$
For $p = 1, ..., K$
    $(f_p, D_p) = Belief(C, Seed_p)$
End For
End Algorithm

4   Conclusions and Future Work

We have presented two tools to reduce the human effort while validating CAD documents that have been automatically generated from a class of complex engineering drawings called Pipping and Instrumentation diagrams (P&IDs).

The first tool detects incorrectly identified components in automatically generated CADs through learning their topology. To do so, we have represented the P&IDs by attributed graphs and we have embedded the local structures of components into vectors. Given each vector, a neural network has been used to predict the identity of the component represented by this vector. The second tool helps the engineer to search groups of connected components that are similar to a specific one. Its uniqueness is the fact that it returns several similar and compact sub-graphs. With the first tool we achieve an average reduction of approximately the 40% of the human effort, keeping an error-free process. With the second tool, we are able to find the queried group of components efficiently in more than the 80% of the cases, even achieving the 100% in some cases.

As a future work, we want to move our system from the laboratory to the industry, thus being in use in the digitisation process of P&ID sheets. These methods could be applied to other kind of industries in which the relational information between the components is available. We believe our methods could drastically reduce the human effort and therefore the economical and temporal cost of this essential task.

Acknowledgements. This project has received funding from Martí-Franquès Research Fellowship Programme of Universitat Rovira i Virgili, by The Data Lab and the Oil & Gas Innovation Centres (Scotland), and by Det Norske Veritas Germanischer Lloyd (DNV GL).

References

1. Arroyo, E., Hoernicke, M., Rodríguez, P., Fay, A.: Automatic derivation of qualitative plant simulation models from legacy piping and instrumentation diagrams. Comput. Chem. Eng. **92**, 112–132 (2016). https://doi.org/10.1016/j.compchemeng.2016.04.040
2. Cordella, L.P., Vento, M.: Symbol recognition in documents: a collection of techniques? Int. J. Doc. Anal. Recogn. **3**(2), 73–88 (2000)
3. Elyan, E., Moreno-García, C.F., Jayne, C.: Symbols classification in engineering drawings. In: International Joint Conference on Neural Networks (IJCNN) (2018). https://doi.org/10.1109/IJCNN.2018.8489087
4. Gibert, J., Valveny, E., Bunke, H.: Graph embedding in vector spaces by node attribute statistics. Pattern Recogn. **45**(9), 3072–3083 (2012)
5. Kang, S.O., Lee, E.B., Baek, H.K.: A digitization and conversion tool for imaged drawings to intelligent piping and instrumentation diagrams (P&ID). Energies **12**(2593), 1–26 (2019). https://doi.org/10.3390/en12132593
6. Moreno-García, C.F., Elyan, E.: Digitisation of assets from the oil & gas industry: challenges and opportunities. In: International Conference on Document Analysis and Recognition (ICDAR), pp. 16–19. No. Workshop on Industrial Applications of Document Analysis and Recognition (WIADAR) (2019). https://doi.org/10.1109/ICDARW.2019.60122
7. Moreno-García, C.F., Elyan, E., Jayne, C.: Heuristics-based detection to improve text/graphics segmentation in complex engineering drawings. In: Engineering Applications of Neural Networks, vol. CCIS 744, pp. 87–98 (2017)
8. Moreno-García, C.F., Elyan, E., Jayne, C.: New trends on digitisation of complex engineering drawings. Neural Comput. Appl. **31**(6), 1695–1712 (2018). https://doi.org/10.1007/s00521-018-3583-1
9. Rahul, R., Paliwal, S., Sharma, M., Vig, L.: Automatic information extraction from piping and instrumentation diagrams. In: International Conference on Pattern Recognition Applications and Methods (ICPRAM), pp. 163–172 (2019). https://doi.org/10.5220/0007376401630172, http://arxiv.org/abs/1901.11383
10. Rantala, M., Niemistö, H., Karhela, T., Sierla, S., Vyatkin, V.: Applying graph matching techniques to enhance reuse of plant design information. Comput. Indus. **107**, 81–98 (2019). https://doi.org/10.1016/j.compind.2019.01.005
11. Rica, E., Álvarez, S., Serratosa, F.: Group of components detection in engineering drawings based on graph matching. Eng. Appl. Artif. Intell. **104**, 104404 (2021)
12. Rica, E., Moreno-García, C.F., Álvarez, S., Serratosa, F.: Reducing human effort in engineering drawing validation. Comput. Ind. **117**, 103198 (2020)
13. Santacruz, P., Serratosa, F.: Error-tolerant graph matching in linear computational cost using an initial small partial matching. Pattern Recogn. Lett. **34**, 1–9 (2018)
14. Tombre, K., et al.: Graphics Recognition: Algorithms and Systems: Second International Workshop, GREC 1997, Nancy, France, 22–23 August 1997, Selected Papers, vol. 2. Springer Science & Business Media, Switzerland (1998)

Refining AttnGAN Using Attention on Attention Network

Naitik Bhise⑩, Adam Krzyzak⁽⊠⁾⑩, and Tien D. Bui⑩

Department of Computer Science and Software Engineering, Concordia University,
Montreal, QC H3G 1M8, Canada
krzyzak@cs.concordia.ca, bui@cse.concordia.ca

Abstract. AttnGAN finds the semantic relation between text and image using an attention network. However, some of the words in the text description remain unattended. We propose a solution called Refined AttnGAN, which contains enhanced attention using Attention on Attention architecture. We apply the mode-seeking function to the network to improve the model diversity. The proposed Architecture is evaluated on the Caltech Birds and Microsoft Coco Dataset. Experimental results demonstrate that our model works very well compared to some state-of-the-art methods.

Keywords: Text-to-image synthesis · Mode-seeking loss function · Attention on attention networks

1 Introduction

Text to Image Synthesis has been around since the research of Generative Adversarial Networks [4]. Generator generates images from a latent random vector while Discriminator compares the generated image with the real image. Conditional GANs use a context vector along with a random vector as a basis for generating images [9]. Text to Image Synthesis uses GANs extensively to generate quality images from text descriptions [10, 17–20]. Bi-directional LSTM is used for the generation of text descriptions from the original captions [11]. Single-stage methods use a single Generator and a Discriminator to process text descriptions to image outputs [10, 11]. Multi-stage methods involve two or more stages for the image generation where each stage uses the output of the previous stage as the input [17–20].

The objective of the text to image synthesis methods is to generate quality images by finding a semantic relation between the image and the text description. Many architectures are proposed to deal with that problem [17–20]. StackGAN [18] introduces conditional augmentation of global sentence vector and spatial replication for generation of next stages. Every word in a sentence has a different level of importance. AttnGAN [17] uses this principle to implement an attention network between the word embedding vectors and image features. Word embeddings vectors are extracted from the Bi-LSTM network along with the global

A. Krzyzak et al. (Eds.): S+SSPR 2022, LNCS 13813, pp. 283–291, 2022.
https://doi.org/10.1007/978-3-031-23028-8_29

sentence vector [11]. DMGAN implements a memory module [16] for finding the semantic relation between text descriptions and image features [20]. Enhanced AttnGAN [3] uses the global sentence vector in another attention network in conjunction with the original attention of AttnGAN. The model generates better evaluations on the fashion dataset [12].

The attention mechanism plays a crucial role in such a system that must capture global dependencies [6]. Attention on Attention (AoA) [6], extends the conventional attention mechanisms by adding another attention. Also, cGANs generate many output modes for an input caption due to multi-modality [9]. But some of the modes are absent from the output distribution causing less diversity. There are many methods for finding the solution to the mode-collapse problem. [8] gives a simple description of the mode-seeking objective function. It maximizes the ratio of the distance between the images generated by two random noise vectors to the distance between the noise vectors.

Our architecture provides an enhanced attention module using the concept of original attnGAN architecture. We apply AoA method to the original attention in the architecture. Caltech Birds [14] and Microsoft Coco [7] datasets are used for method evaluation. We conduct two experiments on these datasets by modifying the loss function in one experiment with the mode-seeking loss. The background section mentions the literature consulted for this research. The method section describes the modules in detail and the construction of the architecture. The results section gives information about the results observed from the experiments and the Conclusion section summarizes the observations. Our objectives for this research can be summarized as follows:

- We try to improve the attention of the AttnGAN network on the word descriptions and to further enrich the image features.
- We attempt to increase the diversity of the output images by using the mode-seeking loss function.
- Our work achieves better performance on both Caltech Birds (CUB) dataset and Microsoft COCO dataset than the original model.

2   Background

Text to Image Synthesis is an important field working with the generation of images from text captions using GANs [4]. Reed [10] constructed the first architecture with a single-stage network generating images of 64X64 resolution. The network finds a semantic relation between the text description and the images. Reed improved the localization constraints by constructing GAWNN [11] architecture. TAC-GAN [2] and PPGN [15] gave images conditioned on a class label and caption. StackGAN [18] brought the multi-stage GAN architectures useful for stabilizing the GAN training. StackGAN-v2 [19] uses conditional loss and colour regularization for identifying the class and the quality of the image generated. AttnGAN [17] makes use of word vectors for providing attention to the image features and improving their text content. Similarly, Dynamic memory GAN [20] uses the memory architecture for text to image synthesis. [1] provided an improvement in the image quality and the FID score [5] after the application of the mode-seeking loss function [8]. Modifications are done in the previous architectures like the e-AttnGAN [3] network which modified the attention of the attention to work on the Fashion dataset [12].

3   Method

In this section, we present a brief discussion of the methods involved in the construction of the Refined AttnGAN architecture and also highlight its components in greater detail. Refined AttnGAN consists of three supplementary modules relative to the AttnGAN.

3.1   AttnGAN

AttnGAN network allows an attention-driven, multi-stage refinement for fine-grained text-to-image generation [17]. It enhances the small subregions of the image by focusing on the relevant words.

The text features are extracted using a bi-directional LSTM encoder. The last two hidden vectors form the word embeddings while sentence embeddings are extracted from the output of the encoder. The attnGAN tries to implement fine-grained word-level information with the help of an attention network. The network consists of a set of hidden vectors that are being processed by the GANs to generate an image. The hidden vectors contain the image-level information like the sentence feature of the single-stage GAN [10]. The hidden vector information is improved with the help of an attention network between the word embeddings and the currently hidden vector.

A new loss function, DAMSM loss, is developed for the training of the generator. The DAMSM loss is designed to learn the attention model in a semi-supervised manner, in which the only supervision is the matching between entire images and whole sentences. The word vectors and sentence vectors are matched with the image and vice versa using the posterior probability function and added to form the complex loss function of the DAMSM loss.

3.2   Global Vector Attention

AttnGAN focuses on the text to image synthesis method with the help of two attention stages. The attention network in AttnGAN focuses on the relationship between the text and the image of the previous stage. The attention was increased by using an extra attention network between the global sentence vector and the image features of the previous stage.

The global sentence vector describes the sentence features generated from the output of the Bi-directional LSTM of AttnGAN. The two attentions are combined with a FILM structure [3] (Fig. 1) .

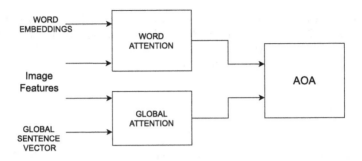

Fig. 1. Refined attention uses this new architecture in place of the word attention in the AttnGAN network in the original network

3.3 Attention on Attention

Attention on Attention is a module that tries to enhance the attention provided by the original attention by adding another sigmoid activation layer between the Query Q and the value output V of an attention network. AoA generates an information vector and a gate vector using two linear transformations conditioned on the current context and the previous attention result. Then it provides the result by applying element-wise multiplication between the inputs g and i.

$$i = W_i^q q + W_i^v \hat{v} + b_i \tag{1}$$

$$g = W_g^q q + W_g^v \hat{v} + b_g \tag{2}$$

$$result = i \odot g \tag{3}$$

3.4 Mode-seeking Function

The mode-seeking function aims at increasing the output modes of the generator distribution which are generally dominated by higher modes. It involves sampling two noise functions through an adversarial network and maximizing the distance between the output. For two noise functions z_1 and z_2 sampled through a GAN network, the mode-seeking loss objective can be found by comparing their respective outputs $I_1 = G(c, z_1)$ and $I_2 = G(c, z_2)$.

$$L_{ms} = \frac{d_I(G(c, z_1), G(c, z_2))}{d_z(z_1, z_2)}$$

where d_I defines the distance metric between the two image outputs and d_z defines the distance metric between the two noise vectors.

3.5 Refined-GAN Architecture

The architecture is constructed by modifying the attention in the AttnGAN. The attention in the AttnGAN can be supplemented with the global vector Attention to have two outputs of the word-level and the sentence-level attention. The two attention are applied to the AoA layer and the resulting output is concatenated with the original image feature vector. The vectors are passed through the stages to form the new image feature vector used for the Image generation of the next Generator.

3.6   Objective Function

The generator objective function for the training of the architecture consists of conditional loss L^G_{cond}, unconditional loss L^G_{uncond} and the DAMSM loss L^G_{DAMSM} which determines the attention similarity between the image and the text descriptions. For the second experiment, the mode-seeking loss L_{ms} is appended to the existing loss function of the generator. The discriminator objective consists of the conditional L^D_{cond} and unconditional loss L^D_{uncond} with respect to the real data as in the original attnGAN network [17].

$$L_G = L^G_{cond} + L^G_{uncond} + L^G_{DAMSM} + L^G_{ms}$$
$$L_D = L^D_{cond} + L^D_{uncond}$$

3.7   Implementation Details

Our architecture is implemented with a batch size of 20 for the CUB birds dataset and 10 for the Microsoft Coco dataset. The parameters of the AttnGAN as well as the DAMSM parameters are set according to the original network [17]. The co-efficient for the mode-seeking loss objective is set to be 1 as suggested in the [1].

4   Results

The result section explains the datasets used for the experiments. A study of the results is presented with the quantitative evaluation in the form of evaluation metric scores. The images are also described qualitatively and a short ablation study is reported.

4.1   Datasets

The experiments are conducted on the Caltech Birds dataset and Microsoft Coco dataset. Caltech Birds is a small dataset of 11,788 images with 200 bird categories. Microsoft COCO dataset includes a training set with 80000 images and a test set with 40000 images. The CUB dataset is processed according to the [18]. The implementation of the models is achieved with the help of the Pytorch library.

Table 1. Evaluation results (FID score) on CUB-200-2011 and Coco dataset with two different architectures of refined attention and refined attention with mode-seeking

Architecture	CUB dataset	Coco dataset
StackGAN++ [19]	35.11	33.88
AttnGAN [17]	23.98	35.49
Refined Attn	23.78	31.68
Refined Attn + mode-seeking	20.73	27.64

4.2 Quantitative Results

We use Frechet Inception Distance and Inception score [13] as our metrics for evaluation of the image quality. FID measures the distance between the synthetic data distribution and the real data distribution. Inception Score measures the output quality and diversity of the generated images (Table 1).

The objective of the first experiment is to incorporate better attention into the network to increase image quality. The second experiment focuses on improving the diversity of the generated images. It can be observed that the FID score has decreased in the case of both datasets. We notice a slight decline in the CUB dataset from 23.98 to 23.78 while the Coco dataset has a substantial reduction from 35.49 to 31.68. The introduction of the mode-seeking loss improves the score even further. The Table 2 compares the FID scores of the two architectures with the old architecture of AttnGAN and StackGAN++.

In the case of Inception Score, the change is not so evident, in the case of the CUB dataset and the inception score actually decreases. The Coco dataset has a slight decline in the inception score with the addition of Refined attention. The Inception metric increases by 2 points for the method with mode-seeking function. The rise is mostly due to the increase in image diversity.

4.3 Qualitative Results

We have seen that there is a considerable change in the FID values due to the Refined attention network. We compare the results through the Fig. 2. There are more instances that the object is well separated from the background and the

Table 2. Evaluation results (Inception score) on CUB-200-2011 and Coco dataset with different architecture variant and one mode-seeking addition.

Architecture	CUB dataset	Coco dataset
StackGAN [19]	3.84 ± 0.06	4.77 ± 0.06
AttnGAN [17]	4.36 ± 0.03	25.89 ± 0.47
Refined Attn	4.32 ± 0.19	24.32 ± 0.58
Refined Attn + mode-seeking	4.33 ± 0.15	26.23 ± 0.42

this larger bird is almost completely black with a white patch and a blue patch on its neck.

this small bird with grayish wingbars and crown, yellowish white belly small beak compared to the body.

the bird has a green breast and abdomen as well as a tiny bill.

this is a bird with a brown belly, blue wings and a large beak.

this fluffy bird is a pretty solid brown with white mixed in to the feathers.

Refined Attention + mode-seeking

Refined Attention

Original AttnGAN

Fig. 2. Birds images for different text descriptions for three architectures from bottom-original AttnGAN, Refined Attention, Refined Attention + mode-seeking loss

object is complete in both cases like the original AttnGAN. It is easier to say that the improved attention has brought about more elaborate features within birds where it concentrates on the small parts of the sentence and thus executes the generation of a better image. The attention effect can be seen in the case of Coco results as the wooden flooring and the home decor clearly in the images. Thus, the attention improvement has caused a better look at the detail of the image and the mode-seeking function has led to the improvement of the image quality by strengthening the inferior modes of the network.

5 Conclusion

In this work, we present a new procedure for modifying the attention network in the original AttnGAN network [17]. The attention network is modified with attention with the global vector and the resulting output is used along with the original attention output in an Attention on Attention network. The mode-seeking function is also applied to the combination to extract new features. We saw that the evaluation metrics improve due to the process and it could be seen in the image samples where the image was well attended as compared to the original network. Thus, the qualitative and quantitative results allow us to conclude the improvement of semantic content due to increased attention and the improvement of diversity due to mode-seeking function.

References

1. Bhise, N., Zhang, Z., Bui, T.D.: Improving text to image generation using mode-seeking function. arXiv preprint arXiv:2008.08976 (2020)
2. Dash, A., Gamboa, J.C.B., Ahmed, S., Liwicki, M., Afzal, M.Z.: TAC-GAN-text conditioned auxiliary classifier generative adversarial network. arXiv preprint arXiv:1703.06412 (2017)
3. Emir Ak, K., Hwee Lim, J., Yew Tham, J., Kassim, A.: Semantically consistent hierarchical text to fashion image synthesis with an enhanced-attentional generative adversarial network. In: Proceedings of the IEEE International Conference on Computer Vision Workshops (2019)
4. Goodfellow, I.J., et al.: Generative adversarial networks. In: Annual Conference on Neural Information Processing Systems (NeurIPS), pp. 2672–2680 (2014)
5. Heusel, M., Ramsauer, H., Unterthiner, T., Nessler, B., Hochreiter, S.: GANs trained by a two time-scale update rule converge to a local Nash equilibrium. In: Advances in Neural Information Processing Systems, pp. 6626–6637 (2017)
6. Huang, L., Wang, W., Chen, J., Wei, X.Y.: Attention on attention for image captioning. In: Proceedings of the IEEE International Conference on Computer Vision, pp. 4634–4643 (2019)
7. Lin, T.-Y., et al.: Microsoft COCO: common objects in context. In: Fleet, D., Pajdla, T., Schiele, B., Tuytelaars, T. (eds.) ECCV 2014. LNCS, vol. 8693, pp. 740–755. Springer, Cham (2014). https://doi.org/10.1007/978-3-319-10602-1_48
8. Mao, Q., Lee, H.Y., Tseng, H.Y., Ma, S., Yang, M.H.: Mode seeking generative adversarial networks for diverse image synthesis. In: Proceedings of the IEEE Conference on Computer Vision and Pattern Recognition, pp. 1429–1437 (2019)
9. Mirza, M., Osindero, S.: Conditional generative adversarial nets. arXiv preprint arXiv:1411.1784 (2014)
10. Reed, S., Akata, Z., Yan, X., Logeswaran, L., Schiele, B., Lee, H.: Generative adversarial text to image synthesis. arXiv preprint arXiv:1605.05396 (2016)
11. Reed, S.E., Akata, Z., Mohan, S., Tenka, S., Schiele, B., Lee, H.: Learning what and where to draw. In: Advances in Neural Information Processing Systems, pp. 217–225 (2016)
12. Rostamzadeh, N., et al.: Fashion-gen: The generative fashion dataset and challenge. arXiv preprint arXiv:1806.08317 (2018)
13. Salimans, T., Goodfellow, I., Zaremba, W., Cheung, V., Radford, A., Chen, X.: Improved techniques for training GANs. In: Advances in Neural Information Processing Systems, pp. 2234–2242 (2016)
14. Wah, C., Branson, S., Welinder, P., Perona, P., Belongie, S.: The CALTECH-UCSD birds-200-2011 dataset (2011)
15. Wang, T.C., Liu, M.Y., Zhu, J.Y., Tao, A., Kautz, J., Catanzaro, B.: High-resolution image synthesis and semantic manipulation with conditional GANs. In: Proceedings of the IEEE Conference on Computer Vision and Pattern Recognition, pp. 8798–8807 (2018)
16. Weston, J., Chopra, S., Bordes, A.: Memory networks. arXiv preprint arXiv:1410.3916 (2014)
17. Xu, T., et al.: AttnGAN: fine-grained text to image generation with attentional generative adversarial networks. In: Proceedings of the IEEE Conference on Computer Vision and Pattern Recognition, pp. 1316–1324 (2018)
18. Zhang, H., et al.: StackGAN: text to photo-realistic image synthesis with stacked generative adversarial networks. In: Proceedings of the IEEE International Conference on Computer Vision, pp. 5907–5915 (2017)

19. Zhang, H., et al.: StackGAN++: realistic image synthesis with stacked generative adversarial networks. IEEE Trans. Pattern Anal. Mach. Intell. **41**(8), 1947–1962 (2018)
20. Zhu, M., Pan, P., Chen, W., Yang, Y.: DM-GAN: dynamic memory generative adversarial networks for text-to-image synthesis. In: Proceedings of the IEEE Conference on Computer Vision and Pattern Recognition, pp. 5802–5810 (2019)

An Autoencoding Method for Detecting Counterfeit Coins

Iman Bavandsavadkouhi$^{(\boxtimes)}$, Saeed Khazaee , and Ching Y. Suen

CENPARMI, Concordia University, Montreal, Canada
{i_bavand,s_khaza,suen}@encs.concordia.ca

Abstract. We use coins in our daily life to pay for bus, metro tickets, vending machines, etc. However, the market for antique and historical coins is another place, where the quality of coins and their genuinity play a significant role. Hence, researchers have considered different methods in coin detection studies. In recent years 2-D and 3-D image processing approaches have been widely used in image-based coin detection. In this paper, we propose a method to detect counterfeit coins based on image content. We employed SIFT, SURF, and MSER to determine the similarity degree of our datasets. Then, we evaluate those descriptors by statistical analysis to see which one is the most effective criterion for counterfeit coin detection. According to experiments, SIFT was selected as the most reliable algorithm for the Danish coin image dataset. Then, we train an autoencoder to find anomalies in the coin images. The trained autoencoder receives a coin image as input and generates a new image. The output image is compared with a basic image using the selected criterion. If the similarity between these two images meets a threshold then the coin is genuine. Most counterfeit coin detection methods require fake data for training. This can be eliminated by our autoencoding-based anomaly method.

Keywords: Counterfeit detection · Coin recognition · Anomaly detection

1 Introduction

A lot of companies, museums, and government agencies are increasing the demand for automatic coin classification system to categorize rare, historical, and ordinary coins. However, a lot of illegal producers have counterfeited precious coins causing major costs and damage to the coin market and society [1]. According to the Counterfeit Monitoring System (CMS) which reported that between 2013 and 2017, 837,910 counterfeit euro coins were confiscated in Europe, with a total value of 1,330,401 Euros. However, in terms of antique coins, there are billions of dollars in the coin counterfeiting business every year throughout the world [2]. CBC News reported on the existence of a counterfeit coin in January 2022. According to this news, Ontario Provincial Police have warned residents to be cautious with their pocket change after discovering a fake two-dollar coin

© The Author(s), under exclusive license to Springer Nature Switzerland AG 2022
A. Krzyzak et al. (Eds.): S+SSPR 2022, LNCS 13813, pp. 292–301, 2022.
https://doi.org/10.1007/978-3-031-23028-8_30

in a store. There are likely to be more fake toonies on the market [3]. As a result of the growing demand for automated techniques to detect counterfeit coins, image-based coin recognition has become critical in counterfeit coin detection [4]. Numerous research projects have been conducted in the field of numismatics to detect counterfeit coins and several interesting methods have been proposed for counterfeit coin detection [4,5]. Usually, in coin detection, most of the automatic fake machine detectors are supplied by the preliminary mechanism for measuring a different aspect of the characteristics of coins [6]. But, when the physical features of fake and genuine coins are identical, these technologies cannot differentiate between them. In recent years, image-based methods for counterfeit coins identification have been extensively studied in the literature [1,4,7]. Researchers have also used image processing techniques to extract features of coin images by analyzing their textures [8,9]. Also, some authors have used edge detection to extract features of coin images for coin detection. Several researchers have used edge detection approaches for feature extraction of coin images in their studies [9,10]. As we know, feature extraction is not an appropriate idea while handling noisy and corrupted images, for instance, rust, dust, or sulfated coin images. In the latest studies in coin detection context, CENPARMI members have produced several novel methods for spotting counterfeit coins. Lately, they transferred 2-D image processing in coin detection into the 3D level. The authors of [3,11] have employed a 3-D scanner to scan and model the number of coins capturing height and depth instead of levels of color. Although 3D scanning is robust against coins of poor quality stated above, the long processing time is still a serious challenge. The possibility of employing an autoencoder for counterfeit anomaly coin detection has not been studied yet and this is still an open field of research. The authors of [12], have utilized an autoencoder for coin detection, while their research focused on coin weights. In this paper, a method is proposed to determine the degree of similarity between coin images in our datasets to detect counterfeit coins. Also, a method is proposed for counterfeit coins detection without the necessity of using fake ones for training, which prevents producing counterfeit coins in the future.

The remainder of this paper is organized as follows. In Sect. 2, we compare the coin images together to select the similarity criterion. In Sect. 3, we design the autoencoder and propose our model for encoding and decoding the images. In Sect. 4, we talk about the procedure of training the model. In Sect. 5, we explain how the proposed model detects counterfeit coins. In Sect. 6, we talk about the experimental results. Finally, we conclude in Sect. 7.

2 Selecting Similarity Criterion

2.1 Comparison of Images

In this section, we produce data from MSER, SIFT, and SURF descriptors for coin images in our datasets through pairwise comparison to achieve a criterion for coin similarity. In this study, Danish 20 Kroner coins dataset that includes four types made in 1990, 1991, 1996, and 2008 is used. As a first step to select similarity criterion, a pairwise comparison is made among all images in our dataset. Accordingly, the first image of our dataset is compared with all other images in the dataset and the second image is compared with all other images. Similarly, we continue this process for all the remaining images. According to our preliminary studies, the value produced from each descriptor is different in scales and it is better to be normalized. The values of all three comparisons from the descriptors are normalized to the range $(0, 1)$ using Eq. 1 where α is the similarity between two images and α' is the normalized value.

$$\alpha' = \frac{\alpha - min(\alpha)}{max(\alpha) - min(\alpha)} \tag{1}$$

Tables 1 demonstrates the values for comparing the similarity of one pair of images in terms of MSER, SIFT, and SURF descriptors for genuine coins. The number of pairs of images depends on the number of datasets in each year. Table 2 shows the number of images and pairs of images in each separate year for both genuine and counterfeit coins. As we mentioned earlier, based on how the pairs of images are selected, the number of pairs of images, n', is obtained using the Eq. 2, where n is the number of images in the dataset.

$$n' = \frac{n(n - 1)}{2} \tag{2}$$

Table 1. The values for comparing the similarity of a pair of images for genuine coins.

Year of Danish coin	MSER	SURF	SIFT
1990	0.1915	0.4135	0.7021
1991	0.3078	1.0000	0.5409
1996	0.1758	0.3648	0.4102
2008	0.3909	0.2456	0.6190

2.2 Selecting a Similarity Function

As mentioned previously, the purpose of this study is to select an algorithm among the mentioned descriptors as the best similarity function. This approach is based on our dataset and will help us in the subsequent steps. The number of comparisons depends on the number of images in our dataset. For this purpose, the result of descriptors in the histogram are compared. Since histograms are used to get the density of the underlying data distribution [13], it is leveraged to find out what values of the similarities occur most often. Figure 1 (a) to (c) represents histograms of the results of each descriptor in 1990s Danish coins. For instance, (a) illustrates the similarity by using the SIFT for 140 pairs of images in the 1990s Danish genuine coins, which is equal to 0.4.

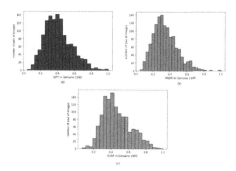

Fig. 1. (a), (b), and (c) histograms of the result of the similarity between pair of images by using SIFT, SURF, and MSER in Danish 1990s genuine coins respectively.

In the next level, Median, Mean, and Variance are used for the comparisons to find the effectiveness of values. Median is used to get a good idea of where a dataset's center is located. In addition, Mean is used to determine the center of our numerical datasets. Variance is used to measure the average degree to which point it differs from the mean. Finally, Standard Deviation is used to determine how measurements for a group are spread out from the average (mean or expected value). Figure 2 illustrates the bar chart of mean, median, std, and variance for SIFT, SURF, and MSER resulting from Danish 1990 genuine coin dataset.

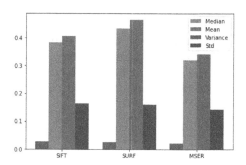

Fig. 2. Bar chart of mean, median, std, and variance for SIFT, SURF, and MSER resulted from Danish 1990s genuine coin dataset.

After using the mentioned descriptors for image comparison, one descriptor is selected as the most reliable one for the rest of the study. Two types of the Danish coin image datasets are compared. According to the result, there are a lot of rates of '1' among the comparison results. According to our investigations, unlike SURF descriptors, there were no miscalculations for SIFT and MSER in this study. According to statistical analysis, the similarity values calculated by the SIFT is more reliable than MSER in all datasets. Thus, SIFT is chosen as the similarity criterion for the next step in this research. Besides, an image that has the highest similarity with the other images is selected as a base or reference image that is supposed to be a good representative of all images. An image that has the highest rate of similarity with other images is selected using the SIFT descriptor. As a result, an image is selected as the basic coin image.

3   Autoencoder Architecture

An autoencoder learns to encode(compress) and decode(reconstruct) unlabeled data in such a way that keeps the most value of input data during reconstruction [14,15]. In other words, when comparing input and output, their difference should be minimum. Here, the amount of difference is known as loss function. In this part, a model is built, leading to the lowest loss function after reconstruction. Indeed, when the system is trained with our dataset, the loss should be reduced during the compress and decompress representation processes [17]. The loss value is used for evaluating our proposed model in the following steps. The autoencoder is constructed with Convolutional Neural Network (CNN) layers to obtain the most commendatory result.

4   Training the Autoencoder

In the first step, the training dataset is built with genuine Danish coins. For training the model, our trainset contains only genuine coins, and we keep the fake samples to test the model.

According to our model, the reconstruction error is relatively low while decompression. In other words, it has tried to reduce the training loss. As a result, the model's decompressed version produces a new image with the most diminutive change from the original (input image). As mentioned earlier, the goal is to build an autoencoder to find anomalous coins that can be applied for counterfeit coin detection.

5   Custom Classifier

This study trains a specific dataset with an autoencoder to have the minimum loss function during the reconstruction. Thus, the measure of loss function is placed in a specific range with a particular threshold. In this case, the intention is to create a machine learning model that learns to reconstruct an image as similar as possible to the original input image. The autoencoder is fed by genuine coin images in the first step. After autoencoder reconstruction, a new image is generated that is expected to be very similar to the original one (input). In this step, the similarity of the selected coin image and the image generated by the autoencoder is measured. As previously mentioned, an autoencoder creates a new coin image after being trained on the original coin images. To calculate the similarity of these two images, SIFT descriptor is used. According to our prior findings in this study, the SIFT descriptor produces the best result between the images in our dataset. A result with a specified threshold is obtained after measuring these two images (selected basic coin image and autoencoder output coin image). Then, to show how the proposed model works, the previously trained autoencoder is tested with a counterfeit coin image. For example, when the model is trained with the genuine Danish coins using the counterfeit Danish coins, the model reconstructs another output. Same as the first step, the similarity of this output is compared with the selected basic image. This similarity is lower than the previous step. Therefore, a defective result is achieved when testing the system with a counterfeit coin. It means that the reconstruction loss is pretty high when testing the system with a counterfeit image coin. As stated earlier, the loss function for genuine coins should be in a specific threshold. Thus, when finding any other loss function outside the specified threshold between the output and the selected coin image, the outcome might be known as anomalous or counterfeit. Figure 3 illustrates a sample of input and output in the proposed method in Danish 1996 coin images.

(a)　　　　　(b)　　　　　(c)　　　　　(d)

Fig. 3. Examples of input and output of autoencoder for the genuine and fake coins after the training process.

(a) Represents a sample of input genuine coin image (trained data) to the system (b) Illustrates the output (reconstructed) coin image by the system when using genuine coin images as input. (c) Demonstrates the counterfeit coin image as input for testing the system. (d) Shows the output (reconstructed) coin image by the system when using counterfeit coin image as input.

6   Experimental Results

In this section, we conducted experiments to evaluate the performance of the proposed method. The hardware used during the test was an iIntel(R) Core(TM) i7-7500U CPU @ 2.70GHz 2.90 GHz, DDR4 16GB RAM, NVIDIA GeForce 940MX with 4 GB Dedicated VRAM. The operating system used was Windows 10–64 bit, and the programming environment was MATLAB 2014 and Google Colaboratory.

6.1 Dataset

In this study, CENPARMI Danish coin dataset [4] that includes Danish 1990, 1991, 1996 and 2008 is used. To train and evaluate the system performance, 214 Danish genuine coin images and 163 Danish counterfeit coin images were used. Table 2 indicates four types of Danish coins that were used in the study for model training and evaluation. It is worth mentioning that the Law Enforcement Office provides all of the coins and all of the coins have been scanned by the 3-D scanner in previous studies [4].

Table 2. The properties of coins and pair of images used in this research.

Dataset	Number of images		Number of pair of images		Training set	Test set	
	Genuine	Fake	Genuine	Fake	Genuine	Genuine	Fake
Danish 1990	51	25	1275	300	50	13	10
Danish 1991	51	14	1275	91	50	14	12
Danish 1996	90	10	4005	45	87	15	9
Danish 2008	22	114	231	6441	20	20	20
All coins	214	163	6786	6877	207	62	51

6.2 Results

For evaluation, the system receives the images as input from the test set that contains previously unseen images of fake and genuine coins. Then, the SIFT descriptor is used to measure similarity between the input and selected basic image. Therefore, the system is first fed with 2008 year genuine coin image for reconstruction by the autoencoder. Then, the similarity between the input and a selected basic coin image is measured by SIFT. As mentioned before, we selected a genuine high-quality coin image as the basic image. According to our investigation, the average similarity between the selected basic coin image and the generated ones calculated by SIFT is 0.96. It is worth mentioning that, all coin images stated in this section are meant to be 2008 year Danish coin images. A counterfeit coin is used for the same year as input to test the system, and then the degree of resemblance is measured. As a result, the SIFT algorithm found an average of 0.78 similarity between generated Danish counterfeit coin image and the selected basic coin image. Here, abnormality can be discovered based on the difference between the results of the SIFT descriptor. This abnormality can realize and detect the fake coins. In other words, a model is learnt by an autoencoder to reconstruct data with the lowest loss function. The model generates an image as output with maximum similarity regarding the input. As a result, if genuine Danish coins are used as input for the model, it should be able to rebuild them with a similarity of 0.96. As a matter of fact, the loss in reconstruction between the input and output image in genuine coins in this year is 0.04. Thus, when the same model is fed with the same kind of coins while the reconstruction loss is more than 0.04 or value of similarity is less than 0.96, the mentioned input can be detected as anomalous or counterfeit coins. Thus, any input that reconstructed any other value of 0.96 could be anomalous. As we mentioned before, counterfeit coins are used as input. As can be seen, the similarity level between the selected basic coin image and output is 0.78. On the other hand, the loss function is 0.22, which is much greater than 0.04. Based on this amount of differentiation, it can be inferred that the input in the second portion was different from the input in the first step. As a result, with the performed model, the genuine Danish coins images could be detected from the counterfeit coin images. Table 3 demonstrates the similarity between the basic genuine image and output image when the system is fed with genuine coin images for the first time. Table 4 illustrates the similarity between the basic image and output image when the system is tested with counterfeit coin images for the first time. Also, the proposed method is compared with four recent studies published in the field of counterfeit coin detection. It should be noted that the data for this comparison was exactly the same as the data used for training and evaluating our proposed method. To this end, the accuracy of methods trained by the Danish coin dataset is computed and compared. According to Table 5, the proposed method using autoencoder outperformed other methods in some cases. Although the accuracy of the proposed model is a bit lower than [11], our proposed model is faster than others in detecting counterfeit coins. It is worth mentioning that our proposed model is trained without the necessity of counterfeit coins.

Table 3. Average of comparison of the generated coin image by autoencoder with the selected basic coin image when we use genuine coin images as input.

Danish coins	Similarity by SIFT
1990	0.98
1991	0.97
1996	0.93
2008	0.96

Table 4. Average of comparison of the generated coin image by autoencoder with the selected basic coin image when we use counterfeit coin image as input.

Danish coins	Similarity by SIFT
1990	0.83
1991	0.81
1996	0.71
2008	0.78

Table 5. Comparing the proposed method with several different previous methods in terms of accuracy

Dataset	[8]	[16]	[1]	[11]	Proposed
Danish 1990	NA	87.2%	93.2%	98.6%	98.0%
Danish 1991	92.1%	86.4%	96.6%	98.0%	97.1%
Danish 1996	96.8%	98.0%	100%	99.8%	99.7%
Danish 2008	95.5%	92.8%	99.6%	99.9%	99.6%

7   Conclusion

In this paper, a new counterfeit coin detection method with an autoencoder was proposed. In the first stage of the study, the results of MSER, SIFT, and SURF in the Danish coin images were examined. Then, all the images in each separate year were compared with the above-mentioned descriptors. Statistical analysis was performed to find the optimal descriptor for producing more reliable similarity for a pair of coin images in our dataset. Thus, SIFT descriptors were selected for our datasets. In the next level of study, we proposed our model by designing an autoencoder with layers of Convolutional Neural Networks, we encoded genuine coin images and reconstruct them from latent. We trained the model to reconstruct the input genuine coin images with the highest accuracy and the lowest loss of function achievable. We measured the similarity between the output of the model and the input images of genuine coins. For the final stage of the research, we utilized a specific range as a criterion. Ultimately, we used counterfeit coin images as input for the previously trained model. As a result, we determined that the range of difference is significantly higher than the specified criteria. This study permits us to train a system to detect counterfeit coins without requiring fake coins for the same one.

Acknowledgement. This research was supported by NSERC, the Natural Sciences and Engineering Research Council of Canada.

References

1. Liu, L., Yue, L., Suen, C. Y.: An image-based approach to detection of fake coins. In: Transactions on Information Forensics and Security, vol. 12, pp. 1227–1239. IEEE (2017)
2. The protection of Euro coins in 2017 (2017). http://ec.europa.eu/info/sites/default/files/
3. Fake Toonies discovered in Hawkesbury, Ont. https://www.cbc.ca/news/canada/ottawa/fake-toonies-hawkesbury-1.6313064
4. Khazaee, S., Sharifi Rad, M., Suen, C.Y.: Detection of counterfeit coins based on modeling and restoration of 3d images. In: Barneva, R.P., Brimkov, V.E., Tavares, J.M.R.S. (eds.) CompIMAGE 2016. LNCS, vol. 10149, pp. 178–193. Springer, Cham (2017). https://doi.org/10.1007/978-3-319-54609-4_13

5. Rácz, A., Héberger, K., Rajkó, R., Elek, J.: Classification of counterfeit coins using multivariate analysis with X-ray diffraction and X-ray fluorescence methods. Forensic Sci. Int. **115**, 129–134 (2001). https://doi.org/10.1016/S0379-0738(00)00309-1
6. Tresanchez, M., Pallejà, T., Teixidó, M., Palacín, J.: Using the optical mouse sensor as a two-Euro counterfeit coin detector. Sensors **9**(9), 7083–7096 (2009)
7. Jain, N., Jain, Neha.: Coin recognition using circular Hough transform. Int. J. Electron. Commun. Comput. Technol. (IJECCT). **2**(3), 2249–7838 (2012)
8. Sun, K., Feng, BY., Atighechian, P., Levesque, S., Sinnott, B., Suen, C. Y.: Detection of counterfeit coins based on shape and letterings features. In: Proceedings of 28th ISCA International Conference on Computer Applications in Industry and Engineering, San Diego, pp. 165–170 (2015)
9. Huber, R., Ramoser, H., Mayer, K., Penz, H., Rubik, M.: Classification of coins using an Eigen space approach. Pattern Recogn. Lett. **26**(1), 61–75 (2005)
10. Van Der Maaten, L., Postma, E.: Towards automatic coin classification. In: Proceedings of the EVA-Vienna, pp. 19–26 (2006)
11. Sharifi Rad, M., Khazaee, S., Suen, C. Y.: Detection of counterfeit coins based on 3d height-map image analysis. In: Expert Systems with Applications. Elsevier (2021). https://doi.org/10.1016/j.eswa.2021.114801
12. Tzu-Hsuan, L., Jehn-Ruey, J.: Anomaly detection with autoencoder and random forest. In: International Computer Symposium (ICS) 2020 (2020). https://doi.org/10.1109/ICS51289.2020.00028
13. Pearson, K.: Contributions to the mathematical theory of evolution. II. Skew variation in homogeneous material. Philos. Trans. R. Soc. London. A. **186**, 343–414 (1895)
14. Goodfellow, I., Bengio, Y., Courville, A.: Deep Learning. MIT Press, Cambridge (1999). http://www.deeplearningbook.org
15. Kramer, M.A.: Nonlinear principal component analysis using auto-associative neural networks. AIChE J. **37**, 233–243 (1991). https://doi.org/10.1002/aic.690370209Citations
16. Semin, K., Lee, S.H., Ro, Y.M.: Image-based coin recognition using rotation-invariant region binary patterns based on gradient magnitudes. J. Vis. Commun. Image Represent. **32**, 217–223 (2015)
17. keras Homepage. http://blog.keras.io/building-autoencoders-in-keras.html

Tarragona Graph Database for Machine Learning Based on Graphs

Elena Rica⬤, Susana Álvarez⬤, and Francesc Serratosa$^{(\boxtimes)}$⬤

Universitat Rovira i Virgili, Tarragona, Spain
`francesc.serratosa@urv.cat`

Abstract. Attributed graphs are commonly used to represent structured objects ranging from images or chemical compounds to social networks, among others. In the last years, graphs have become a powerful tool in machine learning in different fields. For instance, graph generation (i.e. drug design), graph analysis (i.e. marked prediction) or graph matching (i.e. hand written character recognition). When new algorithms in these fields are presented, it is necessary to consciously test them in different situations and applications to evaluate their quality. With this concept in mind, we present a new repository of ten graph datasets, which has the unique characteristic that registers are composed of pairs of classified graphs with their node-to-node ground-truth correspondences. These datasets were previously used in state-of-the-art articles but we have given them a unified format and they are ready to be used for new algorithm-testing purposes.

Keywords: Graph database · Graph matching · Graph generation · Graph analysis

1 Introduction

Attributed graphs have found widespread applications in several research fields of machine learning. This is due to their ability to represent structured objects through unary and binary local entities. Unary entities are represented by graph nodes and binary entities are represented by graph edges. Usually, both, nodes and edges have some attributes that represent local features in the elements they represent. Currently, machine learning techniques based on graphs can be divided in several fields, such as, graph generation, graph analysis or graph matching, among others.

Graphs generation [8] are methods that automatically generate a new graph that has some specific properties given some previous knowledge. For instance, in drug discovery, graphs are generated that represent new compounds that have some physico or chemical properties. Besides, graph analysis [21] is based on deducing or predicting some unknown information in a graph. For instance,

© The Author(s), under exclusive license to Springer Nature Switzerland AG 2022
A. Krzyzak et al. (Eds.): S+SSPR 2022, LNCS 13813, pp. 302–310, 2022.
https://doi.org/10.1007/978-3-031-23028-8_31

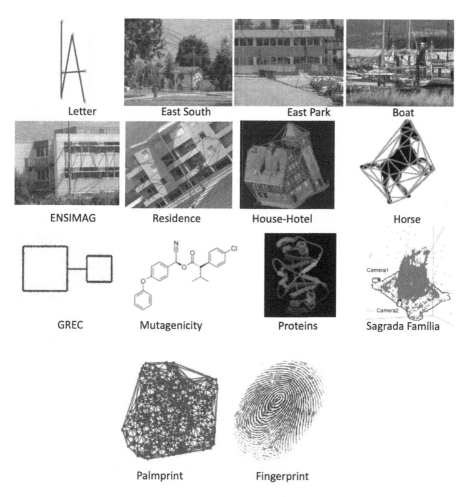

Fig. 1. The first example of each element in the 10 datasets.

given some automatically scanned engineering drawings, the method returns the probabilities that nodes or edges have been improperly scanned. Finally, graph matching [14,16,24,25] gather the methods that given a pair of graphs, return their similarity based on a similarity or dissimilarity measure. Moreover, some of them also return the best mapping between nodes of both graphs (node-to-node correspondence).

In all of these methods, it is crucial to have public databases to analyse the returned results and to compare the new presented methods. In S+SSPR-2008 congress [22] and S+SSPR-2016 congress [19], two papers were published whose aim was to present a graph database for testing purposes. In the second paper, the number of graphs and also the number of nodes of some of the graphs increased with respect to the first one. Moreover, a graph matching contest was reported in [1], which compared and analysed the current algorithms in 2016.

In this paper, we present a new graph repository[1], which incorporates the new needs of current graph methods. It is composed of databases that have been presented previously in some congresses and journals but we have put them together with the same format, and also, we have included some functions to extract the graphs and query the databases and some benchmarks. Note there are other publicly available databases such as OGB[2] but the novelty of our database is its format, which is described in the next section. Figure 1 shows the first example of each element in the datasets.

2   Database Format

The repository is a format-unified compilation of graph databases representing different types of data. Every database of this repository is a collection of registers with the same format. The i-th register is: (G_i, G'_i, f_i, C_i). Both G_i and G'_i are attributed graphs that belong to the same class C_i and they are defined in the same attribute domain. The ground-truth correspondence between the nodes of G_i and the nodes of G'_i, f_i, has been deduced by a human or an another system, which considers other parameters or the raw data. Note, this representation of data has been used in learning methods such as [2,6,7,9,10] because it is useful to analyse graph matching processes from a local point of view.

Datasets are divided in learning, validation and test sets with the aim of being used by machine learning techniques. The learning set is used to learn the parameters of the designed algorithm. Besides, the validation set is used to tune and improve the parameters obtained in the learning step, and the test set is used to test the final quality of the algorithm.

3   Database Description

The repository is divided in 10 datasets, which are called: Letters, Rotation-Zoom, House-Hotel, Horse, Palmprint, Sagrada Familia 3D, Fingerprint, Proteins, Mutagenicy and Grec. Table 1, Table 2 and Table 3 show the main features of these datasets, which are detailed in the next subsections.

---

[1] https://deim.urv.cat/francesc.serratosa/databases/.
[2] https://ls11-www.cs.tu-dortmund.de/staff/morris/graphkerneldatasets.

Table 1. Description of Letter and Rotation Zoom datasets.

Database name		Letter			Rotation zoom				
		Low	Med	High	East_S.	East_P.	Boat	Ens.	Resid
Graphs	Train	750	750	750	4	4	4	4	4
	Validation	750	750	750	3	3	3	3	3
	Test	750	750	750	3	3	3	3	3
Correspondences	Train	37500	37500	37500	16	16	16	16	16
	Validation	37500	37500	37500	9	9	9	9	9
	Test	37500	37500	37500	9	9	9	9	9
Number of classes		15	15	15	1	1	1	1	1
Number of node attributes		2	2	2	64	64	64	64	64
Node attributes' description		(x, y)			SIFT				
Avg. Nodes		4.6	4.6	4.6	50	50	50	50	50
Avg. Edges		6.2	6.4	9	278.2	276.4	278.8	277.7	275.8
Avg. Num. Del./Ins.		0.5	0.5	0.5	32.4	29.5	29.5	31.8	29.2
Max. Nodes		8	9	9	50	50	50	50	50
Max. Edges		12	14	8	282	280	282	284	280
Max. Num. Del./Ins.		4	5	5	50	50	50	50	49

Table 2. Description of House-Hotel and Horse datasets.

Database name		House-hotel	Horse					
			N 10	R 10	S 10	N 90	R 90	S 90
Graphs	Train	71	79	80	80	67	67	67
	Validation	71	60	60	60	67	67	67
	Test	70	60	60	60	66	66	66
Correspondences	Train	2525	80	80	80	67	67	67
	Validation	2525	60	60	60	67	67	67
	Test	2458	60	60	60	66	66	66
Number of classes		2	5	1	1	1	1	1
Number of node attributes		60	60	60	60	60	60	60
Node attributes' description		Content shape						
Avg. Nodes		30	35	35	35	35	35	35
Avg. Edges		154.4	180.7	183.8	185.8	180.7	183.8	185.8
Avg. Num. Del./Ins.		0	0	0	0	0	0	0
Max. Nodes		30	35	35	35	35	35	35
Max. Edges		158	184	184	186	184	184	186
Max. Num. Del./Ins.		0	0	0	0	0	0	0

Table 3. Description of Grec, Mutagenicity, Protein, Sagrada Familia, Palmprint and Fingerprint datasets.

Database		Grec	Mutag.	Protein	Sa. Fa.	Palm.	Finger.
Graphs	Train	10	20	10	136	80	450
	Validation	10	20	10	135	-	450
	Test	20	30	10	135	80	450
Correspondences	Train	100	200	100	18496	320	1350
	Validation	100	200	100	18225	-	1350
	Test	200	300	100	18225	320	1350
Number of classes		12	2	6	1	20	150
Number of node attributes		3	1	3	3	5	4
Node attributes' description		x,y $Type$	$Symbol$	$Amino\ acid$	x,y,z	x,y,A,Q	$x,y,T/B,A$
Avg. Nodes		14.2	40	30	39.4	557.5	20
Avg. Edges		215	1926.7	936.7	456.5	3314.2	100.1
Avg. Num. Del./Ins.		0.4	0.01	3.3	30.1	101.4	0
Max. Nodes		20	70	40	141	1505	20
Max. Edges		380	4830	1560	1918	8962	106
Max. Num. Del./Ins.		5	14	3.3	139	619	0

3.1   Letters

The Letters database was originally presented in [22] and consists of a set of graphs that represent 15 classes of artificially distorted letters of the Latin alphabet. Attributes on nodes are the bidimensional position (x, y) of the junctions and the edges do not have attributes. The ground-truth correspondences between nodes are known due to graphs of each class are generated from an original prototype. There are three variants of the database (Low, Med and High) depending on the distortion degree with respect to the original prototype.

3.2   Rotation Zoom

This database contains graphs representing images that have been extracted from five classes of images of outdoors scenes called Boat, East Park, East South, Ensimag and Resid [19]. In each image, the 50 best salient points were selected. Each node in the correspondent graph represents one of these salient points. The attribute in nodes is a 64-size feature vector obtained by the SIFT extractor [18]. Edges were computed by Delaunay triangulation and do not have attributes. The ground-truth correspondence between nodes of graphs is deduced from the homography between original images. These images have been used in different learning algorithms such as [23].

3.3   House-Hotel

The original CMU House-Hotel database consists of graphs corresponding of images of a toy house and a hotel. One of the first works of machine learning using this database was [7]. Each graph has 30 nodes that correspond to landmark points labelled with Context Shape features [4]. Edges of the graphs were built using Delaunay triangulation and the correspondences were manually set.

3.4 Horse

The elements of this database are based on an original drawing of a horse taken from [6]. This database is separated in three databases (Noise, Shear and Rotate) depending on the transformation that has been applied to artificially generate them. Each graph has 35 nodes that represent hand-marked salient points and the attributes in nodes are the Context Shape features. The edges are unattributed and were built using Delaunay algorithm. Due to distortions were artificially created, the ground-truth correspondences were easily deduced.

3.5 Palmprint

The original images of this database are contained in [13]. This database has images of palmprints of right and left hands of 20 different people. Each person represents a different class in the database. Each node represents a minutia. Attributes in nodes contain the minutiae position, angle, type (termination or bifurcation) and quality (good or poor). Edges do not have attributes and were conformed using Delaunay triangulation. The correspondences between graphs are generated using a greedy matching algorithm based on Hough transform [20].

3.6 Sagrada Familia 3D

This database consists of a set of graphs where each one represents a cloud of 3D points representing Sagrada Familia church of Barcelona [11, 12]. The attributes on nodes are their 3D positions and edges do not have attributes. These 3D points were extracted taking pictures from different positions around the church. A 3D model was built using the whole sequence of 2D images with the Bundler method that also returns the correspondence between the 3D points and the 2D points. With these correspondences, graphs were constructed.

3.7 Fingerprint

The original Fingerprint database used in [3] consists of graphs generated from skeletonised fingerprint images. Nodes represent minutia and their attributes contain the minutiae position, angle and type of minutia (termination or bifurcation). Edges connect nodes that are directly linked. Attributes in the edges are the angle denoting the inclination with respect to the horizontal direction.

3.8   Proteins

This database consists of graphs representing six different classes of proteins originally used in [5]. The graphs are constructed from the Protein Data Bank and labelled with their corresponding enzyme class. Nodes are labelled by their type and their amino acid sequence. Each node is connected to an edge by its three nearest neighbours in space and edges are labelled by their type and length.

3.9   Mutagenicity

The original Mutagenicity database was prepared in [15] and later in [22]. The database presents graphs representing molecules of two classes (mutagen and nonmutagen). The nodes of the graphs are the atoms and the covalent bonds are the edges. The attribute in each node is the number of its chemical symbol and the attribute of each edge is the valence of the linkage.

3.10   Grec

The original Grec database was published in [17]. It consists of graphs representing symbols from architectural and electronic diagrams. Nodes of graphs are intersections, corners or end points of the drawings. Edges are the lines connecting them. Nodes are labelled by their position (x, y) and type of node (intersection, corner and end point) and edges are labelled by three attributes: its type (line or arc), the angle with respect to the horizontal direction and the diameter in case of arcs.

4   Database Interaction

The database is presented as a Matlab variable and also a Python dictionary. Together with the database, a set of functions is available to interact with the databases. To obtain the main features of each database, we provide *"Test_Summarize"* script that gives a list of characteristics of each database separately. The output of this script are the numbers of graphs, correspondences and classes of each database, as well as the maximum and average number of nodes and edges of the graphs of each database.

   We explain below the main functions used by the previous script to obtain these characteristics:

- *GetNumGraphs* returns the total number of different graphs (*"totalGraphs"*) in Learning, Validation and Test set of the specified database "set".
- *GetNodesStatistics* returns the average number of nodes in graphs (*"averageNodes"*) and the maximum number of nodes in graphs (*"maxNodes"*).
- *GetEdgesStatistics* returns the average number of edges in graphs (*"averageEdges"*) and the maximum number (*"maxEdges"*) of each database.
- *GetDeletionsStatistics* returns the average and maximum of deletions in the correspondences (*"averageDeletions"* and *"maxDeletions"* respectively).

- *GetNumPairs* returns the total number of correspondences in each dataset (*"totalPairs"*).
- *GetNumClasses* returns the number of classes of each subset of the database (*"totalClasses"*).
- *GetNumGraphsByClasses* returns the total number of different graphs depending on the type of class (*"totalGbC"*).

5   Conclusions

This paper describes a new graph repository composed of 10 different datasets, which have the main features that each element in the dataset is composed of a pair of graphs, their class and also their best node-to-node correspondence. Moreover, the database and also some functions to use it are publicly available. We believe this repository is going to be useful to test new algorithms in different machine learning fields such as graph generation, graph matching or graph analysis.

References

1. Abu-Aisheh, Z., et al.: Graph edit distance contest: results and future challenges. Pattern Recognit. Lett. **100**, 96–103 (2017). https://doi.org/10.1016/j.patrec.2017. 10.007
2. Algabli, S., Serratosa, F.: Embedding the node-to-node mappings to learn the graph edit distance parameters. Pattern Recogn. Lett. **112**, 353–360 (2018)
3. Algabli, S., Serratosa, F.: Learning graph matching substitution weights based on a linear regression. In: 25th International Conference on Pattern Recognition, ICPR 2020, Virtual Event/Milan, Italy, 10–15 January 2021, pp. 53–58. IEEE (2020). https://doi.org/10.1109/ICPR48806.2021.9412699
4. Belongie, S.J., Mori, G., Malik, J.: Matching with shape contexts. In: Krim, H., Yezzi, A.A. (eds.) Statistics and Analysis of Shapes. Modeling and Simulation in Science, Engineering and Technology, pp. 81–105. Springer, Heidelberg (2006). https://doi.org/10.1007/0-8176-4481-4_4
5. Borgwardt, K.M., Ong, C.S., Schönauer, S., Vishwanathan, S.V.N., Smola, A.J., Kriegel, H.: Protein function prediction via graph kernels. In: Proceedings Thirteenth International Conference on Intelligent Systems for Molecular Biology 2005, Detroit, MI, USA, 25–29 June 2005, pp. 47–56 (2005). https://doi.org/10.1093/bioinformatics/bti1007
6. Bronstein, A.M., Bronstein, M.M., Bruckstein, A.M., Kimmel, R.: Analysis of two-dimensional non-rigid shapes. Int. J. Comput. Vis. **78**(1), 67–88 (2008). https://doi.org/10.1007/s11263-007-0078-4
7. Caetano, T.S., McAuley, J.J., Cheng, L., Le, Q.V., Smola, A.J.: Learning graph matching. IEEE Trans. Pattern Anal. Mach. Intell. **31**(6), 1048–1058 (2009)
8. Cao, N.D., Kipf, T.: MolGAN: an implicit generative model for small molecular graphs. CoRR abs/1805.11973 (2018). http://arxiv.org/abs/1805.11973
9. Conte, D., Serratosa, F.: Interactive online learning for graph matching using active strategies. Knowl. Based Syst. **205**, 106275 (2020). https://doi.org/10.1016/j.knosys.2020.106275

10. Cortés, X., Serratosa, F.: Learning graph-matching edit-costs based on the optimality of the oracle's node correspondences. Pattern Recogn. Lett. **56**, 22–29 (2015)

11. Cortés, X., Serratosa, F.: Cooperative pose estimation of a fleet of robots based on interactive points alignment. Expert Syst. Appl. **45**, 150–160 (2016). https://doi.org/10.1016/j.eswa.2015.09.049

12. Cortés, X., Serratosa, F.: Learning graph matching substitution weights based on the ground truth node correspondence. Int. J. Pattern Recognit. Artif. Intell. **30**(02), 1650005 (2016)

13. Dai, J., Feng, J., Zhou, J.: Robust and efficient ridge-based palmprint matching. IEEE Trans. Pattern Anal. Mach. Intell. **34**(8), 1618–1632 (2011)

14. Garcia-Hernandez, C., Fernández, A., Serratosa, F.: Ligand-based virtual screening using graph edit distance as molecular similarity measure. J. Chem. Inf. Model. **59**(4), 1410–1421 (2019). https://doi.org/10.1021/acs.jcim.8b00820

15. Kazius, J., McGuire, R., Bursi, R.: Derivation and validation of toxicophores for mutagenicity prediction. J. Med. Chem. **48**(1), 312–320 (2005)

16. Koutra, D., Shah, N., Vogelstein, J.T., Gallagher, B., Faloutsos, C.: DeltaCon: principled massive-graph similarity function with attribution. ACM Trans. Knowl. Discov. Data **10**(3), 28:1–28:43 (2016). https://doi.org/10.1145/2824443

17. Liu, W., Lladós, J. (eds.): GREC 2005. LNCS, vol. 3926. Springer, Heidelberg (2006). https://doi.org/10.1007/11767978

18. Lowe, D.G.: Distinctive image features from scale-invariant keypoints. Int. J. Comput. Vis. **60**(2), 91–110 (2004). https://doi.org/10.1023/B:VISI.0000029664.99615.94

19. Moreno-García, C.F., Cortés, X., Serratosa, F.: A graph repository for learning error-tolerant graph matching. In: Robles-Kelly, A., Loog, M., Biggio, B., Escolano, F., Wilson, R. (eds.) S+SSPR 2016. LNCS, vol. 10029, pp. 519–529. Springer, Cham (2016). https://doi.org/10.1007/978-3-319-49055-7_46

20. Ratha, N.K., Karu, K., Chen, S., Jain, A.K.: A real-time matching system for large fingerprint databases. IEEE Trans. Pattern Anal. Mach. Intell. **18**(8), 799–813 (1996). https://doi.org/10.1109/34.531800

21. Rica, E., Álvarez, S., Serratosa, F.: Group of components detection in engineering drawings based on graph matching. Eng. Appl. Artif. Intell. **104**, 104404 (2021). https://doi.org/10.1016/j.engappai.2021.104404

22. Riesen, K., Bunke, H.: IAM graph database repository for graph based pattern recognition and machine learning. In: da Vitoria Lobo, N., et al. (eds.) SSPR /SPR 2008. LNCS, vol. 5342, pp. 287–297. Springer, Heidelberg (2008). https://doi.org/10.1007/978-3-540-89689-0_33

23. Serratosa, F.: A general model to define the substitution, insertion and deletion graph edit costs based on an embedded space. Pattern Recognit. Lett. **138**, 115–122 (2020). https://doi.org/10.1016/j.patrec.2020.07.010

24. Serratosa, F.: Redefining the graph edit distance. SN Comput. Sci. **2**(6), 1–7 (2021). https://doi.org/10.1007/s42979-021-00792-5

25. Solé-Ribalta, A., Serratosa, F.: Graduated assignment algorithm for multiple graph matching based on a common labeling. Int. J. Pattern Recognit. Artif. Intell. **27**(1), 1350001 (2013). https://doi.org/10.1142/S0218001413500018

Human Description in the Wild: Description of the Scene with Ensembles of AI Models

Vincenzo Dentamaro ⬛, Vincenzo Gattulli$^{(\boxtimes)}$ ⬛, Paolo Giglio ⬛,
Donato Impedovo ⬛, and Giuseppe Pirlo ⬛

Department of Computer Science, University of Bari Aldo Moro, Bari, Italy
vincenzo.gattulli@uniba.it

Abstract. Describing an image scene in Natural Language is a very complex procedure for a machine. Many researchers have used Natural Language Processing approaches. In this paper Machine Learning and Computer Vision models will be illustrated with the purpose of describing a picture in the wild. Action Recognition models, Face Recognition with gender and age and Clothing Recognition will be performed in combination with the purpose of generating a textual sentence belonging to natural language describing the scene in the picture. The proposed technique can target multiple domains, specifically useful for preventing cyberbullying situations. In addition, an attempt will be made to exceed for each model the current SoA.

Keywords: Action recognition · Machine learning · Computer vision · Face recognition · Clothing recognition · Ensemble models

1 Introduction

Humans can easily describe the content of an image, whereas, for a machine, it is a highly complex task. Automatic understanding of images is a fundamental problem of Artificial Intelligence, which aims at extracting the essential information from a photograph, capturing the relationships and details of the elements described by it, then enabling a natural language reproduction of its contents. Automatic understanding of image content could affect social relationships, mitigating and preventing bullying and cyberbullying situations. In fact, it might be useful to describe through text the content of images in kids' smartphones. Images in kids' smartphones might contain violent actions that described could be filtered and annotated by keywords. Recent developments in the fields of Machine Learning and Computer Vision, the set of processes that aim to create an approximate model of the real world, have brought tremendous progress in achieving this goal, enabling the development of software architectures that combine artificial intelligence and natural language processing to produce textual descriptions of real-life images. These systems, while generating accurate descriptions of the general context of an image, do not allow attention to be focused on the details of the elements it describes, such as the people present, their clothing, and the details of their faces.

A. Krzyzak et al. (Eds.): S+SSPR 2022, LNCS 13813, pp. 311–322, 2022.
https://doi.org/10.1007/978-3-031-23028-8_32

In this paper, an approach is proposed that aims at describing images *"In the wild,"* considering different contexts and situations in the world, through the combination of different types of AI algorithms working together to create a unique description in natural language. The goal is to develop an AI algorithm that allows a machine to understand the scene shown in images that might contain one or more people, focusing on learning their individuals' faces (understanding gender and age) and body (understanding clothing and color and actions performed).

The following paper is structured as follows: Section 2 will discuss SoA studies that have implemented the models used for Action Recognition, Clothing Recognition and Gender and Age Recognition. Section 3 will analyze the methodology and pipeline that will constitute the final system. Section 4 will analyze SoA datasets useful for training AI models. Section 5 will present observations and results inherent in the specific modules. Finally, Section 6 will conclude the paper.

2   Related Work

This chapter summarize some state-of-the-art approaches useful for the implementation of the final system.

Regarding clothing recognition, some works have been studied: Ge et al. [1] present versatile benchmark (including clothes detection, pose estimation, segmentation, and retrieval). DeepFashion2 contains 801K clothing items and 873K Commercial-Consumer clothes pairs with models Match R-CNN [1]. Sidnev et al. [2] propose Deep-Mark an one-shot approach for clothing detection with CenterNet. The accuracy achieved with DeepFashion2 dataset with clothing detection (0.723 mAP) clothing landmark estimation (0.532 mAP), with neural network one stage [2].

Regarding the recognition of age or gender, the following works have been studied and considered: Eidinger et al. [3] offer a unique data set of face images, labeled for age and gender, acquired by smart-phones with dropout-support vector machine approach. The aim is performing a robust face alignment technique that outperform state-of-the-art [3]. Levi et al. [4] show with deep-convolutional neural networks (CNN) there is a significant increase in performance. The method is evaluated with Audience benchmark for age and gender estimation [4].

Finally, about action recognition, several works have been studied: Khan et al. [5] for the action detection task, perform extensive experiments on Stanford-40 dataset e PASCAL VOC 2012. The approach achieved with Stanford-40 5.7% MAP and with PASCAL VOC 2012 2.1% MAP. The approach outperforms the SoA a MAP of 45.4% on Stanford-40 and 31.4% on PASCAL VOC 2012 [5]. Mohammadi et al. [6] show approach with eight pre-trained CNNs and Transfer Learning technique. In addition is applied an attention mechanism on the CNN output feature maps on Stanford 40 dataset. The Ensemble Learning technique is proposed. Finally, the best configuration achieves 93.17% accuracy [6]. The following chapters will mention these papers but also others with the aim of comparing them with the approaches chosen and implemented in this paper.

3   Methods

The designed system is composed of various AI modules with the purpose of performing optimal scene description (see Fig. 1 (Left)).

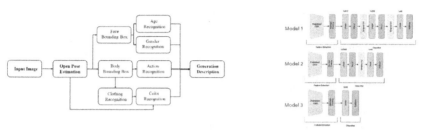

Fig. 1. Pipeline system implemented (Left), Architectures used for classification, respectively Model1, Model2 and Model3 (Right)

The pipeline involves an initial calculation of key points extrapolated from the input image. The key points will identify the positions of the person's body and face (*Open Pose Estimation*). Gender and age recognition is applied to the face (*via Face Bounding Box of OpenPose*), while clothing recognition (*via Body Bounding Box of OpenPose*) and action recognition is applied to the body. The clothing items found, combined with the key points generated by OpenPose, allow recognition and prediction of the colors of each garment worn by the person in the photo. Action recognition is done through the Body Bounding Box of OpenPose to predict the action taken at that moment. Finally, having collected all this information, the final natural language textual description is generated through the composition of other sentences generated by each module. The following sub-sections describe the pipeline modules semantically (see Fig. 1 (Left)).

3.1   Open Pose Estimation

This methodology generates all body and face key points for each person in the image. From these, the coordinates of the points with the least and greatest value on both the horizontal and vertical axes are taken, values corresponding to the coordinates of the upper left and lower right edges of the bounding box. Using key points to approximate the position of the face and body allows not only to perform both steps in one operation (without having to operate additional object detectors and face detectors) but also to have more accurate face bounding boxes that cover only the actual area of the person's face. This choice was also motivated by a lack of need for very accurate bounding boxes but only an approximation of location since this will only be used to increase the speed of clothing recognition (by providing as input the single portion of the image containing the person instead of the full image). For the body bounding box, a 20% increase in width and height is applied. Without this increment, narrower and lower bounding boxes could be generated than the actual position of the person [7].

3.2 Clothing Recognition by Object Detection

The networks used to accomplish these tasks are the YOLOv5-Small architecture (a model of the YOLO architecture with enhancements designed to increase the system's ad-training speed) and the Faster-RCNN architecture with a ResNet50 backbone for classification. Both networks are pre-trained on the *msCOCO* dataset and trained through fine-tuning. The dataset used for this phase is DeepFashion2. A scaling to 256×256 pixels was applied to each image. The networks were trained using the stochastic gradient descent optimization function, with a batch size of 64 elements with 100 epochs.

3.3 Clothing Color Recognition

In this model, techniques are applied to group the RGB values in a portion of an image into different sets. This methodology is called clustering, a form of unsupervised learning that allows an input dataset to be divided into clusters (groups) containing similar elements. The algorithm chosen is K-Means, which, takes as input the list of pixels contained in the image and an N value (number of clusters to be generated) and is able to group the RGB values of the image into N groups each of which will have an associated centroid. The color defined by the centroid of the cluster with the most elements will then be considered the main color of the image. Having extracted the portion of the image containing the clothing item, the following methodology was considered more performant and faster: Apply clustering to the pixels defined by the body key points that fall within the bounding box of the clothing item. Using the body key points made it possible to exclude all those colors that did not belong to the garment, thus generating colors that were much more faithful to the actual color.

3.4 Action Recognition

Several SoA Convolutional Networks were used for the feature extraction phase: ResNet50, VGG19, DenseNet121, SqueezeNet1.1, and InceptionV3, with and without the use of auxiliary outputs. All these networks were used pre-trained on the ImageNet dataset. Three neural network architectures are provided for the classification phase, shown in Fig. 1 (right). The training phase of these networks was carried out using stochastic gradient descent optimizers, and Adam, different types of image augmentation are also used on the dataset. The datasets used in this phase are UCF101 and Stanford40.

3.5 Gender and Age Classification from the Face

The structure of the models used at this stage is like the classification of shares. Following the parameters and classifiers provided in the work of Zhang et al. [8] for gender and age classification using ROR (residual networks of residual networks) architecture, ResNet50, ResNet152 (pre-trained on ImageNet) with batch sizes of 128 and 48 will be used, testing both fine-tuning and transfer learning techniques in addition to Adam and SDG optimizers. The VGG Face model, an architecture based on VGG16 optimized and pre-trained on large-scale face datasets, is also trained to provide more specific features in face classification [9]. The age prediction from the face was performed as a classification task and not regression. Thus, the exact age is not predicted, but a probability distribution over eight different age groupings is predicted, following the standard defined by the *Adience dataset.*

4 Dataset

Different types of datasets had to be adopted in the project's development. The division was done following the format 70% training, 20% validation, and 10% testing, also performing a form of balancing the partitions so that each partition has a similar number of elements for each class in the dataset. *Legend: Configuration 1 (Dataset for pre-trained networks), Configuration 2 (Dataset for Object Detection), Configuration 3 (Dataset for action recognition), Configuration 4 (Dataset for gender and age classification) (See Table 1).*

Table 1. Name, notes, and description of Chosen Datasets in SoA

Conf.	Note	Description	
1	Datasets used for training SoA networks for classification and Object Detection, which will then be further trained by transfer learning and fine tuning	*msCOCO*: Large-scale dataset presenting annotations for Object Detection, segmenting, and description generation. It contains more than 330,000 images with more than 1.5 million object instances, divided into 91 classes. The images contained represent everyday scenes and objects in their natural context. The YOLO family models are pre-trained on this dataset as it represents the state of the art for Object Detection [10]	*ImageNet*: Dataset for image classification constructed based on WordNet. Consisting of 3.2 million images divided into 1000 different classes, this dataset is the largest ever composed for image classification, with the most significant number of classes present [11]

(continued)

Table 1. (*continued*)

Conf.	Note	Description	
2	In the datasets for Object Detection, each image has associated with a set of bounding boxes that define the location of the objects. The format of the annotations follows the standard defined by the *msCOCO* dataset, where each bounding box is described by the upper-left edge, height, and width coordinates	*DeepFashion2*: Dataset used for outfit recognition. They are composed of 491,000 images captured from fashion store websites and outfit-sharing forums. Each of the photos in this dataset contains annotations regarding the clothing worn by the person, with bounding boxes locating each garment and its class annotation among the 13 presents. This dataset is the largest for clothing recognition [1]	
3	This dataset presents a set of images depicting a subject while completing a particular action. The annotation will then be the type of action performed in the image. The state of the art of action recognition is more concentrated on the recognition of actions in the video. For this reason, it was also decided to consider datasets composed of videos	*Stanford40*: Dataset of images of people performing various actions divided into 40 classes. For each class of actions, there are 180 to 300 images for a total of 9532 elements, thus forming the most extensive dataset for still action recognition present in state of the art. That dataset presents a bounding box that encloses the subject that is acting [12]	*UCF101*: The most prominent video dataset presents for action recognition. This dataset is composed of 13,000 videos totaling 27 h of recordings. Each video was retrieved from video-sharing platforms such as YouTube, thus providing user-created videos. This dataset has 101 different classes of videos, making it the most challenging dataset for action recognition given a large number of videos and the diversity of videos [13]. In this case, a single frame was extracted for every 25 frames of each video, thus creating a dataset of about 100,000 images

(*continued*)

Table 1. (*continued*)

Conf.	Note	Description	
4	Of the two datasets chosen, versions containing only the person's face aligned with a head position were used	*UTK Faces:* Large-scale dataset with focus on the wide variety of ages of people present, with faces of people up to 116 years of age. Consisting of 20,000 different images, this dataset presents not only annotations of age and gender but also ethnicity, as well as providing 68 key points for the representation of each face [14]	*Adience:* Faces of people are taken from real pictures of everyday life. It presents a total of 26,580 photos with 2284 different people. In addition to gender, it also presents annotations on age group, with eight different groups [3]

5 Experimentation and Results

This chapter will present the results of experiments carried out on the pipeline modules. The results obtained with all existing SoA models will also be compared.

5.1 Clothing Recognition

Model training was done with batch size 64, fixed grain-size 256×256 images for 100 epochs with Mean Average Precision (mAP) metrics (see Table 2).

Table 2. Clothing recognition results

Model	mAP [0.5]	mAP [0.5 < 0.95]
YOLOv5-Small	0.711	0.544
YOLOv5-Large	0.688	0.454
Faster-RCNN	0.650	0.401
DeepMark2 [2]	0.723	N.a
DF2 Mask-RCNN [1]	0.667	N.a

The best model turns out to be *YOLOV5-Small*. This presents high accuracy values on the recognition of clothing items that appear most frequently within the dataset (0.93 for short-sleeved t-shirts and 0.95 for regular pants) and average values for the other clothing classes. The results are in line with the average accuracy of the present SoA models.

5.2 Actions Classification

Once the networks (*ResNet50, VGG19, DenseNet121, InceptionV3*) are trained, an ensemble of these models is created to increase the overall accuracy of the system. A comparison between the proposed ensemble and benchmarks of other models for action recognition on *Stanford40* (See Table 3):

Table 3. Ensemble neural network models

Model	F1 score
Ensemble	76.60%
Action-specific detectors [5]	75.50%
Attention mechanism ensemble [6]	93.17%

Although the proposed ensemble model is in line with similar architectures for classifying actions on the Stanford40 test set, it performs worse than the best model present at SoA. The latter employs attention mechanisms, a larger image size (512 × 512), a purpose-built neural network for classification, and a more significant number of training epochs. The proposed model, which has only a feature extraction phase and a single fully connected layer for classification, did not achieve the model's results [6]. We now present the performance of the models trained on the UCF101 dataset converted to an image set. At this stage, the VGG19 model was trained with the Model 3 and Model 1 classifier (see Table 4).

Table 4. VGG19 model results with dataset UCF101

VGG19	F1 score
Model 3 Batch 128	68.03%
Model 1 Batch 128	69.49%
Model 1 Batch 128 Data Augmentation	69.80%
Model 1 Batch 128 Data Augmentation Label Smoothing	69.00%
Model 1 Batch 64 Data Augmentation Adam	55.28%
Model 1 Batch 16 Data Augmentation Label Smoothing	**71.89%**

The SoA has no other architectures based on action recognition on the UCF101 dataset converted to an image set, so it was impossible to compare the results obtained with other models in the literature.

5.3   Gender and Age Classification

In a first phase of experimentation, several models were trained by cross dataset technique. UTK Faces is used entirely for training, (with division 80% training and 20% validation) and then model evaluation is performed on all elements of the *Adience* dataset. The best result was from the ResNet50 Transfer Learning ADAM model with 75% F1-score. Then additional models are trained by 5-fold cross validation on the *Adience dataset* (See Table 5).

Table 5. 5-fold cross validation con dataset Adience with Data Augmentation

Model (Gender)	1	2	3	4	5	Average
ResNet152 Batch 48	83%	88%	82%	85%	85%	84.60%
ResNet50 Batch 128	82%	85%	82%	84%	83%	83.2%
VGG Face Batch 16	**90%**	**91%**	**88%**	**92%**	**89%**	**90.00%**
VGG Face Batch 64	85%	90%	85%	89%	87%	87.20%

The model with the best performance was used for training the network for classification into age groups. It is evaluated in two modes, accuracy in predicting the exact age group and 1-off accuracy. A prediction is also considered exact if the age group before or after the true value is predicted (see Table 6).

Table 6. VGG Face Batch 16 model with Data Augmentation for Exact Age and Age 1-off

Model	1	2	3	4	5	Average
VGG Face Batch 16 - Data Augmentation (Exact Age)	69%	55%	62%	52%	60%	59.60%
VGG Face Batch 16 - Data Augmentation (Age 1-off)	**97%**	**93%**	**95%**	**94%**	**95%**	**94.80%**

The best model turns out to be VGG Face trained with batch 16 and image augmentation of the training dataset. This model is now compared with some of the present SoA models (see Table 7).

Table 7. Comparison of different SoA models

Model	Gender	Age	Age 1-off
Our Model	**90.00%**	**59.60%**	**94.80%**
LBP + FPLBP + Dropout 0.8 [3]	77.80%	45.10%	79.50%

(continued)

Table 7. (*continued*)

Model	Gender	Age	Age 1-off
Levi, G. & Hanasser [4]	86.80%	50.70%	86.80%
Whole Component [15]	89.60%	54.30%	87.60%
ROR Architecture [8]	90.87%	61.78%	92.15%
ROR Architecture IMDB-Wiki Fine Tuning [8]	93.24%	66.74%	97.38%

The proposed model thus has higher accuracy than most state-of-the-art architectures, exceeded only by the model presented by A. Ke Zhang et al., which uses a ROR architecture with two fine-tuning steps, first on the IMDB-Wiki dataset and then on the Adience dataset with fivefold cross-validation. The model generates values that deviate only slightly from the true value, thus allowing it to create ages that accurately approximate the person's true age. Some practical examples of the output of the final architecture, in which the models were used, are presented (see Fig. 2). Final sentence generation uses standard generated sentences with dynamic terms based on the output of the various components.

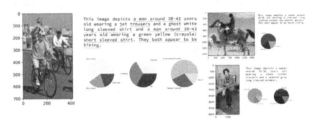

Fig. 2. Examples description Final system

6 Conclusion

This paper illustrated a design strategy with AI models to perform scene descriptions. The final architecture exploits the models: YOLOv5-Small for clothing recognition, VGG-Face trained with batch 16 for gender and age recognition, and the VGG19 model trained with batch 16 and label smoothing on the UCF101 dataset for action recognition. Action recognition using models trained on UCF101 does not present the expected results on images as very different from those in the video frames of the dataset used. Using images captured from videos with very different qualities led the system to suboptimal performance. The modular structure of the proposed architecture allows different patterns to be easily added for additional detail recognition to provide more accurate descriptions. The use of body key points can be used to provide an estimate of the person's position, such as "sitting," "standing," etc. Facial key points can be used to derive the person's eye shape or color to describe them even more accurately. Finally, the use of OpenPose and

YOLO for body and face position estimation and clothing recognition allows for very high inference rates, with the possibility of applying the proposed architecture to video formats, such as video streams from video surveillance cameras.

Acknowledgments. This work is supported by the Italian Ministry of Education, University and Research within the PRIN2017 - BullyBuster project - A framework for bullying and cyberbullying action detection by computer vision and artificial intelligence methods and algorithms.

References

1. Ge, Y., Zhang, R., Wang, X., Tang, X., Luo, P.: DeepFashion2: a versatile benchmark for detection, pose estimation, segmentation and re-identification of clothing images. In: 2019 IEEE/CVF Conference on Computer Vision and Pattern Recognition (CVPR), 2019-June, pp. 5332–5340 (2019). https://doi.org/10.1109/CVPR.2019.00548
2. Sidnev, A., Trushkov, A., Kazakov, M., Korolev, I., Sorokin, V.: DeepMark: one-shot clothing detection. In: Proceedings - 2019 International Conference on Computer Vision Workshop, ICCVW 2019, pp. 3201–3204 (2019). https://doi.org/10.1109/ICCVW.2019.00399
3. Eidinger, E., Enbar, R., Hassner, T.: Age and gender estimation of unfiltered faces. IEEE Trans. Inf. Forensics Secur. **9**, 2170–2179 (2014). https://doi.org/10.1109/TIFS.2014.2359646
4. Levi, G., Hassncer, T.: Age and gender classification using convolutional neural networks. In: IEEE Computer Society Conference on Computer Vision and Pattern Recognition Workshops, 2015-October, pp. 34–42 (2015). https://doi.org/10.1109/CVPRW.2015.7301352
5. Khan, F.S., Xu, J., van de Weijer, J., Bagdanov, A.D., Anwer, R.M., Lopez, A.M.: Recognizing actions through action-specific person detection. IEEE Trans. Image Process. **24**, 4422–4432 (2015). https://doi.org/10.1109/TIP.2015.2465147
6. Mohammadi, S., Majelan, S.G., Shokouhi, S.B.: Ensembles of deep neural networks for action recognition in still images. In: 2019 9th International Conference on Computer and Knowledge Engineering, ICCKE 2019, pp. 315–318 (2020). https://doi.org/10.1109/ICCKE4 8569.2019.8965014
7. Cao, Z., Hidalgo, G., Simon, T., Wei, S.E., Sheikh, Y.: OpenPose: realtime multi-person 2D pose estimation using part affinity fields. IEEE Trans. Pattern Anal. Mach. Intell. **43**, 172–186 (2021). https://doi.org/10.1109/TPAMI.2019.2929257
8. Zhang, K., et al.: Age group and gender estimation in the wild with deep RoR architecture. IEEE Access **5**, 22492–22503 (2017). https://doi.org/10.1109/ACCESS.2017.2761849
9. Parkhi, O.M., Vedaldi, A., Zisserman, A.: Deep Face Recognition. In: Proceedings of the British Machine Vision Conference (BMVC), pp. 41.1–41.12 (2015). https://doi.org/10.5244/C.29.41
10. Lin, T.-Y., et al.: Microsoft COCO: common objects in context. In: Fleet, D., Pajdla, T., Schiele, B., Tuytelaars, T. (eds.) ECCV 2014. LNCS, vol. 8693, pp. 740–755. Springer, Cham (2014). https://doi.org/10.1007/978-3-319-10602-1_48
11. Deng, J., Dong, W., Socher, R., Li, L.J., Li, K., Fei-Fei, L.: ImageNet: a large-scale hierarchical image database. In: 2009 IEEE Conference on Computer Vision and Pattern Recognition, pp. 248–255 (2009). https://doi.org/10.1109/CVPRW.2009.5206848
12. Yao, B., Jiang, X., Khosla, A., Lin, A.L., Guibas, L., Fei-Fei, L.: Human action recognition by learning bases of action attributes and parts. In: Proceedings of the IEEE International Conference on Computer Vision, pp. 1331–1338 (2011). https://doi.org/10.1109/ICCV.2011.6126386

13. Soomro, K., Roshan Zamir, A., Shah, M.: UCF101: A Dataset of 101 Human Actions Classes From Videos in The Wild (2012)
14. Zhang, Z., Song, Y., Qi, H.: Age progression/regression by conditional adversarial autoencoder. In: IEEE Conference on Computer Vision and Pattern Recognition (CVPR 2017) (2017)
15. Huang, C.T., Chen, Y., Lin, R., Kuo, C.C.J.: Age/gender classification with whole-component convolutional neural networks (WC-CNN). In: Proceedings - 9th Asia-Pacific Signal and Information Processing Association Annual Summit and Conference, APSIPA ASC 2017, 2018-February, pp. 1282–1285 (2018). https://doi.org/10.1109/APSIPA.2017.8282221

Author Index

Printed in the United States
by Baker & Taylor Publisher Services